AF356604

Plant Signaling, Behavior and Communication

Plant Signaling, Behavior and Communication

Editors

Frantisek Baluska
Gustavo Maia Souza

Basel • Beijing • Wuhan • Barcelona • Belgrade • Novi Sad • Cluj • Manchester

Editors
Frantisek Baluska
University of Bonn
Bonn
Germany

Gustavo Maia Souza
Federal University of Pelotas
Pelotas
Brazil

Editorial Office
MDPI AG
Grosspeteranlage 5
4052 Basel, Switzerland

This is a reprint of articles from the Special Issue published online in the open access journal *Plants* (ISSN 2223-7747) (available at: https://www.mdpi.com/journal/plants/special_issues/plant_signaling_behavior_communication).

For citation purposes, cite each article independently as indicated on the article page online and as indicated below:

Lastname, A.A.; Lastname, B.B. Article Title. *Journal Name* **Year**, *Volume Number*, Page Range.

ISBN 978-3-7258-1617-0 (Hbk)
ISBN 978-3-7258-1618-7 (PDF)
doi.org/10.3390/books978-3-7258-1618-7

Contents

About the Editors

Frantisek Baluska

Frantisek Baluska was Group Leader at the IZMB, University of Bonn. He is one of the leading scientists in the fields of cell biology, cytoskeletons, polarity, and plant sensory biology. He has published more than 200 peer reviewed papers. Web of Science has scored 327 of his entries with H-Index 67. In order to foster this new sensory and behavioural view of plants and their roots, he has cofounded together with Taylor & Francis two scientific journals: *Plant Signaling & Behavior* and *Communicative and Integrative Biology*. He is the editor of the book series 'Signaling and Communication in Plants' at the Springer Verlag. He also cofounded the Society of Plant Signaling & Behavior.

Gustavo Maia Souza

Gustavo Maia Souza is a plant ecophysiologist researching the complex relationships between plants and their environments, especially under stressful conditions, examining plants as cognitive agents. He has developed the concept of the "plant electrome" to help to uncover the internal signaling network through which a modular system emerges with wholeness and individuality. He is the group leader of the Laboratory of Cognition and Plant Electrophysiology at the Federal University of Pelotas, Brazil, also acting as coordinator of the Postgraduate Program in Plant Physiology. His more than 100 per reviewed papers are mostly devoted to plant stress behavior under water deficit conditions, plant cognition, philosophy, and the theory of plant ecophysiology. Currently, he is one of the Editors-in-Chief of *Plant, Signaling & Behavior*.

Editorial

Plant Signaling, Behavior and Communication

Frantisek Baluska [1,*] and Gustavo Maia Souza [2,*]

[1]   Institute of Cellular and Molecular Botany (IZMB), University of Bonn, 53115 Bonn, Germany
[2]   Department of Botany, Institute of Biology, Federal University of Pelotas, Pelotas 96160, RS, Brazil
*   Correspondence: baluska@uni-bonn.de (F.B.); gumaia.gms@gmail.com (G.M.S.)

Being sessile organisms that need to effectively explore space (above and below ground) and acquire resources through growth, plants must simultaneously consider multiple possibilities and wisely balance the energy they spend on growth with the benefits for survival. Unlike animals whose body structure is set early in their development, plants have a modular architecture that grows indefinitely from meristems, which are versatile tissues found throughout the plant, particularly at shoot and root apexes. This growth results in new branches, shoots, leaves, and roots throughout a plant's life, facilitating resource acquisition. Additionally, these modules are more than mere building blocks; they allow plants to perceive and locally react to environmental stimuli, thus enabling a range of appropriate responses to different signals and threats. This modularity is not indicative of isolated units but rather a cohesive network that allows for self-organization across multiple scales—from molecular to cellular levels and from individual organisms to entire communities—creating a complex, integrated signaling system within the plant [1,2].

Emerging areas in plant sciences are increasingly focusing on plant signaling, communication, cognition, and behavior. New research has shown that plants are far more intricate and engaged in their interactions with both living and non-living environments. This Special Issue was focused on the unique sensory systems of plants, including the detection and transmission of signals, the gathering and processing of sensory information related to actively adapting to stress, and the dynamics of communication among plants and their surroundings, including other plants and other living beings. Therefore, it is not surprising that a large amount of evidence has been accumulated showcasing astonishing cognitive plant abilities, such as their ability to accurately find resources, to make decisions, and to communicate with each other about their "findings".

Herein, the exquisite dynamic behavioral capacities of plants that are embedded in an ever-changing environment are well illustrated by Pavlovic et al. [3], evidencing the accurate and selective way by which the carnivorous plant *Drosera capensis* senses and behaves with regard to different stimuli. Costa et al. [4] document the abilities of plants to respond to different cues that, presented as single or combined stimuli, engender different systemic signals (both electrical and hydraulic). The changes in the dynamic of different systemic signals, particularly the bioelectrical ones, induce different responses at modules distant from the local stimulated tissues, highlighting plants' capacities to integrate and to coordinate responses systemically. Such ability to integrate signals is also demonstrated at the tissue level when collective stomatal behavior changes under different external cues (light and drought), affecting photosynthetic efficiencies—as showed in the mathematical simulations by Sukhova et al. [5]. Exemplifying the plethora of sophisticated mechanisms used by plants to sense their environment, Yamashita and Baluska [6] propose that plant eye-like ocelli, which allow plant-specific kind of vision [7,8] and which evolved from the algal ocelloids, are part of complex plant sensory systems and guide cognition-based plant behavior, such as the mimicking of diverse host plants by woody vine *Boquila trifoliata* [9–11] and root light escape tropism [12].

How plants sense the environment and then behave is not only a matter of perceiving and processing different external cues. The area where plants are integrated can also

Citation: Baluska, F.; Souza, G.M. Plant Signaling, Behavior and Communication. *Plants* **2024**, *13*, 1132. https://doi.org/10.3390/plants13081132

Received: 3 April 2024
Accepted: 11 April 2024
Published: 18 April 2024

affect their ability to explore available resources. Mahal at al. [13] showed the ability of wheat plants to explore nutrients, varying both in their amount and distribution in the soil, exhibiting not only belowground responses but also aboveground changes; however, in this case, the influence of the area in which the plants grew was not clear. On the other hand, Chautá and Kessler [14] discussed how plants respond to changes in light quality and exposure to chemicals released by neighboring plants (volatile organic compounds, VOCs). The study found that these factors strongly interact and influence on the production of secondary metabolites, both volatile and non-volatile, in plants, affecting how plants detect and respond to VOCs emitted by other plants. The findings indicate that plants can integrate various environmental cues to modulate their chemical outputs, which in turn can affect the interactions within plant populations and communities.

Moreover, the ability of plants to communicate "stress calls" to other ones is well illustrated by Falik and Novoplansky [15], who report that drought cuing and relayed cuing is observed in intra- and interspecific neighbor combinations, but their strength depends on plant identity and position. Accordingly, Midzi et al. [16] highlight the ecological relevance of such interactions under various environmental stresses and the growing understanding of the mechanisms involved and the significance of VOC-mediated inter-plant interactions under both biotic and abiotic stresses. As an interesting example of inter-plant interactions, Le Ding et al. [17] show that intraspecific kin recognition may facilitate cooperation between genetically related biotypes to compete with interspecific rice, offering many potential implications and applications in paddy systems.

Facing a constantly changing environment and, at the same time, interacting with other plant species through a sophisticated communication system may require that, at some point, plants are challenged to make choices to ensure survival. Lee et al. [18] bring such exciting possibilities, suggesting that decision making is especially relevant to the issue of plant intelligence as it is commonly taken to be characteristic of cognition. As a matter of evidence, Wang et al. [19] present the case of pea plants searching for support and likely making choices between different possibilities, showing distinct preference for their support. Interestingly, the dynamics of pea plants' climbing movements show distinct kinematic traits allowing for automatic classification using machine learning methods, as illustrated by Wang et al. [20]. Making decisions is considered a high-level cognitive capacity of living beings, surpassing the abilities of sensing and responding. In a complex and demanding environment, it is likely that making choices would demand the ability to attend specific cues more relevant for plant survival. In this direction, Parise et al. [21] have proposed that plants can show states of attention when facing specific challenges co-occurring with different cues. They suggest that the phenomenon of attention in plants would be reflected in their electrophysiological activity, which can be analyzed by investigating the potential existence of different band frequencies (including low, delta, theta, mu, alpha, beta, and gamma waves) using a protocol adapted from neuroscientific research.

Conflicts of Interest: The authors declare no conflict of interest.

References

1. De Kroon, H.; Visser, E.J.W.; Huber, H.; Hutchings, M.J. A modular concept of plant foraging behaviour: The interplay between local responses and systemic control. *Plant Cell Environ.* **2009**, *32*, 704–712. [CrossRef] [PubMed]
2. Lüttge, U. Integrative emergence in contrast to separating modularity in plant biology: Views on systems biology with information, signals and memory at scalar levels from molecules to the biosphere. *Theor. Exp. Plant Physiol.* **2021**, *33*, 1–13. [CrossRef]
3. Pavlovic, A.; Vrobel, O.; Tarkowski, P. Water cannot activate traps of the carnivorous sundew plant Drosera capensis: On the trail of Darwin's 150-years-old mystery. *Plants* **2023**, *12*, 1820. [CrossRef] [PubMed]
4. Costa, Á.V.L.; Oliveira, T.F.d.C.; Posso, D.A.; Reissig, G.N.; Parise, A.G.; Barros, W.S.; Souza, G.M. Systemic signals induced by single and combined abiotic stimuli in common bean plants. *Plants* **2023**, *12*, 924. [CrossRef] [PubMed]
5. Sukhova, E.; Ratnitsyna, D.; Gromova, E.; Sukhov, V. Development of two-dimensional model of photosynthesis in plant leaves and analysis of induction of spatial heterogeneity of CO_2 assimilation rate under action of excess light and drought. *Plants* **2022**, *11*, 3285. [CrossRef] [PubMed]
6. Yamashita, F.; Baluška, F. Algal ocelloids and plant ocelli. *Plants* **2023**, *12*, 61. [CrossRef] [PubMed]
7. Haberlandt, G. *Die Lichtsinnesorgane der Laubblätter*; Engelmann: Leipzig, Germany, 1905.

8. Wager, H. The perception of light in plants. *Ann. Bot.* **1909**, *23*, 459–489. [CrossRef]
9. Baluška, F.; Mancuso, S. Vision in plants via plant-specific ocelli? *Trends Plant Sci.* **2016**, *21*, 727–730. [CrossRef] [PubMed]
10. Mancuso, S.; Baluška, F. Plant ocelli for visually guided plant behavior. *Trends Plant Sci.* **2017**, *22*, 5–6. [CrossRef] [PubMed]
11. White, J.; Yamashita, F. *Boquila trifoliolata* mimics leaves of an artificial plastic host plant. *Plant Signal. Behav.* **2022**, *17*, 1977530. [CrossRef] [PubMed]
12. Mo, M.; Yokawa, K.; Wan, Y.; Baluška, F. How and why do root apices sense light under the soil surface? *Front. Plant Sci.* **2015**, *6*, 775. [CrossRef]
13. Mahal, H.F.; Barber-Cross, T.; Brown, C.; Spaner, D.; Cahill, J.F., Jr. Changes in the amount and distribution of soil nutrients and neighbours have differential impacts on root and shoot architecture in wheat (*Triticum aestivum*). *Plants* **2023**, *12*, 2527. [CrossRef] [PubMed]
14. Chautá, A.; Kessler, A. metabolic integration of spectral and chemical cues mediating plant responses to competitors and herbivores. *Plants* **2022**, *11*, 2768. [CrossRef] [PubMed]
15. Falik, O.; Novoplansky, A. Interspecific drought cuing in plants. *Plants* **2023**, *12*, 1200. [CrossRef] [PubMed]
16. Midzi, J.; Jeffery, D.W.; Baumann, U.; Rogiers, S.; Tyerman, S.D.; Pagay, V. Stress-induced volatile emissions and signalling in inter-plant communication. *Plants* **2022**, *11*, 2566. [CrossRef] [PubMed]
17. Le Ding, L.; Zhao, H.-H.; Li, H.-Y.; Yang, X.-F.; Kong, C.-H. Kin recognition in an herbicide-resistant barnyardgrass (*Echinochloa crus-galli* L.) biotype. *Plants* **2023**, *12*, 1498. [CrossRef] [PubMed]
18. Lee, J.; Segundo-Ortin, M.; Calvo, P. Decision making in plants: A rooted perspective. *Plants* **2023**, *12*, 1799. [CrossRef] [PubMed]
19. Wang, Q.; Guerra, S.; Bonato, B.; Simonetti, V.; Bulgheroni, M.; Castiello, U. Decision-making underlying support-searching in pea plants. *Plants* **2023**, *12*, 1597. [CrossRef] [PubMed]
20. Wang, Q.; Barbariol, T.; Susto, G.A.; Bonato, B.; Guerra, S.; Castiello, U. Classifying circumnutation in pea plants via supervised machine learning. *Plants* **2023**, *12*, 965. [CrossRef] [PubMed]
21. Parise, A.G.; Oliveira, T.F.d.C.; Debono, M.-W.; Souza, G.M. The Electrome of a parasitic plant in a putative state of attention increases the energy of low band frequency waves: A comparative study with neural systems. *Plants* **2023**, *12*, 2005. [CrossRef] [PubMed]

Review

Stress-Induced Volatile Emissions and Signalling in Inter-Plant Communication

Joanah Midzi [1,2], David W. Jeffery [1,2], Ute Baumann [1], Suzy Rogiers [2,3], Stephen D. Tyerman [1,2] and Vinay Pagay [1,2,*]

[1] School of Agriculture, Food and Wine, The University of Adelaide, Glen Osmond, SA 5064, Australia
[2] Australian Research Council Training Centre for Innovative Wine Production, Urrbrae, SA 5064, Australia
[3] New South Wales Department of Primary Industries, Wollongbar, NSW 2477, Australia
* Correspondence: vinay.pagay@adelaide.edu.au

Abstract: The sessile plant has developed mechanisms to survive the "rough and tumble" of its natural surroundings, aided by its evolved innate immune system. Precise perception and rapid response to stress stimuli confer a fitness edge to the plant against its competitors, guaranteeing greater chances of survival and productivity. Plants can "eavesdrop" on volatile chemical cues from their stressed neighbours and have adapted to use these airborne signals to prepare for impending danger without having to experience the actual stress themselves. The role of volatile organic compounds (VOCs) in plant–plant communication has gained significant attention over the past decade, particularly with regard to the potential of VOCs to prime non-stressed plants for more robust defence responses to future stress challenges. The ecological relevance of such interactions under various environmental stresses has been much debated, and there is a nascent understanding of the mechanisms involved. This review discusses the significance of VOC-mediated inter-plant interactions under both biotic and abiotic stresses and highlights the potential to manipulate outcomes in agricultural systems for sustainable crop protection via enhanced defence. The need to integrate physiological, biochemical, and molecular approaches in understanding the underlying mechanisms and signalling pathways involved in volatile signalling is emphasised.

Keywords: abiotic stress; biotic stress; plant–plant interactions; priming; stress signalling; VOCs; volatile-mediated signalling

Citation: Midzi, J.; Jeffery, D.W.; Baumann, U.; Rogiers, S.; Tyerman, S.D.; Pagay, V. Stress-Induced Volatile Emissions and Signalling in Inter-Plant Communication. *Plants* **2022**, *11*, 2566. https://doi.org/10.3390/plants11192566

Academic Editors: Frantisek Baluska and Gustavo Maia Souza

Received: 21 August 2022
Accepted: 14 September 2022
Published: 29 September 2022

Publisher's Note: MDPI stays neutral with regard to jurisdictional claims in published maps and institutional affiliations.

1. Introduction

Climate change has exacerbated the multifarious effects of environmental stresses on crop growth and development, thereby compromising sustainable agricultural productivity worldwide. Biotic stresses in the form of insects, bacteria, viruses, fungi, nematodes, arachnids, and weeds account for over 30% of losses from the annual global food production capacity, or approximately US$500 billion [1]. Abiotic stresses including drought, extreme temperatures, and salinity are major yield-limiting factors of economically important food crops globally [2]. Of these stresses, drought is one of the most significant, given that its frequency and severity has been forecasted to increase due to climate change [3]. Greater effort has therefore been directed towards the implementation of sustainable agricultural management and drought mitigation strategies in major crop-growing regions worldwide.

To compensate for their immobile nature, plants acclimatise to various environmental stresses with an array of complex molecular, physiological, and biochemical adaptations, which ultimately allow them to survive and even maintain potential rates of growth [4]. These adaptations may be expressed constitutively, or in many cases, are activated only in the presence of stressors. The evolved innate immunity or basal resistance of a plant is regulated by an intricate network of endogenous signalling molecules, receptor proteins, and transcriptional regulators. In response to environmental stress, plants exhibit an upregulation of the expression of stress-related genes encoding for defence proteins, such as trypsin protease inhibitors (PI) and pathogenesis-related (PR) proteins, which act against herbivory and pathogen attack, respectively. There is also an elevated production of secondary metabolites including osmoprotectants and toxins with deterrent/antifeedant activity [5–13]. Timely defence responses to biotic and abiotic stresses offer a fitness benefit to plants that largely depends on the capacity to quickly and accurately recognise external stress stimuli [14,15]. Upon stress perception, the plant will activate various defence-signalling pathways, leading to subsequent gene expression [15,16].

Among the secondary metabolites, phytochemicals in the form of volatile organic compounds (VOCs) have been identified as chemical-signalling molecules involved in both intra- and inter-plant communication, providing a fitness benefit to both the emitter and neighbouring receiver plants [17]. Plants respond to various biotic and abiotic stresses by emitting VOCs that fall into various compound classes, namely: terpenoids, benzenoids/phenylpropanoids, and fatty acid derivatives [17–25]. Recent studies in various crop species have explored an intriguing possibility of developing stress tolerance in plants through their exposure to VOCs from stressed neighbours, a process referred to as "priming" [5–13,26–30]. In the classical study by Karban et al. [27], volatile-mediated airborne signalling between native tobacco (*Nicotiana attenuata*) and sagebrush (*Artemisia tridentata*) was demonstrated. The sagebrush plants were clipped experimentally to mimic insect damage, and the emitted VOCs induced herbivore resistance in the neighbouring tobacco plants.

A follow-up study by Kessler et al. [31] reported an upregulation of herbivore-regulated genes in the receiver tobacco plants, but with no evidence of direct elicitation of defensive secondary metabolites. Interestingly, following post-challenge with *Manduca sexta* caterpillars, the receiver tobacco plants had an accelerated production of PI proteins, which was not evident in plants not previously exposed to the clipped sagebrush volatiles [31]. This study demonstrated how volatile-mediated plant–plant interaction primes defence responses in non-stressed receiver plants, inducing a faster and stronger response to a real stress. The production of defensive metabolites such as (Z)-3-hexenyl-vicianoside [12] and genes of defence-related enzymes including PI, threonine deaminase, and α-dioxygenase [31] have been shown to increase as a result of VOC exposure, ultimately conferring tolerance or resistance in non-stressed plants. Overall, however, the signalling pathways and mechanisms in volatile-mediated plant–plant interactions are vaguely understood and have been sparsely investigated in stress physiology studies.

This review offers a comprehensive synopsis of volatile-mediated inter-plant communication, referencing studies published from 2000–2021 and using Google Scholar and Web of Science as the main academic search engines. It explores the range of VOCs that elicit various stress responses in non-stressed, neighbouring receiver plants in the face of both biotic and abiotic stresses. Understanding how plants can prepare for impending stresses by recognising VOC signals from stressed plants without themselves having to experience the actual stress is a noble endeavour towards the development of crops that are more tolerant against environmental stresses. Furthermore, the identification of specific receptors, transcription factors (TFs), and other regulatory proteins involved in volatile-mediated signalling, as well as the characterisation of the various defence-signalling pathways induced after VOC perception, is necessary for a full appreciation of airborne signalling between plants under stress. The integration of physiological, metabolome, and transcriptome analyses in volatile-mediated signalling and defence priming will be discussed.

2. Plant Volatile Organic Compounds

Plants emit and respond to a wide range of VOCs, which are generally lipophilic and of adequate volatility to be released into the atmosphere from the liquid phase [22]. Up to 10% of the carbon fixed during photosynthesis can be lost through complex volatile plumes [32]. Beyond terpenoids, which constitute the most complex group, other major classes of VOCs include fatty acid catabolites, aromatic compounds, and amino acid derivatives as products of the shikimic acid pathway [17–24]. Volatiles can either be emitted constitutively [33] or induced in response to factors associated with abiotic stress [21,22] or biotic stress [34]. VOCs have relatively high vapour pressure, low molecular weight, and low polarity properties, accounting for their high volatility [35]. The rate of emission from plant tissue into the atmosphere has been suggested to depend on the volatility, solubility, and diffusivity of the particular volatile compound rather than its rate of synthesis or other physiological mechanisms [36]. Nonetheless, the ratio of compounds in a constitutively emitted VOC bouquet is considered to be dependent on species taxonomy and varies greatly within species [37].

VOC Biosynthesis

About 1700 VOCs have been characterised in different plant volatile blends, where they convey information about the plant identity and physiological condition [38]. The biosynthesis of VOCs occurs in the leaf mesophyll tissues, particularly in the palisade mesophyll cells, prior to their release either via inflicted wounds or the stomata on leaves and stems, which regulate emission rates by the opening and closing of their apertures [33,36,39]. The secretory cells of glandular trichomes are also active sites of VOC production [40]. In non-chlorophyll-containing tissues such as flowers and roots, VOC biosynthesis has been reported to occur in specialised crenulated epidermal cells, whose close proximity to the atmosphere or rhizosphere ensures immediate release [32,40]. Roots produce volatiles in epidermal cells and release them into the rhizosphere to mediate below-ground plant–plant interactions, thereby influencing the diversity of soil microbial communities [32,41,42].

Damage to the plant tissues triggers hydrolytic cleavage of complex membrane lipids by lipases to produce free polyunsaturated fatty acids such as the C_{18} linoleic and linolenic acids [43,44]. Following peroxidation of the fatty acids by lipoxygenase (LOX) enzymes, subsequent cleavage of lipid hydroperoxide by hydroperoxide lyase (HPL) will give rise to a suite of C_6 aldehydes, alcohols and esters—collectively known as green leaf volatiles (GLVs) (Figure 1, plastid) [7,8,34,45]. In the presence of allene oxide synthase (AOS), the lipid hydroperoxide is rechannelled to the jasmonic acid (JA) pathway for JA synthesis [44]. Terpenoids form the largest and most diverse group of organic compounds synthesised via two distinct metabolic pathways: the cystolic mevalonic acid (MVA) pathway and the plastidic 2-C-methyl-D-erythritol 4-phosphate (MEP) pathway. All terpenoids are derived from C_5 isoprene precursors, isopentenyl pyrophosphate (IPP), and dimethylallyl diphosphate (DMADP), with reactions catalysed by various terpene synthases [46] (Figure 1, plastid). Aromatic compounds (in terms of chemistry rather than any potential aroma) are derived from L-phenylalanine and include products from the chain-shortening of *trans*-cinnamic acid and structures derived from lignin biosynthesis that form benzenoids [19,47] (Figure 1, plastid). Additionally, amino acid methionine produces 1-aminocyclopropane-1-carboxylic acid (ACC), which is enzymatically oxidised to form ethylene [48] (Figure 1, cytosol). Methanol is also one of the most abundant VOCs found in plants and has been reported to serve as a signalling molecule in intra- and inter-plant communication [49,50]. It is produced through the methylation of the cell wall constituent pectin during plant growth and senescence. The high pH of herbivore salivary secretions is reported to also activate cell-wall pectin methylesterases, leading to copious amounts of methanol being released (Figure 1, cell wall) [40].

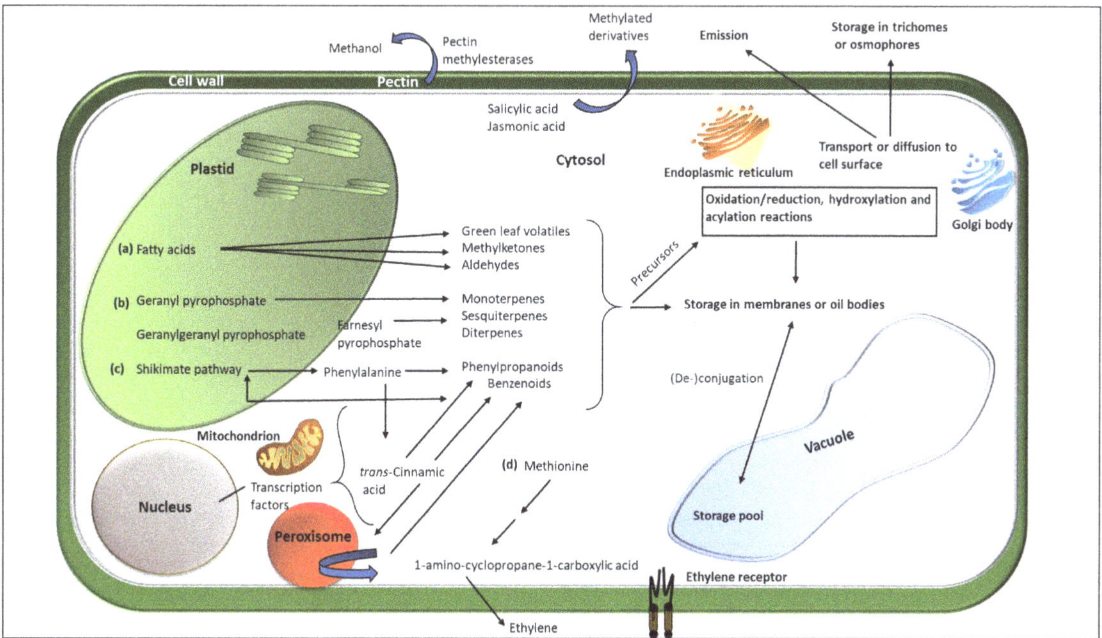

Figure 1. Biosynthetic sites, storage, and transport of major plant volatile organic compounds. The four major biosynthetic classes, (**a**) fatty acids, (**b**) terpenoids, (**c**) benzenoids, and (**d**) amino acids, are represented in the figure. Modification reactions including redox reactions, hydroxylation, acylation, and degradative reactions occur mainly in the cell cytosol. Some reactions may also occur in the membrane-bound sub-cellular compartments such as plastids, mitochondria, and endoplasmic reticulum. Inactive forms of the VOCs such as glycosides are stored in the vacuole, ducts, and other extracellular compartments such as trichomes. Transport of the volatiles is by passive diffusion and possibly via vesicular trafficking by the endoplasmic reticulum, golgi apparatus, and the *trans*-golgi network. (Modified from Pichersky et al. [51]).

3. Plant Volatile Storage, Transport, and Emission

Considering that many of the VOCs produced by plants are likely to be lethal to the plant itself at high concentrations, several self-resistance mechanisms are employed by plants during VOC storage, transport, and emission to avoid self-toxicity (Figure 1). These include vacuolar sequestration, vesicle transport, extracellular excretion, extracellular biosynthesis, and storage of VOCs in cells as inactive non-volatile glycoside precursors [52,53]. Storage of VOCs occurs in various extracellular storage organs, including glandular trichomes, ducts, and laticifers, as well as the sub-cellular membrane-bound compartments such as vacuoles, plastids, mitochondria, and the endoplasmic reticulum (ER) [40,51,54].

Stored VOCs are released after deconjugation of their precursors upon mixing with lytic enzymes after the rupturing of storage organs through mechanical damage [40]. Studies have shown that specific components from herbivore salivary secretions such as β-glucosidases, glucose oxidases, and volicitin act as elicitors of HIPV emission during herbivory to release toxic aglycones of the glycoside conjugates, which have antibiotic and/or antixenotic effects on the herbivores [42,55–59]. When gut regurgitant of cabbage caterpillar *Pieris brassicae* was used in a study by Mattiacci et al. [60] to treat artificially damaged cabbages, a release of volatile blends similar to those observed from herbivore-damaged plants was noted. Parallel results were observed when β-glucosidase from bitter almonds was used to treat undamaged *P. lunatus* [61]. Interestingly, when lima bean plants were treated with JA solution, a similar VOC blend was emitted. It was postulated that the lytic enzymes in herbivore saliva hydrolyse cell structures into oligosaccharides and pectin, which in turn act as first signals leading to gene activation and de novo metabolite synthesis via internal signal transduction pathways such as the JA pathway [62,63].

Despite the presumption that VOC movement from the site of synthesis into the atmosphere occurs via passive diffusion, Widhalm et al. [32] questioned this supposition and highlighted the possibility of high barrier resistance in the movement of lipophilic VOCs across the cytosol, plasma membrane, aqueous cell wall, and cuticle. Movement via diffusion alone is likely to be too slow to account for the high emission rates observed during stress responses. Based on Fick's first law of diffusion, it was determined that VOCs had to accumulate to toxic levels internally before they could be emitted at the observed emission rates, so it has been suggested that more active trafficking mechanisms may be involved [32]. Plausible mechanisms involve vesicular trafficking associated with the endoplasmic reticulum and Golgi apparatus [32,64], soluble carrier proteins [65], plasma-membrane localised transporters such as ATP-Binding Cassette (ABC) transporters [66,67], and small carrier proteins in the cell apoplast, such as lipid transfer proteins (LTPs) [68,69]. From these studies, it can be hypothesised that the same mechanisms that facilitate volatile emission from cells into the atmosphere may also be involved in VOC recognition and uptake [70].

Apart from VOC emissions from ruptured specialised storage structures such as glandular trichomes and ducts, emissions from plants are predominantly via stomata in leaves or directly from the epidermal cells in tissues lacking stomata, such as flower petals and roots. Emission of VOCs via the cuticle is an alternative pathway for some monoterpenoids. This has been calculated to contribute only 10–20% of the total monoterpenoid emissions, and therefore would not account for the high emission rates observed when stomatal conductance is low [71]. Stomata provide a low resistance pathway for volatile emissions, but they may decrease the efflux of VOCs from plants during stomatal closure resulting from various stressors [72,73].

The magnitude of stress-induced VOC emissions in a plant depends on its stress tolerance as well as the severity, timing, and duration of that particular stressor [21]. Unlike storage emissions, de novo emissions are generally more sensitive to stress and have been suggested to be the primary source of emitted isoprenoids [74,75]. Constitutive emissions can be altered through immediate stress responses and acclimation after stress recovery [76,77]. Volatiles including LOX-pathway products [78] and methanol [79] are induced during early stress (mild stress), reflecting signal activation at the membrane level and in cell walls [80]. Later stress responses include emissions of specialised isoprenoids [22]. Induced volatile phytohormones including MeJA, MeSA, and ET mediate stress signal perception, transduction, and propagation, leading to the activation of gene expression of defence-related genes, including enzymes involved in VOC synthesis [21]. The lifetime of these induced VOC emissions is finite following stress exposure. LOX products and MeSA have been shown to last for 15h after ozone exposure [79] and five to seven days under heat stress [81], and isoprenoid emissions continued for two days after herbivory [82].

Physicochemical and physiological factors limit VOC emission rates in plants, with the extent of control varying with the specific volatiles stored in the leaf [36]. Physicochemical constraints affect the emission of the synthesised VOCs to the ambient air by limiting their volatility and diffusion [36]. Leaf anatomy, specifically the presence of specialised VOC storage structures such as glands and resin ducts, has also been shown to influence emission rates [83]. Large diffusion resistances of various VOC storage pools exist between the intercellular spaces and the ambient air as a result of layers of epithelial and sclerenchyma cells lining the storage structures [74], unless rupturing of the pools occurs. Biochemical and molecular control over VOC synthesis in response to physiological factors such as light [84], temperature, drought [75,83], and high carbon dioxide concentration [83,85] has been shown to relate to the availability of immediate VOC precursors as well as on the activity rate of flux-controlling enzymes.

The biosynthesis of signalling VOCs depends on the plastidic pool of intermediate products of photosynthesis, including glyceraldehyde 3-phosphate (G3P) and erythrose 4-phosphate (Ery4P), as well as sufficient supply of phosphoenol pyruvate (PEP) from glycolysis [74] (Figure 2). During mild stress, the photosynthetic machinery is affected such that the carbon assimilation rate is significantly decreased [86]. Emission rates of most VOCs are expected to be reduced under such conditions [84]. However, an increase in emissions after stress exposure has been observed during drought [76,87], heat [88], salinity [89], and ozone stress [90], indicating the acclimation of VOC synthesis under stress conditions and the plant's ability to maintain high emission rates after stress relief.

Mild water stress conditions, like most abiotic stressors, typically reduce stomatal conductance, thereby reducing intercellular CO_2 concentration [86] and increasing leaf temperature because of constrained leaf transpiration [91]. The temperature response of VOC emissions is a function of increased enzyme activity and substrate availability [92], such that an increase in temperature reduces photosynthetic metabolites and energy, subsequently reducing de novo VOC synthesis and emission [93]. Although the reduction in stomatal conductance limits CO_2 supply to the Calvin–Benson cycle of photosynthesis, photosynthetic electron transport is not inhibited, as the triggering of photorespiration may effectively supply CO_2 and phosphoglycerate (G3P) needed to drive the cycle [21] (Figure 2). Therefore, ATP and NADPH remain available for the reduction in carbon reserves in the form of starch and sugar for isoprenoid synthesis [21]. Typically, sustained moderate or strong rapid drought stress will eventually lead to a significant reduction in VOC emissions, with augmented emissions after rewatering [76,87]. However, prolonged water stress can lead to accelerated leaf senescence and retardation in key terminal enzyme activity, leading to low emission rates during and after stress [84,94–96].

Mild heat stress enhances the activities of flux-controlling enzymes involved in the terpenoid synthesis pathway at the expense of carbon fixation enzymes. Subsequently, emission increases as the VOC synthesis pathway becomes more competitive for carbon and photosynthetic electrons than carbon fixation [88,94]. Increases in temperature also affect photorespiration, which can indirectly regulate PEP levels available to form intermediate substrates for VOC biosynthesis [97]. Light intensity affects the amount of the terpenoid precursor G3P available from photosynthesis, as well as the energetic co-factors ATP and NADPH required for its chemical reduction [98].

The diffusion of individual VOCs is influenced by the width of the stomatal opening, the leaf anatomy, and compound molar size [36]. The volatility of a specific VOC is determined by equilibrium partitioning between the intercellular airspaces and the ambient air, which then affect the diffusion gradient [36,74]. Changes in stomatal conductance can differentially affect the emission rates of various VOCs under stress conditions [39]. The emission of methanol [99], linalool [39], acetic acid [100], and acetaldehyde [101] were vulnerable to stomatal conductance, whereas several monoterpenoids such as limonene, *trans*-β-ocimene [39], isoprene [102], and α-pinene [103] were emitted independently of stomatal regulation. This independence from stomatal conductance has been associated with the large Henry's Law constant (*H*; gas/liquid phase partition coefficient) of the monoterpenoids, which implies that they can maintain a high intercellular partial pressure for a given liquid phase concentration. An increase in the diffusion gradient from the intercellular space to the external atmosphere may compensate, at least partially, for the reduced stomatal conductance, in order to control VOC emission rates [39,102,104].

Following an earlier hypothesis that VOC emissions in plants may display hormetic responses to environmental stressors [105], studies involving herbivory [106,107], drought [108,109], CO_2 concentration, light, temperature [92,110], and ozone [105,111,112] have also provided evidence that VOC emission is biphasic by time and dose in response to the stressors, thus indicating hormesis [113,114]. Hormesis is a biphasic stress dose response that depicts adaptive responses of organisms to low-dose stresses whereby they can improve their tolerance to severe stress challenges, whereas higher doses of the stresses will have negative effects on organisms [115] (Figure 2). The occurrence of hormesis of

VOC emission in plants suggests an evolutionary adaptation that acts to maintain fitness in a changing environment in the context of enhancing intra- and interplant communication, defence priming, protection, and defence functions [114].

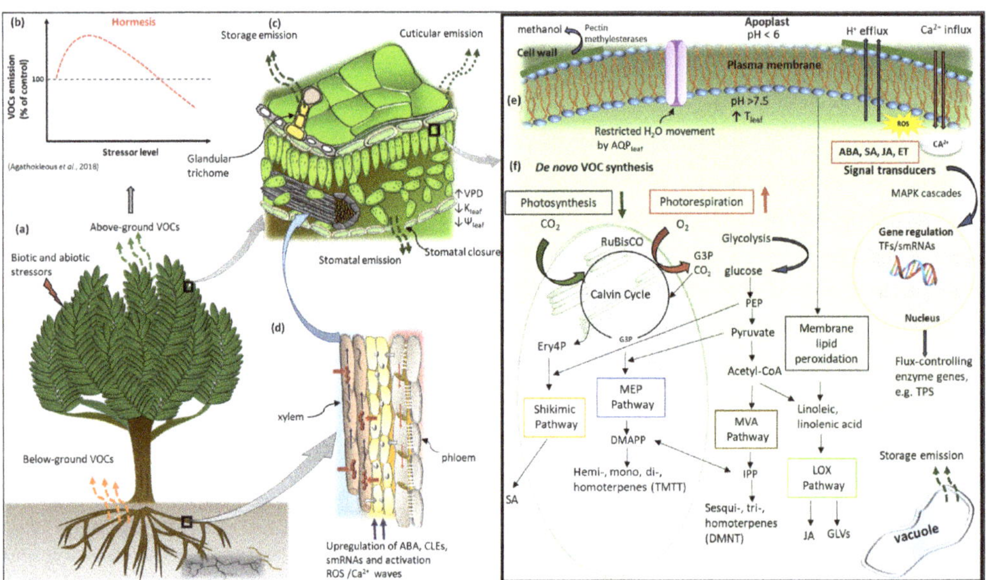

Figure 2. A simplified schematic of the interactions among major VOC biosynthesis pathways in response to stress. (**a**) VOC emissions respond to biotic and abiotic stresses from both below and aboveground sources of a plant. (**b**) A generalised VOC emission response to stress is given, which depicts a hormetic-like biphasic pattern. (**c**) Increased stress results in decreased intercellular CO_2 concentration favouring photorespiration over photosynthesis, which is the source of VOC precursors. (**d**) Various root-to-shoot signalling molecules that are involved in stress signalling. (**e**) Membrane depolarisation activities involving ROS accumulation and Ca^{2+} influx as well as stress-induced phytohormones, e.g., ABA, JA, SA, and ET, activate defence signal transduction pathways that trigger gene expression of flux-controlling enzymes, e.g., TPS genes. Stressors affecting leaf temperature (T_{leaf}) influence the activity of the enzymes. (**f**) De novo VOC synthesis in the chloroplast and cytosol involves VOC precursors such as G3P (for the MEP pathway) and Ery4P (for the shikimic pathway), PEP from glycolysis, as well as its downstream metabolites pyruvate and acetyl-CoA that are involved in the Shikimic, MVA, and LOX biosynthesis pathways. Note that the sizes of organelles are not drawn to scale. Abbreviations: ABA, abscisic acid; JA, jasmonic acid; SA—salicylic acid; ET, ethylene; TPS, terpene synthase; G3P, glyceraldehyde 3-phosphate; MEP, 2-C-methyl-D-erythritol 4-phosphate; Ery4P, erythrose 4-phosphate; PEP, phosphoenol pyruvate; MVA, mevalonic acid; LOX, lipoxygenase; CLE25, CLAVATA3/EMBRYO-SURROUNDING REGION-RELATED 25; smRNA, small RNAs [114].

4. Volatile Uptake, Perception, and Signalling

The adsorption of VOCs on the surface of the leaves as well as their solubilisation and enzymatic detoxification within the leaves are what drive the uptake, deposition, and storage of VOCs [116]. VOCs' chemical reactivity, physicochemical properties, and involvement in leaf metabolism all play a role in how much they react in leaves and can potentially be absorbed by vegetation [116]. Though stomatal uptake is a significant sink for VOCs in leaves, VOCs can adsorb directly onto the cuticle as a result of gas-phase deposition [117–120]. The cuticle opens up as a route to and from the leaf tissues for organic compounds with substantial lipophilicity and low vapour pressure in addition to the stomatal route [118]. The degree of stomatal conductance along with the combination of VOC physicochemical properties will determine whether VOCs are released or absorbed through the cuticle, the stomatal pores, or both.

The direction of the concentration gradient between the leaf intercellular air space and the surrounding air determines whether a certain VOC is released or absorbed at any particular time. At very low internal concentrations, the uptake rate is controlled essentially by stomatal conductance as well as the ambient concentration of the VOC [116]. A more significant uptake of monoterpenes was observed when the ambient air concentration was higher versus when it was lower [121]. The thickness and conductivity of the leaf boundary layer also have a bearing on VOC foliar uptake and deposition, which itself is influenced by leaf morphology and wind speed [119]. Higher uptake rates have been noted in plants with thinner leaves and greater surface area per unit dry mass [121]. Overall, it is anticipated that temperature, VOC and cuticle physicochemical properties, and cuticular structure would all play major roles in regulating the deposition and release of compounds from leaf surfaces [116].

Once deposited on the plant's surface, stress-induced VOCs may act as resistance elicitors perceivable by the plant's evolved and highly sophisticated surveillance system. Subsequently, unique molecular mechanisms of several defence-signalling pathways are initiated [122,123], thereby conferring some degree of immunity in distal regions of the same plant or a neighbouring plant, thus acting as long-distance damage-associated molecular patterns (DAMPs) [124,125]. Elicitation by the volatile DAMPs ranges from early signalling cascades (including Ca^{2+} influxes, oxidative burst, and MAPK activation) to phenotypic defence responses [126]. Methanol emissions in plants, just like GLVs, are reliably associated with injury and have been suggested to act as wound signals in plants [49]. Studies have shown that methanol activated oxidative burst and MAPK cascades in *Festuca arundinacea* (tall fescue) grass and tomato [127]. Isoprene was shown in *Arabidopsis* to induce signal transduction networks associated with stress response and plant growth regulators, such as JA, SA, and ET, which promote defence [128,129].

Although the molecular basis for VOC perception and signalling is vaguely understood, it is likely that VOC signals are first perceived at the membranes, making the transmembrane pathway a more plausible route for signal generation and transduction. On the other hand, a plant's response could be simply evoked by volatiles that might be deposited on the plasma membrane due to their lipophilic or amphiphilic nature, which could probably distort the membrane organisation [130]. Subsequent VOC metabolisation or VOC recognition in the cytoplasm of those VOCs that manage to pass through the membrane is also a possibility, as observed in studies related to the uptake of ^{13}C- labelled GLVs in plants [131,132]. Unlike most VOCs, ethylene perception and signal transduction have been well-established through extensive studies using *Arabidopsis* [133–140]. First suggested by Burg and Burg [133], this gaseous phytohormone binds reversibly to its receptor via a protein-bound transition metal co-factor such as copper. Compelling evidence from biochemical data has shown that the ethylene signal is perceived at the membrane of the endoplasmic reticulum by a family of receptors, including ETHYLENE RESPONSE1 (ETR1), ETHYLENE RESPONSE SENSOR (ERS1), ETR2, ETHYLENE IN-SENSITIVE 4 (EIN4), and (ERS2), which repress ethylene through the negative regulator CONSTITUTIVE RESPONSE1 (CTR1). Downstream and secondary ethylene responsive gene expression are positively regulated by EIN2 and EIN3 [137,139,141–143].

Efforts to understand the molecular aspects of VOC perception and signalling have involved studies in exogenous applications of VOCs to plants. In these studies, VOC-binding proteins involved in gene transcriptional regulation have been identified [70]. A transcriptional co-repressor, TOPLESS-like protein (TPL), was shown in tobacco to bind to β-caryophyllene, a sesquiterpene, to initiate the expression of stress response gene *NtOsmotin*, a pathogenesis-related protein [70]. The study suggested significant specificity of the receptors for individual VOCs as well as selectivity in inducing gene expression in plants. Though the involvement of membrane receptors in the VOC-sensing mechanism cannot be ruled out, it is clear that some nuclear proteins may act as VOC receptors in plants.

Documented studies on mechanisms by which plants perceive various stimuli such as light [144,145], sound [146–149], and touch [150] have shown uncanny similarities with those exhibited in mammalian systems. It is also probable that the mechanisms involved in how plants "smell" odours follow the same paradigm as that observed in mammals—excluding the existence of a nervous system. In mammals, the olfactory system has been described as discriminatory, in that each volatile compound is perceived to have a distinct odour, and even subtle changes in odourant structure or concentration can potentially alter the VOC's "code", thereby shifting the perception of its odour quality [151]. A similar discriminatory behaviour has also been observed in the parasitic dodder plant (*C. pentagona*), which locates its host plants based on its perception of their unique volatile blends [152] and is repelled by specific VOC blends of its non-hosts. In mammals, a combinatorial receptor coding scheme is used by the olfactory system, where specific combinations of odourant receptors (ORs) recognise specific volatiles. Studies by Malnic et al. [151] have shown that even subtle alterations in the volatile compound or changes in its concentration may result in changes in its "code", thereby affecting the perceived quality of the odour. Whether plants follow a similar model for their VOC receptors remains unclear, but it is worth considering in future endeavours to understand the mechanisms of volatile perception and signalling.

5. Volatile-Mediated Intra- and Inter-Plant Communication

Plants have evolved unique communication mechanisms between various organs for the enhancement of their development and providing resistance to stress. Upon perception of an environmental stress, a plant will activate several long-distance signals to trigger systemic stress responses with the aid of the plant's vasculature (Figure 3). Environmental information can be conveyed from the roots to the aerial parts of the plant, and vice-versa, via well-orchestrated cell-to-cell and/or long-distance signals. This root-to-shoot/shoot-to-root signalling is crucial in the activation of defence responses in the target tissues, which allow for plant adaptation to both biotic and abiotic stresses at the whole-plant level. These possibly interconnected signals, which include electrical, hydraulic, and chemical signals, differ in their mode of action as well as propagation speed [153,154]. As a consequence of multiple stresses, these signals tend to overlap in their presence and speed, making it difficult to associate a particular response to a specific stress stimulus [154,155].

Unlike most chemical signals, VOCs are long-distance signals that can induce systemic stress responses in distant plant organs that have little or no distinct vascular connection with parts of the stressed plant [169,170]. The role of volatiles in intra- and inter-specific communication, their chemical diversity, and mode of action is well-documented [20,171–179]. They also serve as signalling molecules that are involved in intra- and inter-plant communication, providing a fitness benefit to the "emitter" and modifying behaviour in neighbouring "receiver" plants [17]. Plants are reported to be able to distinguish between the different volatile blends [152]. As individual VOCs are not species-specific, plants formulate unique messages by adjusting the individual components of a volatile bouquet [180], which in turn induce specific responses in the receiver. For instance, the parasitic dodder plant (*Cuscuta pentagonia*) uses specific VOC blends as foraging cues to locate its host plants [152]. The presence of (*Z*)-3-hexenyl acetate in the constitutive VOC blend of wheat, a non-host to the dodder plant, was reported to be absent in tomato, suggesting its role as a repellent to the dodder [152].

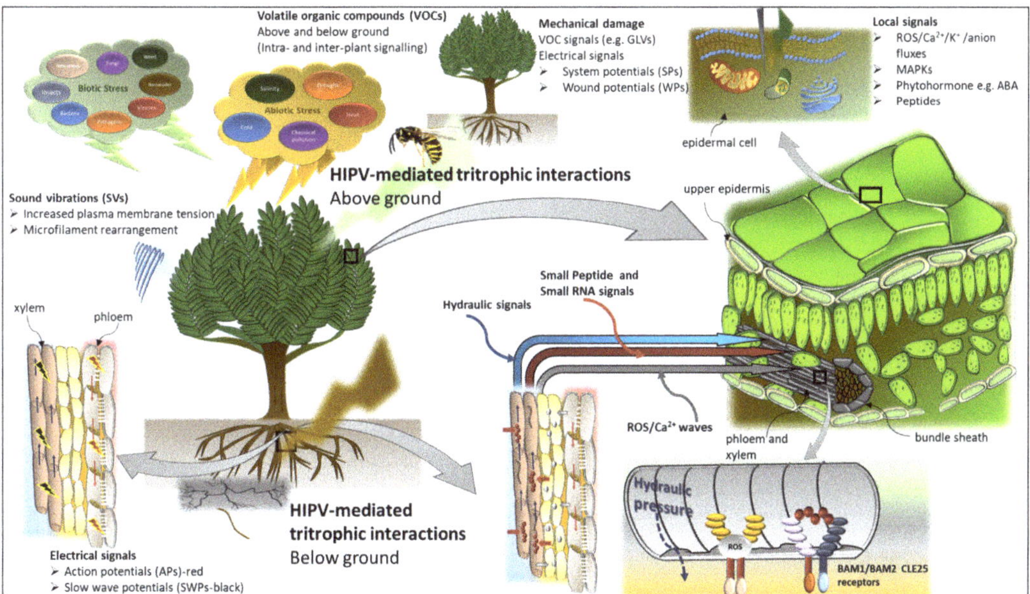

Figure 3. Overview of long distance signals in plants. Long-distance signals include electrical, hydraulic, and chemical signals. Electrical signals found in plants include: slow wave potentials (SWPs) propagated in the functional xylem; action potentials (APs) initiated in the phloem; system potentials (SPs) propagated in the apoplast following mechanical perturbations or wounding; and wound potentials (WPs) through changes in cell turgor leading to plasma membrane depolarisation. Hydraulic signals involve changes in turgor pressure, mass flow, and pressure waves. SWPs are closely linked to hydraulic signals as a result of cavitation events or changes in turgor [154–156]. Chemical signals can be classified as: (i) secondary messengers including reactive oxygen species (ROS), inositol triphosphate (IP), Ca^{2+}, K^+, and anion fluxes; (ii) signalling cascade chemicals including mitogen-activated protein kinases (MAPKs); and (iii) chemical response signals including phytohormones and volatile organic compounds (VOCs). Herbivore-induced plant volatiles (HIPVs) have been well-documented involving both above- and below-ground biocontrol of herbivores by insect predators and parasitoids, such as the yellow jacket wasp and the parasitic wasp, respectively [157,158]. Recent advanced analyses have elucidated the role of various mobile molecules including small peptides [155,159,160] and small RNAs (small interfering RNA and micro RNAs) in long-distance systemic signalling [159,161–163]. Stomatal closure is one of the initial responses to osmotic stress to prevent hydraulic failure [164] and is regulated by abscisic acid (ABA) [165,166]. During water stress, the root-to-shoot communication is mediated by a small mobile root-derived peptide, CLAVATA3/EMBRYO-SURROUNDING REGION-RELATED 25 (CLE25), which then triggers ABA accumulation in the leaves via BARELY ANY MERISTEM1 (BAM1) and BAM2 receptors in the leaf vascular bundles [159]. Sound vibrations (SVs) or acoustic signals produced by both biotic and abiotic stresses act as stimuli capable of priming plants for future stress challenges and as long-range signals that activate plant-signalling pathways [148]. Leaf vibrations caused by herbivore chewing [148,149] and the "clicking" sound produced by the collapsing water column (cavitation) in the xylem [167] have been demonstrated to trigger systemic responses in distal regions of the plant. The perturbation of the plasma membrane by SVs is characterised by a sequence of molecular episodes including cell wall modification and microfilament rearrangement in plant cells [147,149,168].

Volatile blend composition has a bearing on the specificity of volatile-mediated plant–plant communication in congeneric species such that "eavesdropping" by other species can be avoided [27,180,181]. In such cases, competing plant heterospecies may not be able to "decode" the chemical stress signals, which may otherwise have served to prepare them for a potential threat [179,182].

The allelopathic effect of VOCs, particularly terpenoids, has also been employed as part of the plant's defence against its competing neighbours by remotely suppressing the competitors' growth [183]. The cellular mode of action that underlies the allelopathic effect of terpenoids, such as eucalyptol, β-myrcene, camphor, camphene, menthol, and α- and β-pinene, occurs via cytoskeletal disruption, with microtubules being the primary targets [184–188]. Essential oils containing these terpenoids have been used for weed control as well as biological control of pests [188,189], facilitated by the specific nature of allelopathic compounds [190–192].

Under biotic stress, plants respond to insect and pathogen attacks by emitting herbivore-induced plant volatiles (HIPVs) and pathogen-induced volatiles (PIPVs), respectively [30,193–198]. Such volatiles may be exploited by neighbouring hetero- or conspecifics to either benefit or compromise the fitness of the emitter plant. The HIPVs/PIPVs can either benefit the emitter plant by attracting pollinators and natural enemies of its herbivores, or they can negatively affect the plant by acting as foraging cues to parasitic plants and as host location cues by herbivores [40,152]. Interestingly, in some HIPV-mediated plant interactions, receiver plants have been shown to adsorb VOCs from their stressed neighbours and process them into their lethal derivatives, which are detrimental to herbivore growth and survival [12]. In some instances, the receiver plant will re-release VOCs that have been adsorbed from the source plant against insect pests, in what is known as associational resistance (AR) [199].

Abiotic factors such as high light intensity [200], nutrient availability [201], salinity [11], temperature [29,72,73,104,202], wind and UV radiation [203,204], ozone exposure [205], and mechanical damage [5] have also been shown to induce VOC emissions in plants. In response to abiotic stress, VOCs offer protection against high-temperature exposure in addition to improving the thermo-tolerance of photosynthetic tissues [22,206–210]. To alleviate the negative effects of oxidative stress induced in response to environmental stressors, plant VOCs can stabilise and protect cellular membranes by quenching ROS species or by altering ROS signalling [211,212]. The role and mechanism of action of stress-induced VOCs in curbing the detrimental effects of biotic and abiotic stresses is summarised in Figure 4.

The potential role of stress-induced VOCs in priming neighbouring plants for enhanced stress tolerance has been explored [5–13,27,29,30,213], with only a few studies focusing on abiotic stress. Priming is an adaptive strategy characterised by an enhanced responsiveness of defence mechanisms to stress as a result of a previous stress challenges. This involves subtle physiological, molecular, and epigenetic alterations in the plant leading to increased stress resistance and/or tolerance [214]. GLVs comprise an important chemical group in the HIPV/PIPV plume emitted in response to mechanical damage, herbivory, or pathogen attack. The priming effect of GLVs has been characterised by an increase in defence-related gene expression and an augmented production of secondary metabolites with antixenotic or antibiotic effects on the biotic stressors [20,193,215,216]. In studies that involved priming for salinity stress, a significant increase in salt tolerance was observed in *Arabidopsis* and lima beans (*Phaseolus lunatus*) plants, independent of ABA and salinity stress-signalling pathways) [11,13]. An increase in photosynthetic rate and relative growth rate was observed in the plants previously exposed to VOCs from salt-stressed plants. In a similar study by Landi et al. [28], enhanced photosynthetic activity and reproductive success was also observed in *Ocimum basilicum*. In this particular study, the observed VOC profile differences of the emitter and the receiver plants were thought to suggest the ability of the receiver plants to propagate the volatile signal to their own neighbours [28]. Beyond these examples, Table 1 gives a brief account of a selection of studies on VOC-mediated priming in plant–plant communication under biotic and abiotic stresses.

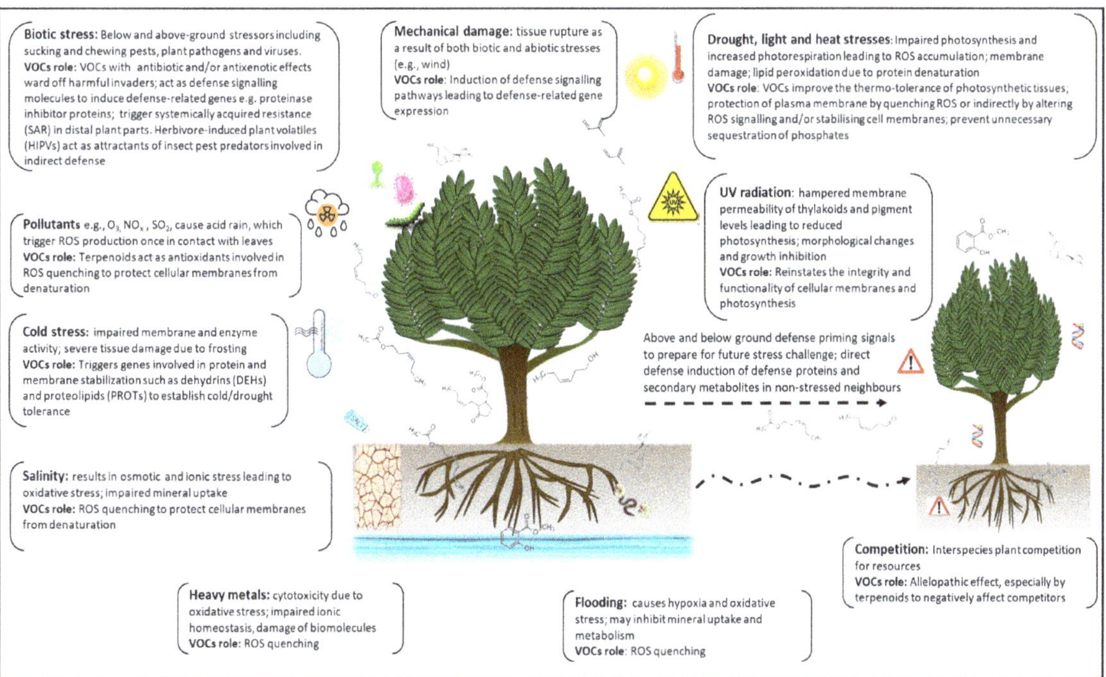

Figure 4. Role and mechanism of action of stress-induced volatile organic compounds (VOCs). In response to both biotic and abiotic stresses, VOCs induce various defence-signalling pathways and are involved in intra- and inter-plant priming against future stress challenges. The antixenotic and antibiotic effects of VOCs facilitate direct and indirect defence responses against biotic stressors. Both abiotic and biotic stress induce the production of reactive oxygen species (ROS) whose negative oxidative effects are curbed by VOCs through ROS quenching and cell membrane stabilisation. Thermo-tolerance of photosynthetic tissue is also facilitated by some VOCs under high temperature conditions. (Modified from Vickers et al. [208]).

Apart from the GLVs that are released in response to herbivory and mechanical damage, terpenoids including monoterpenes, sesquiterpenes, and hemiterpenes have also been identified and associated with defence priming in receiver plants. Typical VOCs include monoterpenes β-ocimene, 1,8-cineole, linalool, α- and β-pinene; sesquiterpenes β-caryophyllene, (*E*)-α-bergamotene and (*E*)-β-farnesene; hemiterpenes (*E*)-4,8-dimethylnona-1,3,7-triene (DMNT), and 4,8,12-trimethyltrideca-1,3,7,11-tetraene (TMTT). Phenylpropenes including eugenol and methyl eugenol were detected in salt-stressed plant emissions.

Table 1. Volatile-mediated priming in plant–plant communication under biotic and abiotic stress conditions.

Plant System	Stress Stimulus	Emitter VOCs Identified	Priming Effect on Receiver	References
Lima bean (*Phaseolus lunatus*)	Herbivory: Red spidermite (*Tetranychus urticae*)	β-Ocimene (*E*)-4,8-Dimethylnona-1,3,7-triene (DMNT) 4,8,12-Trimethyltrideca-1,3,7,11-tetraene (TMTT) Linalool (*E*)-2-Hexenal	• Increase in ethylene, JA • Upregulation of ethylene biosynthesis genes • *S*-adenosylmethionine (SAM), 1-aminocyclopropane-1-carboxylic acid oxidase	[7]
Maize (*Zea mays*)	Mechanical damage Herbivory: Beet armyworm (*Spodoptera exigua*)	GLVs Terpenoids	• Rapid JA production on post challenge with *S. exigua* regurgitant and mechanical damage	[8]
Maize (*Zea mays*)	Herbivory: Cotton leaf worm (*Spodoptera littoralis*)	GLVs Linalool (*E*)-4,8-Dimethyl-1,3-7-nonatriene (DMNT) Phenethyl acetate Indole Geranyl acetate (*E*)-β-Caryophyllene (*E*)-α-Bergamotene (*E*)-β-Farnesene β-Sesquiphellandrene	• Direct response: • reduction in caterpillar development and feeding on plant • Indirect response: enhanced aromatic and terpenoid emissions resulting in increased attraction of parasitic wasp, *Cotesia marginiventris* • Increase in JA-inducible genes	[9]
Tomato (*Solanum lycopersicum*)	Herbivory: Tobacco cutworm (*Spodoptera litura*)	(*Z*)-3-Hexen-1-ol	• Absorbed (*Z*)-3-hexen-1-ol processed to its glycoside, (*Z*)-3-hexenyl vicianoside, reducing development and longevity of the cutworms	[12]
Sagebrush (*Artemisia tridentata*) and wild tobacco (*Nicotiana. attenuata*)	Mechanical damage (clipping)	(*Z*)-3-Hexenal (*Z*)-3-Hexen-1-ol (*Z*)-3-Hexanyl acetate Methacrolein (3*R*,7*S*)-Methyl jasmonate	• Enhanced herbivore defence • Increased proteinase inhibitor (PI) • Increased polyphenol oxidase (PPO)	[5]

Table 1. *Cont.*

Plant System	Stress Stimulus	Emitter VOCs Identified	Priming Effect on Receiver	References
Sagebrush (*Artemisia tridentata*) and wild tobacco (*N. attenuata*)	Mechanical damage	β-Pinene 1,8-Cineole (*E*)-Ocimene *p*-Cymene Camphor Linalool β-Phellandrene Artemisole (*Z*)-3-Hexenal (*Z*)-3-Hexen-1-ol (*Z*)-3-Hexanyl acetate	• Enhanced defence against herbivore *Manduca sexta* • Increase in trypsin PI • Upregulation of threonine deaminase and pathogen-inducible α-dioxygenase (PIOX_NICAT) • Downregulation of tobacco NtPII10 photosystem II protein and ribulose 1,5-bisphosphate carboxylase subunit	[31]
Arabidopsis thaliana	Salinity	VOCs unknown	• Increased salt tolerance, independent of ABA and salinity stress-signalling pathways	[11]
Basil (*Ocimum basilicum*)	Salinity	Bornyl acetate Eugenol *cis*-α-Bergamotene Methyl eugenol 1,8-Cineole	• Higher reproductive success due to early flowering and senescence and higher seed yield • VOC fingerprint that overlapped, for most compounds, with that of emitters	[28]
Broad bean (*Vicia faba*)	Salinity	VOCs unknown	• Increased salt tolerance • Increased stomatal conductance, photosynthetic rate, relative growth	[13]
Tea (*Camellia sinensis*)	Cold	Geraniol Linalool Nerolidol MeSA	• Higher Fv/Fm values in receivers • Increased cold tolerance • Upregulation of dehydration-responsive elements (DRE)-binding proteins	[29]

6. Priming: The Cost of Defence

The concept of priming, an indication of hormesis, has gained interest in plant stress physiology due to its feasibility and efficiency, as well as being a cost- and resource-effective tool [8,214]. Priming has been associated with induced resistance (IR) whereby a plant is sensitised for a more robust basal defence response to subsequent attacks [217]. This immunological memory can last throughout the life cycle of a plant and is possibly heritable [214]. IR thus puts the plant in an alert state for future stress challenges and confers broad-spectrum protection against various environmental stresses [214].

To ascertain that effective defence priming has been achieved, more robust defence responses with a low fitness cost must be observed. In addition, a superior response in the presence of the actual stress challenge must be evident to imply that the plant has a memory of the stress stimulus [218]. In the context of priming using VOCs, a receiver plant must be able to associate a particular volatile plume to a specific stress in order to affect corresponding and specific defence responses. The priming of defence responses in receiver plants due to volatile signals may depend on the plant's receptiveness to different concentrations of specific volatile blends or individual compounds, the duration of exposure, and the effective distance from the source plant before the VOC bouquet is diluted in the air to inactive concentrations [42,180,219,220]. In the latter case, as the VOCs are released into the atmosphere, eddy currents dilute them to concentrations that are adequate to prime neighbouring plants for impending stresses [217,221–223]. Since the concentrations of airborne VOC signals required to induce priming are generally lower than those in which a full defence response is elicited [223], low fitness costs are likely to be incurred by the emitter plants. Although it has been presumed that priming does not alter plant metabolic pathways or initiate defence-related gene expression until actual exposure to the stress [11,218,221,224,225], a study by Balmer et al. [226] has shown that direct changes for enhanced tolerance can in fact be triggered with negligible fitness costs in the receiver plants. Priming may, therefore, offer an ecological fitness benefit to plants that are able to launch faster and stronger defence responses when the impending stress arrives [31,227].

Besides the ability to prime or induce defence-related responses in receiver plants, VOC exposure may have the potential to affect nutrient uptake and photosynthesis, thereby modulating primary and secondary metabolisms. This would subsequently lead to a change in the overall growth and development of the plants [228], as observed in VOC-exposed broad bean plants that reduced their photosynthetic rates in response to VOCs whilst they achieved increased salt resilience [13]. A reduction in photosynthesis was also shown in sweet basil plants only after subsequent exposure to salt stress, and they were observed to have early flowering, a greater seed set, and early senescence [28]. A downside of priming has also been observed in studies reporting the downregulation of specific resistance pathways, potentially leading to a reduced defence ability to unrelated stressors that the plant may not have been primed for [214].

This has been observed in transgenerational priming, in which inherited defence hormone crosstalk seemed to compromise the primed progeny's defence response against stresses the parents had not been primed for [214]. Using the study by Luna et al. [229] as an example, it was observed that JA-dependent defence responses were compromised in salicylic acid (SA)-primed *Arabidopsis* progeny, leading to significant susceptibility to necrotrophic fungi [229]. This was accounted for by the probable existence of crosstalk between the JA and SA pathways.

Continued metabolic investment in the primed state in the absence of stress has a low fitness cost to a plant, but it may compromise growth and yield [214,230]. In a study carried out by Crisp et al. [230], the growth and yield of *Arabidopsis* plants previously primed by exposure to drought or high light intensity was significantly reduced in one group of plants, whereas another was observed to recover to their pre-stressed state after rewatering. The latter group was deemed to have "forgotten" about the primed state in contrast to the former, which maintained the hardened/primed state whereby defence responses remained activated. Considering that stresses are usually transitory and often repetitive in a given environment, plants that are able to balance the formation of stress memories for protection against future stress challenges with resetting for normal growth and yield may have developed an evolutionary adaptation strategy that allows them to compete effectively against their con- and hetero-specifics [230]. Where priming is observed to be maladaptive in transgenerational inheritance, plants would be better off resetting their memory and "forgetting" previous stresses to avoid compromising their growth and survival [230].

Overall, priming offers an opportunity for the development and implementation of sustainable crop protection strategies for Smart Agriculture [231]. Whereas studies on volatile-mediated priming of defence responses against biotic stresses in plants have been well-documented [5,6,8,27,31,169,172,222,232], they exist only sparsely for abiotic stresses [11,13,233]. Therefore, the need to explore opportunities that VOC-mediated priming could offer against the various abiotic stressors is greatly emphasised.

7. Exploiting VOC-Mediated Signalling for Future Sustainable Agriculture

7.1. The Ecological Aspects: A Community Perspective

In nature, a plant may respond to a myriad of signals from its neighbours—above and below ground—and/or under different environmental conditions [234]. Despite their sessile nature, plants are able to modulate their phenotype in response to a plethora of environmental stresses and to interactions with various members of their associated community. In these communities, plants are unlikely to experience single stress factors sequentially but rather multiple stresses simultaneously [157,234] while also interacting with both con- and hetero-species [235]. The co-occurrence of multiple stresses may have some additive effects, although in some cases, only one stress takes precedence. This means that plant response due to several stress combinations cannot be accurately extrapolated based on responses to individual stress factors [236], suggesting an interaction effect of the combined stresses on plant response. This too will result in a diverse mix of VOCs being emitted by the plant. In the context of VOC-mediated interactions, the cocktail of stress-induced VOCs emitted by the individual plants contributes significantly towards the complex transcriptional responses observed in the neighbouring plants [237]. It is plausible that receiver plants acquire a stronger basal resistance resulting from the stress-specific VOC blend, rendering them more resilient and fit in the face of diverse future stressors as they remain exposed to the different volatile blends over time.

Studies on volatile-mediated plant–plant communication have indicated its ecological and functional relevance, although most work has understandably been carried out in controlled environments such as glasshouses and growth chambers due to the complexities of building a robust experimental design as well as incorporating adequate controls [238]. Evidence from studies on the volatile-mediated priming of non-stressed plants under various environmental conditions inspires the exploration of further opportunities to induce tolerance in crops against a broad spectrum of stresses. However, this should include moving on from studies carried out under controlled environmental conditions to those that better represent that which occurs in the natural environment. Normal agricultural management activities such as pruning, hedging, and crop harvesting, as well as key phenological events such as flowering and fruit set, are also potential factors that induce VOC emission in plants. The questions of how the neighbouring con- and hetero-specifics respond to the resultant volatile bouquet released by the multi-stressed plant and whether they will be primed for broader tolerance require exploration. Additional considerations include the examination of the effect of wind eddies on the airborne signals, because these may dilute the volatile bouquet to concentrations that may not effectively prime the intended receiver plants as mentioned earlier; the effective distance between the source and receiver plants in different agricultural systems; and diurnal changes in humidity and temperature, which will impact the volatility of the VOCs. These considerations will be relevant in ascertaining the feasibility of plant–plant VOC priming for enhanced stress resilience.

7.2. Technologies for Exploring Plant VOC Signalling Interactions

Attempting to address both the mechanistic and functional-level questions concerning inter-plant VOC signalling, several advances in the fields of metabolomics, volatilomics, transcriptomics, and bioinformatics offer an opportunity for a more holistic understanding of VOC-induced defence responses in plants. Significant progress in the development of VOC collection and analysis technologies that are sensitive and relatively inexpensive has been observed over the past decade [239–241]. These advances include gas chromatography-mass spectrometry (GC-MS) and static and dynamic headspace (SHS, DHS) techniques.

The headspace VOC sampling methods involve non-solvent sorptive extraction techniques, which include headspace solid phase microextraction (HS-SPME) and purge-and-trap headspace (P&T-HS) for static and dynamic extraction, respectively [239,242,243], using collection chambers (refer to Figure 4 in the review by Tholl et al. [239]). Live plants can be analysed with such techniques to provide a more representative volatile profile than with traditional methods such as solvent extraction or steam distillation. It is important, however, to consider changes that may occur in the plant's micro-environment while in the headspace collection chambers. Increases in humidity [39] and temperature, as well as reduced light intensity in the chambers [244,245] are likely to affect transpiration rates and, subsequently, VOC composition and emission rates. Additionally, given the potential for the scalping of VOCs by hydrophobic materials, their adsorption by the lining of collection chambers or sampling bags may require the use of non-reactive materials such as fluorinated ethylene propylene (FEP) or polytetrafluoroethylene (PTFE), which have been shown to be effective in plant VOC sampling [210,246,247].

Innovations that allow for VOC sampling under natural environments in which plants grow have emerged in the past decade. Trends include direct analysis in real time (DART), which is an ionisation technology for rapid non-contact analyte detection on solid or liquid surfaces [248]. Portable devices suitable for both laboratory and field sampling include a portable gas chromatograph (GC) with photoionisation detector (GC-PID) such as the FROG-4000 VOC analyser. Volatile analysis using these devices can be undertaken in water, soil, and air [249]. Micro-versions of these detectors, μGC–PID, are currently being developed and have been demonstrated to facilitate rapid vapour analysis in the field [250]. The FlavourSpec is another portable analytical device consisting of GC coupled to an ion mobility spectrometer (IMS), which can analyse headspace volatiles of solid and liquid samples without prior treatment [243,251]. Additionally, another very cost-effective field sampling technique involving minimum headspace and organism manipulation involves the use of polydimethylsiloxane (PDMS) in the form of silicone tubing. This technique involves using small pieces of the tubing for headspace sampling, combined with thermal desorption (TD)-GC-MS analysis [252]. GC-independent methods such as proton-transfer reaction-mass spectrometry (PTR-MS) have also been relevant in both field and laboratory real-time VOC monitoring [253]. A high-resolution version of the PTR-MS uses a time-of-flight mass spectrometer (PTR-TOF-MS) that is capable of rapid measurement of VOCs at ultra-low concentrations with a resolution of isobaric ions [254]. A brief highlight of the various practical and novel technologies being employed in the metabolite profiling of VOCs is given in this review (Table 2); however, more detailed and extensive reviews have been documented.

Table 2. Practical technologies used in plant volatile metabolite profiling.

Methodology	Real Time Yes/No	Portability Yes/No	References
Static headspace solid-phase microextraction device (SHS-SPME) with conventional gas chromatography-mass spectrometry (GC-MS)	No	No	[239–243]
Dynamic purge- and -trap headspace (P&T-HS) with conventional GC-MS	No	No	[239,242,243]
Polydimethylsiloxane (PDMS) silicone tubing coupled with thermal desorption GC-MS	No	No	[252]
Portable gas chromatograph (GC) with photoionisation detector (GC-PID)	No	Yes	[249]
Micro gas chromatograph (GC) with photoionisation detector (µGC–PID)	No	Yes	[250]
GC coupled to an ion mobility spectrometer (IMS), e.g., FlavourSpec	No	Yes	[243,251]
Direct analysis in real time (DART) mass spectrometry	Yes	No	[248]
Proton-transfer reaction-mass spectrometry (PTR-MS)	Yes	No	[253]
Proton-transfer-reaction time-of-flight mass spectrometry (PTR-TOF-MS)	Yes	No	[254]

With the advent of high-performance computing and statistical analysis packages such as R, there is an opportunity to mine and analyse the data collected from one or a combination of the above-mentioned methodologies to unveil relationships between VOC signals and plant physiology that would allow for a systems-based understanding of plant communication. Furthermore, data fusion approaches [255] that combine real-time VOC signalling measurements, e.g., PTR-MS, with high temporal resolution phenotyping data such as plant growth, spectral responses, and water and nutrient status are available. Modern "omics" methods such as metabolomics, volatilomics, proteomics, and transcriptomics methods all generate vast quantities of data requiring advanced analytics such as chemometrics, machine learning, and deep learning techniques for statistical analyses and prediction of plant responses to biotic and abiotic stresses. For example, Principal Component Analysis (PCA) is a multivariate statistical method to group variables and highlight their relative contributions to other variables based on variance. In contrast, machine learning algorithms based on supervised learning are available, which provide extensive and reliable datasets and can offer significant advantages over traditional chemometrics methods with regards to the prediction of VOC responses,; Random Forest is one example [256,257].

Though the use of physiological data in the evaluation of VOC signalling is relevant under both biotic and abiotic stresses, the impact of the signalling may be overlooked or assumed to be absent if the parameters selected in the study are influenced by several other environmental factors. For instance, in trying to investigate VOC signalling under abiotic stresses such as drought, it may be reasonable to consider observing changes in transpiration rate in the receiver plant. However, transpiration rate may not give a clear indication of signalling due to factors such as glasshouse lighting and stomatal oscillations [258], which depend on diurnal variations. Whether or not the receiver plant is responding to VOCs or to its inherent circadian rhythm may not be well-defined. Other physiological parameters such as ROS scavenging, stimulated root growth, and xylem hydraulic changes may be more difficult to measure directly. A more reliable option may be to investigate the subtle changes that could occur at the transcript level, which would offer an opportunity to understand the immediate responses of the receiver plants to VOC perception. Targeted gene expression using quantitative PCR (qPCR) is relevant in this regard and has been employed in several VOC-signalling studies under biotic [31,259] and abiotic stresses [11]. However, only a few putative genes can be investigated at a time, and those may not represent all the genes involved in VOC signalling. Investigating the gene expression of rate-limiting genes involved in GLV and terpenoid biosynthesis, as well as the various phytohormone signalling pathways, may be important for understanding VOC signalling in plants.

As opposed to qPCR analysis, which determines gene expression levels of a few pre-selected gene transcripts, next generation sequencing (NGS) technologies such as RNA-seq provide in-depth knowledge of a myriad of genes involved in various metabolic pathways and the complex interactions of plants during stress [260]. Gene expression profiling and transcriptome analyses are functional genomics approaches that have facilitated the identification of two broad classes of genes associated with stress responses. One category encompasses genes encoding the proteins involved with cellular osmotic homeostasis and stress protection, and the other consists of TFs and kinases involved in the regulation of stress transduction and gene expression in conjunction with various phytohormones [261]. NGS reveals transcript information, allowing for the identification of unique genes and single nucleotide variants, as well as allele-specific gene expression [261]. The platform offers an opportunity to investigate possible VOC protein receptors, TFs, and regulatory proteins involved in volatile signalling under various environmental stresses, which could prove invaluable in future studies.

To complement gene expression studies, the use of genetically modified plants possessing the ability to either emit or perceive specific components of the wild-type volatile bouquet will be useful in understanding plant–plant VOC signalling [222]. The use of mutant "deaf" receiver plants with functionally impaired VOC receptors for specific volatile substances [141] and "mute" emitter lines with silenced genes of enzymes needed for VOC synthesis [262–267] may be useful in investigating the possible receptors involved in VOC signalling. Confirmation studies involving the use of synthetic analogues of the VOCs could then be undertaken in conjunction with the "mute" emitter plant to ascertain whether the response by receiver is restored after a specific VOC has been made absent in the "mute" emitter.

Efforts directed towards understanding VOC signalling pathways and defence priming can greatly benefit from the integration of physiological, metabolome, proteome, and transcriptome analyses (Figure 5), aided by the technological advancements in those individual fields witnessed over the past decade.

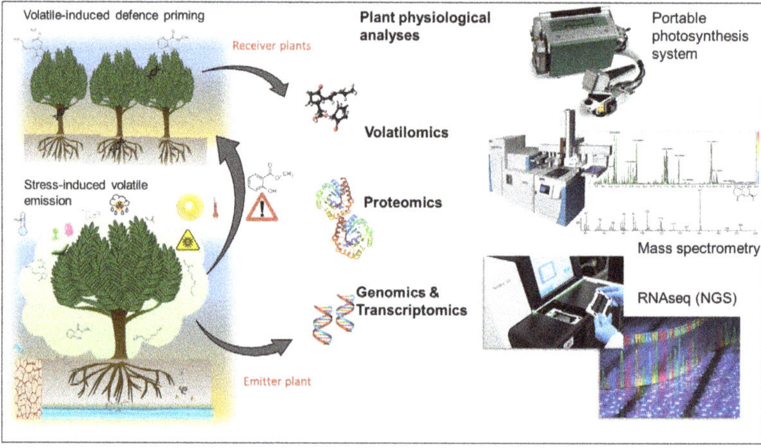

Figure 5. Integrating physiological, metabolome, proteome, and transcriptome analyses in understanding plant volatile signalling.

8. Conclusions and Future Perspectives

The increasing burden of climate change has exacerbated the effects of both biotic and abiotic stresses, thus posing a threat to global agricultural production. The employment of VOCs to enhance plant resilience to stress offers an eco-sustainable strategy for Smart Agricultural practices. The wider application of both natural and synthetic VOCs in most agricultural systems has focused on controlling insect pests by the VOCs acting as herbivore repellents or as attractants of their natural enemies, or on combining volatiles and pheromones for tailored herbivore trapping. The role of plant volatiles in intra- and inter-specific communication, the chemical diversity of VOCs, and their mode of action have been relatively well-documented. Studies have indicated the ecological and functional relevance of VOC-mediated communication, including their priming effect in non-stressed neighbouring plants for augmented defence responses following future stress challenges. Interestingly, most studies to date have mainly focused on biotic stress, with only a minority addressing several abiotic stresses. Moreover, information on the relevant signalling pathways and mechanisms involved in volatile-mediated plant–plant interactions remains nascent and has been sparsely studied in stress physiology studies.

Efforts directed towards metabolite VOC profiling of plant volatile blends emitted under various environmental stresses, the identification of the specific VOC receptors, and TFs and other regulatory proteins involved in the signalling pathways are still required. Future endeavours will benefit greatly from technological advances in the fields of plant physiology, metabolomics, transcriptomics, and bioinformatics that have emerged over the past decade. Indeed, the integration of physiological, metabolomic, and transcriptomic analyses in volatile-mediated signalling and defence priming studies will contribute to achieving a more holistic understanding of VOC-induced defence responses in plants under both biotic and abiotic stresses. As most studies have been carried out under controlled environments such as glasshouses and growth chambers, it would be worthwhile to integrate such studies with those that consider more realistic environments in which plants grow. Plants in their natural environments are part of a community of constantly interacting hetero- and conspecifics and are prone to multiple stresses occurring simultaneously. Investigations into the responses by neighbouring plants to the resultant VOC emissions will provide more conclusive results regarding the feasibility of implementing VOC-mediated priming not only against biotic stresses, but also against various abiotic stresses in agricultural systems. If plants are able to manipulate responses in their neighbours against biotic stresses via VOCs, it is plausible that they could also "warn" them against impending abiotic stresses. The potential of integrating plant–plant VOC-mediated defence priming into existing plant protection strategies against a myriad of abiotic stresses appears convincing and would potentially assist in overcoming major yield-limiting factors that contribute to over 70% of crop yield losses worldwide. As indicated in this review, however, additional studies, are still required in the face of growing pressures on ecosystems and the need to drastically improve the sustainability of agricultural production.

Author Contributions: Author contributions are as follows: Conceptualisation, investigation, visualisation, writing—original draft preparation, J.M.; supervision, writing—review and editing, D.W.J., U.B., S.R., S.D.T. and V.P; Conceptualisation, project administration, funding acquisition, writing—review and editing, V.P. All authors have read and agreed to the published version of the manuscript.

Funding: This work was supported by the Australian Research Council Training Centre for Innovative Wine Production (www.ARCwinecentre.org.au; project number IC170100008), funded by the Australian Government with additional support from Wine Australia, Waite Research Institute and industry partners. The University of Adelaide is a member of the Wine Innovation Cluster. J.M. recognises the support of a Wine Australia supplementary PhD scholarship (WA Ph2104).

Institutional Review Board Statement: Not applicable.

Informed Consent Statement: Not applicable.

Data Availability Statement: Not applicable.

Conflicts of Interest: The authors declare no conflict of interest.

References

1. Oerke, E.C. Crop losses to pests. *J. Agric. Sci.* **2006**, *144*, 31–43. [CrossRef]
2. Mantri, N.; Patade, V.; Penna, S.; Ford, R.; Pang, E. Abiotic stress responses in plants—Present and future. In *Abiotic Stress Responses in Plants: Metabolism, Productivity and Sustainability*; Springer: New York, NY, USA, 2012; pp. 1–19.
3. IPCC. *Climate change 2014: Impacts, Adaptation, and Vulnerability. Part A: Global and Sectoral Aspects. Contribution of Working Group II to the Fifth Assessment Report of the Intergovernmental Panel on Climate Change*; Field, C.B., Barros, V.R., Dokken, D.J., Mach, K.J., Mastrandrea, M.D., Bilir, T.E., Chatterjee, M., Ebi, K.L., Estrada, Y.O., Genova, R.C., et al., Eds.; Cambridge University Press: Cambridge, UK; New York, NY, USA, 2014; p. 1132.
4. Bohnert, H.J.; Gong, Q.; Li, P.; Ma, S. Unraveling abiotic stress tolerance mechanisms–getting genomics going. *Curr. Opin. Plant Biol.* **2006**, *9*, 180–188. [CrossRef] [PubMed]
5. Karban, R.; Baldwin, I.T.; Baxter, K.J.; Laue, G.; Felton, G.W. Communication between plants: Induced resistance in wild tobacco plants following clipping of neighboring sagebrush. *Oecologia* **2000**, *125*, 66–71. [CrossRef] [PubMed]
6. Tscharntke, T.; Thiessen, S.; Dolch, R.; Boland, W. Herbivory, induced resistance, and interplant signal transfer in *Alnus glutinosa*. *Biochem. Syst. Ecol.* **2001**, *29*, 1025–1047. [CrossRef]
7. Arimura, G.-I.; Ozawa, R.; Nishioka, T.; Boland, W.; Koch, T.; Kühnemann, F.; Takabayashi, J. Herbivore-induced volatiles induce the emission of ethylene in neighboring lima bean plants. *Plant J.* **2002**, *29*, 87–98. [CrossRef] [PubMed]
8. Engelberth, J.; Alborn, H.T.; Schmelz, E.A.; Tumlinson, J.H. Airborne signals prime plants against insect herbivore attack. *Proc. Natl. Acad. Sci. USA* **2004**, *101*, 1781. [CrossRef]
9. Ton, J.; D'Alessandro, M.; Jourdie, V.; Jakab, G.; Karlen, D.; Held, M.; Mauch-Mani, B.; Turlings, T.C.J. Priming by airborne signals boosts direct and indirect resistance in maize. *Plant J.* **2007**, *49*, 16–26. [CrossRef]
10. Kessler, A.; Halitschke, R. Specificity and complexity: The impact of herbivore-induced plant responses on arthropod community structure. *Curr. Opin. Plant* **2007**, *10*, 409–414. [CrossRef]
11. Lee, K.; Seo, P.J. Airborne signals from salt-stressed *Arabidopsis* plants trigger salinity tolerance in neighboring plants. *Plant Signal. Behav.* **2014**, *9*, e28392. [CrossRef]
12. Sugimoto, K.; Matsui, K.; Iijima, Y.; Akakabe, Y.; Muramoto, S.; Ozawa, R.; Uefune, M.; Sasaki, R.; Alamgir, K.M.; Akitake, S.; et al. Intake and transformation to a glycoside of (Z)-3-hexenol from infested neighbors reveals a mode of plant odor reception and defense. *Proc. Natl. Acad. Sci. USA* **2014**, *111*, 7144–7149. [CrossRef]
13. Caparrotta, S.; Boni, S.; Taiti, C.; Palm, E.; Mancuso, S.; Pandolfi, C. Induction of priming by salt stress in neighbouring plants. *Environ. Exp. Bot.* **2018**, *147*, 261–270. [CrossRef]
14. Zhang, J.; Nguyen, H.T.; Blum, A. Genetic analysis of osmotic adjustment in crop plants. *J. Exp. Bot.* **1999**, *50*, 291–302. [CrossRef]
15. Saijo, Y.; Loo, E.P.-I. Plant immunity in signal integration between biotic and abiotic stress responses. *New Phytol.* **2020**, *225*, 87–104. [CrossRef] [PubMed]
16. Lata, C.; Prasad, M. Role of DREBs in regulation of abiotic stress responses in plants. *J. Exp. Bot.* **2011**, *62*, 4731–4748. [CrossRef] [PubMed]
17. Heil, M.; Silva Bueno, J.C. Within-plant signaling by volatiles leads to induction and priming of an indirect plant defense in nature. *Proc. Natl. Acad. Sci. USA* **2007**, *104*, 5467. [CrossRef] [PubMed]
18. Shute, C.C. Chemical transmitter systems in the brain. *Mod. Trends. Neurol.* **1975**, *6*, 183–203.
19. Dudareva, N.; Pichersky, E.; Gershenzon, J. Biochemistry of plant volatiles. *Plant Physiol.* **2004**, *135*, 1893–1902. [CrossRef]
20. Dudareva, N.; Negre, F.; Nagegowda, D.A.; Orlova, I. Plant volatiles: Recent advances and future perspectives. *Crit. Rev. Plant Sci.* **2006**, *25*, 417–440. [CrossRef]
21. Niinemets, Ü.; Arneth, A.; Kuhn, U.; Monson, R.K.; Peñuelas, J.; Staudt, M. The emission factor of volatile isoprenoids: Stress, acclimation, and developmental responses. *Biogeosciences* **2010**, *7*, 2203–2223. [CrossRef]
22. Loreto, F.; Schnitzler, J.P. Abiotic stresses and induced BVOCs. *Trends Plant Sci.* **2010**, *15*, 154–166. [CrossRef]
23. Weingart, G.; Kluger, B.; Forneck, A.; Krska, R.; Schuhmacher, R. Establishment and application of a metabolomics workflow for identification and profiling of volatiles from leaves of *Vitis vinifera* by HS-SPME-GC-MS. *Phytochem. Anal.* **2012**, *23*, 345–358. [CrossRef]
24. Niinemets, Ü.; Kännaste, A.; Copolovici, L. Quantitative patterns between plant volatile emissions induced by biotic stresses and the degree of damage. *Front. Plant Sci.* **2013**, *4*, 262. [CrossRef] [PubMed]
25. Zhou, S.; Jander, G. Molecular ecology of plant volatiles in interactions with insect herbivores. *J. Exp. Bot.* **2021**, *73*, 449–462. [CrossRef]
26. Farmer, E.E.; Ryan, C.A. Interplant communication: Airborne methyl jasmonate induces synthesis of proteinase inhibitors in plant leaves. *Proc. Natl. Acad. Sci. USA* **1990**, *87*, 7713. [CrossRef]

27. Karban, R.; Huntzinger, M.; McCall, A.C. The specificity of eavesdropping on sagebrush by other plants. *Ecology* **2004**, *85*, 1846–1852. [CrossRef]

28. Landi, M.; Araniti, F.; Flamini, G.; Piccolo, E.L.; Trivellini, A.; Abenavoli, M.R.; Guidi, L. "Help is in the air": Volatiles from salt-stressed plants increase the reproductive success of receivers under salinity. *Planta* **2020**, *251*, 48. [CrossRef]

29. Zhao, M.; Wang, L.; Wang, J.; Jin, J.; Zhang, N.; Lei, L.; Gao, T.; Jing, T.; Zhang, S.; Wu, Y.; et al. Induction of priming by cold stress via inducible volatile cues in neighboring tea plants. *J. Integr. Plant Biol.* **2020**, *62*, 1461–1468. [CrossRef] [PubMed]

30. Lazazzara, V.; Avesani, S.; Robatscher, P.; Oberhuber, M.; Pertot, I.; Schuhmacher, R.; Perazzolli, M. Biogenic volatile organic compounds in the grapevine response to pathogens, beneficial microorganisms, resistance inducers and abiotic factors. *J. Exp. Bot.* **2021**, *73*, 529–554. [CrossRef]

31. Kessler, A.; Halitschke, R.; Diezel, C.; Baldwin, I. Priming of plant defense responses in nature by airborne signaling between *Artemisia tridentata* and *Nicotiana attenuata*. *Oecologia* **2006**, *148*, 280–292. [CrossRef]

32. Widhalm, J.R.; Jaini, R.; Morgan, J.A.; Dudareva, N. Rethinking how volatiles are released from plant cells. *Trends Plant Sci.* **2015**, *20*, 545–550. [CrossRef]

33. Kesselmeier, J.; Staudt, M. Biogenic volatile organic compounds (VOC): An overview on emission, physiology and ecology. *J. Atmos. Chem.* **1999**, *33*, 23–88. [CrossRef]

34. Paré, P.W.; Tumlinson, J.H. Plant volatiles as a defense against insect herbivores. *Plant Physiol.* **1999**, *121*, 325. [CrossRef] [PubMed]

35. Piechulla, B.; Degenhardt, J. The emerging importance of microbial volatile organic compounds. *Plant Cell Environ.* **2014**, *37*, 811–812. [CrossRef] [PubMed]

36. Niinemets, Ü.; Loreto, F.; Reichstein, M. Physiological and physicochemical controls on foliar volatile organic compound emissions. *Trends Plant Sci.* **2004**, *9*, 180–186. [CrossRef] [PubMed]

37. Bruce, T.J.; Wadhams, L.J.; Woodcock, C.M. Insect host location: A volatile situation. *Trends Plant Sci.* **2005**, *10*, 269–274. [CrossRef] [PubMed]

38. Knudsen, J.T.; Eriksson, R.; Gershenzon, J.; Ståhl, B. Diversity and distribution of floral scent. *Bot. Rev.* **2006**, *72*, 1. [CrossRef]

39. Niinemets, Ü.; Reichstein, M. Controls on the emission of plant volatiles through stomata: Differential sensitivity of emission rates to stomatal closure explained. *J. Geophys. Res. Atmos.* **2003**, *108*, 4211. [CrossRef]

40. Baldwin, I.T. Plant volatiles. *Curr. Biol.* **2010**, *20*, R392–R397. [CrossRef]

41. Jassbi, A.R.; Zamanizadehnajari, S.; Baldwin, I.T. Phytotoxic volatiles in the roots and shoots of *Artemisia tridentata* as detected by headspace solid-phase microextraction and gas chromatographic-mass spectrometry analysis. *J. Chem. Ecol.* **2010**, *36*, 1398–1407. [CrossRef]

42. Heil, M.; Karban, R. Explaining evolution of plant communication by airborne signals. *Trends Ecol. Evol.* **2010**, *25*, 137–144. [CrossRef] [PubMed]

43. Gigot, C.; Ongena, M.; Fauconnier, M.-L.; Wathelet, J.-P.; Jardin, P.D.; Thonart, P. The lipoxygenase metabolic pathway in plants: Potential for industrial production of natural green leaf volatiles. *Biotechnol. Agron. Soc. Environ.* **2010**, *14*, 451–460.

44. Nakashima, A.; von Reuss, S.H.; Tasaka, H.; Nomura, M.; Mochizuki, S.; Iijima, Y.; Aoki, K.; Shibata, D.; Boland, W.; Takabayashi, J.; et al. Traumatin- and dinortraumatin-containing galactolipids in *Arabidopsis*: Their formation in tissue-disrupted leaves as counterparts of green leaf volatiles. *J. Biol. Chem.* **2013**, *288*, 26078–26088. [CrossRef] [PubMed]

45. Yan, S.; Cheng, H.; Yang, H.; Yuan, H.; Zhang, J.; Chi, D. Effects of plant volatiles on the EAG response and behavior of the grey tiger longicorn, *Xylotrechus rusticus* (L.) Coleoptera: *Cerambycidae*. *Kun Chong Xue Bao* **2006**, *49*, 759–767.

46. Wu, W.; Tran, W.; Taatjes, C.A.; Alonso-Gutierrez, J.; Lee, T.S.; Gladden, J.M. Rapid discovery and functional characterization of terpene synthases from four *Endophytic xylariaceae*. *PLoS ONE* **2016**, *11*, e0146983. [CrossRef] [PubMed]

47. Boatright, J.; Negre, F.; Chen, X.; Kish, C.M.; Wood, B.; Peel, G.; Orlova, I.; Gang, D.; Rhodes, D.; Dudareva, N. Understanding in vivo benzenoid metabolism in petunia petal tissue. *Plant Physiol.* **2004**, *135*, 1993–2011. [CrossRef]

48. Rudus, I.; Sasiak, M.; Kępczyński, J. Regulation of ethylene biosynthesis at the level of 1-aminocyclopropane-1-carboxylate oxidase (ACO) gene. *Acta Physiol. Plant* **2013**, *35*, 295–307. [CrossRef]

49. Dorokhov, Y.L.; Komarova, T.V.; Petrunia, I.V.; Frolova, O.Y.; Pozdyshev, D.V.; Gleba, Y.Y. Airborne signals from a wounded leaf facilitate viral spreading and induce antibacterial resistance in neighboring plants. *PLoS Pathog.* **2012**, *8*, e1002640. [CrossRef]

50. Komarova, T.; Sheshukova, E.; Dorokhov, Y. Cell wall methanol as a signal in plant immunity. *Front. Plant Sci.* **2014**, *5*, 101. [CrossRef]

51. Pichersky, E.; Noel, J.P.; Dudareva, N. Biosynthesis of plant volatiles: Nature's diversity and ingenuity. *Science* **2006**, *311*, 808–811. [CrossRef]

52. Jerkovic, I.; Mastelic, J. Composition of free and glycosidically bound volatiles of *Mentha aquatica* L. *Croat. Chem. Acta* **2001**, *74*, 431–439.

53. Sirikantaramas, S.; Yamazaki, M.; Saito, K. Mechanisms of resistance to self-produced toxic secondary metabolites in plants. *Phytochem. Rev.* **2007**, *7*, 467. [CrossRef]

54. Maffei, M.E. Sites of synthesis, biochemistry and functional role of plant volatiles. *S. Afr. J. Bot.* **2010**, *76*, 612–631. [CrossRef]

55. Mithöfer, A.; Wanner, G.; Boland, W. Effects of feeding *Spodoptera littoralis* on lima bean leaves: Continuous mechanical wounding resembling insect feeding is sufficient to elicit herbivory-related volatile emission. *Plant Physiol.* **2005**, *137*, 1160–1168. [CrossRef] [PubMed]

56. Bonaventure, G.; VanDoorn, A.; Baldwin, I.T. Herbivore-associated elicitors: FAC signaling and metabolism. *Trends Plant Sci.* **2011**, *16*, 294–299. [CrossRef] [PubMed]

57. Louis, J.; Luthe, D.S.; Felton, G.W. Salivary signals of European corn borer induce indirect defenses in tomato. *Plant Signal. Behav.* **2013**, *8*, e27318. [CrossRef] [PubMed]

58. Turlings, T.C.J.; Erb, M. tritrophic interactions mediated by herbivore-induced plant volatiles: Mechanisms, ecological relevance, and application potential. *Annu. Rev. Entomol.* **2018**, *63*, 433–452. [CrossRef] [PubMed]

59. Arce, C.M.; Besomi, G.; Glauser, G.; Turlings, T.C.J. Caterpillar-induced volatile emissions in cotton: The relative importance of damage and insect-derived factors. *Front. Plant Sci.* **2021**, *12*, 709858.

60. Mattiacci, L.; Dicke, M.; Posthumus, M.A. beta-Glucosidase: An elicitor of herbivore-induced plant odour that attracts host-searching parasitic wasps. *Proc. Natl. Acad. Sci. USA* **1995**, *92*, 2036–2040. [CrossRef]

61. Hopke, J.; Donath, J.; Blechert, S.; Boland, W. Herbivore-induced volatiles: The emission of acyclic homoterpenes from leaves of *Phaseolus lunatus* and *Zea mays* can be triggered by a β-glucosidase and jasmonic acid. *FEBS Lett.* **1994**, *352*, 146–150. [CrossRef]

62. Doares, S.H.; Syrovets, T.; Weiler, E.W.; Ryan, C.A. Oligogalacturonides and chitosan activate plant defensive genes through the octadecanoid pathway. *Proc. Natl. Acad. Sci. USA* **1995**, *92*, 4095–4098. [CrossRef]

63. Bergey, D.R.; Orozco-Cardenas, M.; de Moura, D.S.; Ryan, C.A. A wound- and systemin-inducible polygalacturonase in tomato leaves. *Proc. Natl. Acad. Sci. USA* **1999**, *96*, 1756. [CrossRef]

64. McFarlane, H.E.; Watanabe, Y.; Yang, W.; Huang, Y.; Ohlrogge, J.; Samuels, A.L. Golgi- and trans-Golgi network-mediated vesicle trafficking is required for wax secretion from epidermal cells. *Plant Physiol.* **2014**, *164*, 1250–1260. [CrossRef] [PubMed]

65. Samuels, L.; McFarlane, H.E. Plant cell wall secretion and lipid traffic at membrane contact sites of the cell cortex. *Protoplasma* **2012**, *249* (Suppl. 1), S19–S23. [CrossRef] [PubMed]

66. McFarlane, H.E.; Shin, J.J.; Bird, D.A.; Samuels, A.L. *Arabidopsis* ABCG transporters, which are required for export of diverse cuticular lipids, dimerize in different combinations. *Plant Cell* **2010**, *22*, 3066–3075. [CrossRef] [PubMed]

67. Adebesin, F.; Widhalm, J.R.; Boachon, B.; Lefèvre, F.; Pierman, B.; Lynch, J.H.; Alam, I.; Junqueira, B.; Benke, R.; Ray, S.; et al. Emission of volatile organic compounds from petunia flowers is facilitated by an ABC transporter. *Science* **2017**, *356*, 1386–1388. [CrossRef]

68. Debono, A.; Yeats, T.H.; Rose, J.K.; Bird, D.; Jetter, R.; Kunst, L.; Samuels, L. *Arabidopsis* LTPG is a glycosylphosphatidylinositol-anchored lipid transfer protein required for export of lipids to the plant surface. *Plant Cell* **2009**, *21*, 1230–1238. [CrossRef]

69. Choi, Y.E.; Lim, S.; Kim, H.J.; Han, J.Y.; Lee, M.H.; Yang, Y.; Kim, J.A.; Kim, Y.S. Tobacco NtLTP1, a glandular-specific lipid transfer protein, is required for lipid secretion from glandular trichomes. *Plant J.* **2012**, *70*, 480–491. [CrossRef]

70. Nagashima, A.; Higaki, T.; Koeduka, T.; Ishigami, K.; Hosokawa, S.; Watanabe, H.; Matsui, K.; Hasezawa, S.; Touhara, K. Transcriptional regulators involved in responses to volatile organic compounds in plants. *J. Biol. Chem.* **2019**, *294*, 2256–2266. [CrossRef]

71. Schurmann, W. Emission of biosynthesized monoterpenes from needles of Norway spruce. *Naturwissenschaften* **1993**, *80*, 276–278. [CrossRef]

72. Llusià, J.; Peñuelas, J. Changes in terpene content and emission in potted Mediterranean woody plants under severe drought. *Can. J. Bot.* **1998**, *76*, 1366–1373.

73. Bourtsoukidis, E.; Schneider, H.; Radacki, D.; Schütz, S.; Hakola, H.; Hellén, H.; Noe, S.; Mölder, I.; Ammer, C.; Bonn, B. Impact of flooding and drought conditions on the emission of volatile organic compounds of *Quercus robur* and *Prunus serotina*. *Trees* **2013**, *28*, 193–204. [CrossRef]

74. Niinemets, U.L.; Reichstein, M.; Staudt, M.; Seufert, G.N.; Tenhunen, J.D. Stomatal Constraints May Affect Emission of Oxygenated Monoterpenoids from the Foliage of Pinus pinea. *Plant Physiol.* **2002**, *130*, 1371–1385. [CrossRef] [PubMed]

75. Copolovici, L.; Niinemets, Ü. Environmental Impacts on Plant Volatile Emission. In *Deciphering Chemical Language of Plant Communication*; Blande, J.D., Glinwood, R., Eds.; Springer International Publishing: Cham, Switzerland, 2016; pp. 35–59.

76. Peñuelas, J.; Filella, I.; Seco, R.; Llusià, J. Increase in isoprene and monoterpene emissions after re-watering of droughted *Quercus ilex* seedlings. *Biol. Plant.* **2009**, *53*, 351–354. [CrossRef]

77. Velikova, V.; Várkonyi, Z.; Szabó, M.; Maslenkova, L.; Nogues, I.; Kovács, L.; Peeva, V.; Busheva, M.; Garab, G.; Sharkey, T.D.; et al. Increased thermostability of thylakoid membranes in isoprene-emitting leaves probed with three biophysical techniques. *Plant Physiol.* **2011**, *157*, 905. [CrossRef]

78. Brilli, F.; Hörtnagl, L.; Bamberger, I.; Schnitzhofer, R.; Ruuskanen, T.M.; Hansel, A.; Loreto, F.; Wohlfahrt, G. Qualitative and quantitative characterization of volatile organic compound emissions from cut grass. *Environ. Sci. Technol.* **2012**, *46*, 3859–3865. [CrossRef]

79. Beauchamp, J.; Wisthaler, A.; Hansel, A.; Kleist, E.; Miebach, M.; Niinemets, Ü.; Schurr, U.L.I.; Wildt, J. Ozone induced emissions of biogenic VOC from tobacco: Relationships between ozone uptake and emission of LOX products. *Plant Cell Environ.* **2005**, *28*, 1334–1343. [CrossRef]

80. Grote, R.; Samson, R.; Alonso, R.; Amorim, J.H.; Cariñanos, P.; Churkina, G.; Fares, S.; Thiec, D.L.; Niinemets, Ü.; Mikkelsen, T.N.; et al. Functional traits of urban trees: Air pollution mitigation potential. *Front. Ecol. Environ.* **2016**, *14*, 543–550. [CrossRef]

81. Karl, T.; Guenther, A.; Turnipseed, A.; Patton, E.G.; Jardine, K. Chemical sensing of plant stress at the ecosystem scale. *Biogeosciences* **2008**, *5*, 1287–1294. [CrossRef]

82. Staudt, M.; Lhoutellier, L. Volatile organic compound emission from holm oak infested by gypsy moth larvae: Evidence for distinct responses in damaged and undamaged leaves. *Tree Physiol.* **2007**, *27*, 1433–1440. [CrossRef]
83. Possell, M.; Loreto, F. The Role of volatile organic compounds in plant resistance to abiotic stresses: Responses and mechanisms. In *Biology, Controls and Models of Tree Volatile Organic Compound Emissions*; Niinemets, Ü., Monson, R.K., Eds.; Springer: Dordrecht, The Netherlands, 2013; pp. 209–235.
84. Niinemets, Ü.; Sun, Z. How light, temperature, and measurement and growth [CO_2] interactively control isoprene emission in hybrid aspen. *J. Exp. Bot.* **2015**, *66*, 841–851. [CrossRef]
85. Li, Z.; Sharkey, T.D. Molecular and pathway controls on biogenic volatile organic compound emissions. In *Biology, Controls and Models of Tree Volatile Organic Compound Emissions*; Niinemets, Ü., Monson, R.K., Eds.; Springer: Dordrecht, The Netherlands, 2013; pp. 119–151.
86. Chaves, M.M.; Flexas, J.; Pinheiro, C. Photosynthesis under drought and salt stress: Regulation mechanisms from whole plant to cell. *Ann. Bot.* **2009**, *103*, 551–560. [CrossRef] [PubMed]
87. Pegoraro, E.; Rey, A.; Bobich, E.G.; Barron-Gafford, G.; Grieve, K.A.; Malhi, Y.; Murthy, R. Effect of elevated CO_2 concentration and vapour pressure deficit on isoprene emission from leaves of *Populus deltoides* during drought. *Funct. Plant Biol.* **2004**, *31*, 1137–1147. [CrossRef] [PubMed]
88. Velikova, V.; Loreto, F. On the relationship between isoprene emission and thermotolerance in *Phragmites australis* leaves exposed to high temperatures and during the recovery from a heat stress. *Plant Cell Environ.* **2005**, *28*, 318–327. [CrossRef]
89. Loreto, F.; Delfine, S. Emission of isoprene from salt-stressed *Eucalyptus globulus* leaves. *Plant Physiol.* **2000**, *123*, 1605–1610. [CrossRef] [PubMed]
90. Loreto, F.; Pinelli, P.; Manes, F.; Kollist, H. Impact of ozone on monoterpene emissions and evidence for an isoprene-like antioxidant action of monoterpenes emitted by *Quercus ilex* leaves. *Tree Physiol.* **2004**, *24*, 361–367. [CrossRef]
91. Grote, R.; Monson, R.K.; Niinemets, Ü. Leaf-level models of constitutive and stress-driven volatile organic compound emissions. In *Biology, Controls and Models of Tree Volatile Organic Compound Emissions*; Niinemets, Ü., Monson, R.K., Eds.; Springer: Dordrecht, The Netherlands, 2013; pp. 315–355.
92. Rasulov, B.; Hüve, K.; Bichele, I.; Laisk, A.; Niinemets, Ü. Temperature response of isoprene emission in vivo reflects a combined effect of substrate limitations and isoprene synthase activity: A Kinetic Analysis. *Plant Physiol.* **2010**, *154*, 1558–1570. [CrossRef]
93. Singsaas, E.L.; Sharkey, T.D. The effects of high temperature on isoprene synthesis in oak leaves. *Plant Cell Environ.* **2000**, *23*, 751–757. [CrossRef]
94. Loreto, F.; Barta, C.; Brilli, F.; Nogues, I. On the induction of volatile organic compound emissions by plants as consequence of wounding or fluctuations of light and temperature. *Plant Cell Environ.* **2006**, *29*, 1820–1828. [CrossRef]
95. Brilli, F.; Barta, C.; Fortunati, A.; Lerdau, M.; Loreto, F.; Centritto, M. Response of isoprene emission and carbon metabolism to drought in white poplar (*Populus alba*) saplings. *New Phytol.* **2007**, *175*, 244–254. [CrossRef]
96. Grote, R.; Lavoir, A.-V.; Rambal, S.; Staudt, M.; Zimmer, I.; Schnitzler, J.-P. Modelling the drought impact on monoterpene fluxes from an evergreen Mediterranean forest canopy. *Oecologia* **2009**, *160*, 213–223. [CrossRef]
97. Rosenstiel, T.N.; Fisher, A.J.; Fall, R.; Monson, R.K. Differential accumulation of dimethylallyl diphosphate in leaves and needles of isoprene- and methylbutenol-emitting and nonemitting species. *Plant Physiol.* **2002**, *129*, 1276–1284. [CrossRef] [PubMed]
98. Sharkey, T.D.; Yeh, S. Isoprene emission from plants. *Annu. Rev. Plant Physiol. Plant Mol. Biol.* **2001**, *52*, 407–436. [CrossRef] [PubMed]
99. Nemecek-Marshall, M.; MacDonald, R.C.; Franzen, J.J.; Wojciechowski, C.L.; Fall, R. Methanol emission from leaves (enzymatic detection of gas-phase methanol and relation of methanol fluxes to stomatal conductance and leaf development). *Plant Physiol.* **1995**, *108*, 1359–1368. [CrossRef] [PubMed]
100. Gabriel, R.; Schäfer, L.; Gerlach, C.; Rausch, T.; Kesselmeier, J. Factors controlling the emissions of volatile organic acids from leaves of *Quercus ilex* L. (Holm oak). *Atmos. Environ.* **1999**, *33*, 1347–1355. [CrossRef]
101. Kreuzwieser, J.; Harren, F.J.M.; Laarhoven, L.J.J.; Boamfa, I.; te Lintel-Hekkert, S.; Scheerer, U.; Hüglin, C.; Rennenberg, H. Acetaldehyde emission by the leaves of trees—Correlation with physiological and environmental parameters. *Physiol. Plant* **2001**, *113*, 41–49. [CrossRef]
102. Fall, R.; Monson, R.K. Isoprene emission rate and intercellular isoprene concentration as influenced by stomatal distribution and conductance. *Plant Physiol.* **1992**, *100*, 987. [CrossRef]
103. Loreto, F. Influence of environmental factors and air composition on the emission of alpha-pinene from *Quercus ilex* leaves. *Plant Physiol.* **1996**, *110*, 267–275. [CrossRef]
104. Sharkey, T.D.; Loreto, F. Water stress, temperature, and light effects on the capacity for isoprene emission and photosynthesis of kudzu leaves. *Oecologia* **1993**, *95*, 328–333. [CrossRef]
105. Calfapietra, C.; Fares, S.; Loreto, F. Volatile organic compounds from Italian vegetation and their interaction with ozone. *Environ. Pollut.* **2009**, *157*, 1478–1486. [CrossRef]
106. Piesik, D.; Delaney, K.J.; Wenda-Piesik, A.; Sendel, S.; Tabaka, P.; Buszewski, B. *Meligethes aeneus* pollen-feeding suppresses, and oviposition induces, *Brassica napus* volatiles: Beetle attraction/repellence to lilac aldehydes and veratrole. *Chemoecology* **2013**, *23*, 241–250. [CrossRef]
107. Harren, F.J.M.; Cristescu, S.M. Online, real-time detection of volatile emissions from plant tissue. *AoB Plants* **2013**, *5*, plt003. [CrossRef] [PubMed]

108. Jubany-Marí, T.; Prinsen, E.; Munné-Bosch, S.; Alegre, L. The timing of methyl jasmonate, hydrogen peroxide and ascorbate accumulation during water deficit and subsequent recovery in the Mediterranean shrub *Cistus albidus* L. *Environ. Exp. Bot.* **2010**, *69*, 47–55. [CrossRef]

109. Copolovici, L.; Kännaste, A.; Remmel, T.; Niinemets, Ü. Volatile organic compound emissions from *Alnus glutinosa* under interacting drought and herbivory stresses. *Environ. Exp. Bot.* **2014**, *100*, 55–63. [CrossRef] [PubMed]

110. Rasulov, B.; Hüve, K.; Laisk, A.; Niinemets, Ü. Induction of a longer term component of isoprene release in darkened aspen leaves: Origin and regulation under different environmental conditions. *Plant Physiol.* **2011**, *156*, 816–831. [CrossRef] [PubMed]

111. Yuan, X.; Calatayud, V.; Gao, F.; Fares, S.; Paoletti, E.; Tian, Y.; Feng, Z. Interaction of drought and ozone exposure on isoprene emission from extensively cultivated poplar. *Plant Cell Environ.* **2016**, *39*, 2276–2287. [CrossRef] [PubMed]

112. Tani, A.; Ohno, T.; Saito, T.; Ito, S.; Yonekura, T.; Miwa, M. Effects of ozone on isoprene emission from two major *Quercus* species native to East Asia. *J. Agric. Meteorol.* **2017**, *73*, 195–202. [CrossRef]

113. Calabrese, E.J.; Mattson, M.P. How does hormesis impact biology, toxicology, and medicine? *NPJ Aging Mech. Dis.* **2017**, *3*, 13. [CrossRef]

114. Agathokleous, E.; Kitao, M.; Calabrese, E.J. Emission of volatile organic compounds from plants shows a biphasic pattern within an hormetic context. *Environ. Pollut.* **2018**, *239*, 318–321. [CrossRef]

115. Jiang, Y.; Ye, J.; Li, S.; Niinemets, Ü. Methyl jasmonate-induced emission of biogenic volatiles is biphasic in cucumber: A high-resolution analysis of dose dependence. *J. Exp. Bot.* **2017**, *68*, 4679–4694. [CrossRef]

116. Niinemets, Ü.; Fares, S.; Harley, P.; Jardine, K.J. Bidirectional exchange of biogenic volatiles with vegetation: Emission sources, reactions, breakdown and deposition. *Plant Cell Environ.* **2014**, *37*, 1790–1809. [CrossRef]

117. Bakker, M.I. Atmospheric Deposition of Semivolatile Organic Compounds to Plants. Ph.D. Thesis, Utrecht University Repository, Utrecht, The Netherlands, 2000.

118. Riederer, M.; Daiß, A.; Gilbert, N.; Köhle, H. Semi-volatile organic compounds at the leaf/atmosphere interface: Numerical simulation of dispersal and foliar uptake. *J. Exp. Bot.* **2002**, *53*, 1815–1823. [CrossRef] [PubMed]

119. Keyte, I.; Wild, E.; Dent, J.; Jones, K.C. Investigating the foliar uptake and within-leaf migration of phenanthrene by moss (*Hypnum Cupressiforme*) using two-photon excitation microscopy with autofluorescence. *Environ. Sci. Technol.* **2009**, *43*, 5755–5761. [CrossRef] [PubMed]

120. Burkhardt, J.; Pariyar, S. Particulate pollutants are capable to 'degrade' epicuticular waxes and to decrease the drought tolerance of Scots pine (*Pinus sylvestris* L.). *Environ. Pollut.* **2014**, *184*, 659–667. [CrossRef]

121. Noe, S.M.; Copolovici, L.; Niinemets, Ü.; Vaino, E. Foliar limonene uptake scales positively with leaf lipid content: "Non-emitting" species absorb and release monoterpenes. *Plant Biol.* **2007**, *9*, e79–e86. [CrossRef] [PubMed]

122. Tör, M.; Lotze, M.T.; Holton, N. Receptor-mediated signalling in plants: Molecular patterns and programmes. *J. Exp. Bot.* **2009**, *60*, 3645–3654. [CrossRef]

123. He, Y.; Zhou, J.; Shan, L.; Meng, X. Plant cell surface receptor-mediated signaling—A common theme amid diversity. *J. Cell Sci.* **2018**, *131*, jcs209353. [CrossRef]

124. Choi, H.W.; Klessig, D.F. DAMPs, MAMPs, and NAMPs in plant innate immunity. *BMC Plant Biol.* **2016**, *16*, 232. [CrossRef]

125. Quintana-Rodriguez, E.; Duran-Flores, D.; Heil, M.; Camacho-Coronel, X. Damage-associated molecular patterns (DAMPs) as future plant vaccines that protect crops from pests. *Sci. Hortic.* **2018**, *237*, 207–220. [CrossRef]

126. Gust, A.A.; Pruitt, R.; Nürnberger, T. Sensing danger: Key to activating plant immunity. *Trends Plant Sci.* **2017**, *22*, 779–791. [CrossRef]

127. Hann, C.T.; Bequette, C.J.; Dombrowski, J.E.; Stratmann, J.W. Methanol and ethanol modulate responses to danger- and microbe-associated molecular patterns. *Front. Plant Sci.* **2014**, *5*, 550. [CrossRef]

128. Harvey, C.M.; Sharkey, T.D. Exogenous isoprene modulates gene expression in unstressed *Arabidopsis thaliana* plants. *Plant Cell Environ.* **2016**, *39*, 1251–1263. [CrossRef] [PubMed]

129. Zuo, Z.; Weraduwage, S.M.; Lantz, A.T.; Sanchez, L.M.; Weise, S.E.; Wang, J.; Childs, K.L.; Sharkey, T.D. Isoprene acts as a signaling molecule in gene networks important for stress responses and plant growth. *Plant Physiol.* **2019**, *180*, 124–152. [CrossRef] [PubMed]

130. Tuteja, N.; Sopory, S.K. Chemical signaling under abiotic stress environment in plants. *Plant Signal. Behav.* **2008**, *3*, 525–536. [CrossRef] [PubMed]

131. Connor, E.C.; Rott, A.S.; Zeder, M.; Jüttner, F.; Dorn, S. ^{13}C-labelling patterns of green leaf volatiles indicating different dynamics of precursors in *Brassica* leaves. *Phytochemistry* **2008**, *69*, 1304–1312. [CrossRef] [PubMed]

132. Farag, M.A.; Paré, P.W. C_6-Green leaf volatiles trigger local and systemic VOC emissions in tomato. *Phytochemistry* **2002**, *61*, 545–554. [CrossRef]

133. Burg, S.P.; Burg, E.A. Molecular requirements for the biological activity of ethylene. *Plant Physiol.* **1967**, *42*, 144–152. [CrossRef]

134. Bleecker, A.B.; Schaller, G.E. The Mechanism of ethylene perception. *Plant Physiol.* **1996**, *111*, 653–660. [CrossRef]

135. Ecker, J.R. The ethylene signal transduction pathway in plants. *Science* **1995**, *268*, 667–675. [CrossRef]

136. Chen, Q.G.; Bleecker, A.B. Analysis of ethylene signal-transduction kinetics associated with seedling-growth response and chitinase induction in wild-type and mutant *Arabidopsis*. *Plant Physiol.* **1995**, *108*, 597–607. [CrossRef]

137. Chang, C.; Kwok, S.F.; Bleecker, A.B.; Meyerowitz, E.M. *Arabidopsis* ethylene-response gene ETR1: Similarity of product to two-component regulators. *Science* **1993**, *262*, 539–544. [CrossRef]

138. Guzmán, P.; Ecker, J.R. Exploiting the triple response of *Arabidopsis* to identify ethylene-related mutants. *Plant Cell* **1990**, *2*, 513–523. [PubMed]
139. Ju, C.; Chang, C. Mechanistic insights in ethylene perception and signal transduction. *Plant Physiol.* **2015**, *169*, 85–95. [CrossRef] [PubMed]
140. Wang, F.; Cui, X.; Sun, Y.; Dong, C.H. Ethylene signaling and regulation in plant growth and stress responses. *Plant Cell Rep.* **2013**, *32*, 1099–1109. [CrossRef] [PubMed]
141. Hua, J.; Chang, C.; Sun, Q.; Meyerowitz, E.M. Ethylene insensitivity conferred by *Arabidopsis* ERS gene. *Science* **1995**, *269*, 1712. [CrossRef]
142. Sakai, H.; Hua, J.; Chen, Q.G.; Chang, C.; Medrano, L.J.; Bleecker, A.B.; Meyerowitz, E.M. ETR2 is an ETR1-like gene involved in ethylene signaling in *Arabidopsis*. *Proc. Natl. Acad. Sci. USA* **1998**, *95*, 5812–5817. [CrossRef]
143. Hall, B.P.; Shakeel, S.N.; Schaller, G.E. Ethylene Receptors: Ethylene perception and signal transduction. *J. Plant Growth Regul.* **2007**, *26*, 118–130. [CrossRef]
144. Chory, J. Light signal transduction: An infinite spectrum of possibilities. *Plant J.* **2010**, *61*, 982–991. [CrossRef]
145. Chamovitz, D. Rooted in sensation: Sight. *New Sci.* **2012**, *215*, 35. [CrossRef]
146. Chamovitz, D. Rooted in sensation: Hearing. *New Sci.* **2012**, *215*, 37. [CrossRef]
147. Gagliano, M. Green symphonies: A call for studies on acoustic communication in plants. *Behav. Ecol.* **2013**, *24*, 789–796. [CrossRef]
148. Appel, H.M.; Cocroft, R.B. Plants respond to leaf vibrations caused by insect herbivore chewing. *Oecologia* **2014**, *175*, 1257–1266. [CrossRef] [PubMed]
149. Mishra, R.C.; Ghosh, R.; Bae, H. Plant acoustics: In the search of a sound mechanism for sound signaling in plants. *J. Exp. Bot.* **2016**, *67*, 4483–4494. [CrossRef] [PubMed]
150. Volkov, A.G.; Adesina, T.; Jovanov, E. Closing of venus flytrap by electrical stimulation of motor cells. *Plant Signal. Behav.* **2007**, *2*, 139–145. [CrossRef] [PubMed]
151. Malnic, B.; Hirono, J.; Sato, T.; Buck, L.B. Combinatorial receptor codes for odours. *Cell* **1999**, *96*, 713–723. [CrossRef]
152. Mescher, M.C.; Runyon, J.B.; De Moraes, C.M. Plant host finding by parasitic plants: A new perspective on plant to plant communication. *Plant Signal. Behav.* **2006**, *1*, 284–286. [CrossRef]
153. Tiwari, V.; Patel, M.K.; Chaturvedi, A.K.; Mishra, A.; Jha, B. Functional characterisation of the Tau Class Glutathione-S-Transferases Gene (SbGSTU) Promoter of *Salicornia brachiata* under salinity and osmotic stress. *PLoS ONE* **2016**, *11*, e0148494. [CrossRef]
154. Huber, A.E.; Bauerle, T.L. Long-distance plant signaling pathways in response to multiple stressors: The gap in knowledge. *J. Exp. Bot.* **2016**, *67*, 2063–2079. [CrossRef]
155. Johns, S.; Hagihara, T.; Toyota, M.; Gilroy, S. The fast and the furious: Rapid long-range signaling in plants. *Plant Physiol.* **2021**, *185*, 694–706. [CrossRef]
156. Malone, M. Wound-induced hydraulic signals and stimulus transmission in *Mimosa pudica* L. *New Phytol.* **1994**, *128*, 49–56. [CrossRef]
157. Dicke, M.; van Loon, J.J.; Soler, R. Chemical complexity of volatiles from plants induced by multiple attack. *Nat. Chem. Biol.* **2009**, *5*, 317–324. [CrossRef]
158. Dicke, M.; Baldwin, I.T. The evolutionary context for herbivore-induced plant volatiles: Beyond the 'cry for help'. *Trends Plant Sci.* **2010**, *15*, 167–175. [CrossRef] [PubMed]
159. Takahashi, F.; Suzuki, T.; Osakabe, Y.; Betsuyaku, S.; Kondo, Y.; Dohmae, N.; Fukuda, H.; Yamaguchi-Shinozaki, K.; Shinozaki, K. A small peptide modulates stomatal control via abscisic acid in long-distance signalling. *Nature* **2018**, *556*, 235–238. [CrossRef] [PubMed]
160. Takahashi, F.; Shinozaki, K. Long-distance signaling in plant stress response. *Curr. Opin. Plant* **2019**, *47*, 106–111. [CrossRef] [PubMed]
161. Tang, G.; Reinhart, B.J.; Bartel, D.P.; Zamore, P.D. A biochemical framework for RNA silencing in plants. *Genes Dev* **2003**, *17*, 49–63. [CrossRef]
162. Yoo, B.-C.; Kragler, F.; Varkonyi-Gasic, E.; Haywood, V.; Archer-Evans, S.; Lee, Y.M.; Lough, T.J.; Lucas, W.J. A systemic small RNA signaling system in plants. *Plant Cell* **2004**, *16*, 1979–2000. [CrossRef]
163. Kondhare, K.R.; Patil, N.S.; Banerjee, A.K. A historical overview of long-distance signalling in plants. *J. Exp. Bot.* **2021**, *72*, 4218–4236. [CrossRef]
164. Charrier, G.; Delzon, S.; Domec, J.-C.; Zhang, L.; Delmas, C.E.L.; Merlin, I.; Corso, D.; King, A.; Ojeda, H.; Ollat, N.; et al. Drought will not leave your glass empty: Low risk of hydraulic failure revealed by long-term drought observations in world's top wine regions. *Sci. Adv.* **2018**, *4*, eaao6969. [CrossRef]
165. Schachtman, D.P.; Goodger, J.Q. Chemical root to shoot signaling under drought. *Trends Plant Sci.* **2008**, *13*, 281–287. [CrossRef]
166. Cramer, G.R.; Urano, K.; Delrot, S.; Pezzotti, M.; Shinozaki, K. Effects of abiotic stress on plants: A systems biology perspective. *BMC Plant Biol.* **2011**, *11*, 163. [CrossRef]
167. Zweifel, R.; Zeugin, F. Ultrasonic acoustic emissions in drought-stressed trees—More than signals from cavitation? *New Phytol.* **2008**, *179*, 1070–1079. [CrossRef]
168. Gagliano, M.; Renton, M.; Duvdevani, N.; Timmins, M.; Mancuso, S. Acoustic and magnetic communication in plants: Is it possible? *Plant Signal. Behav.* **2012**, *7*, 1346–1348. [CrossRef] [PubMed]

169. Frost, C.J.; Appel, H.M.; Carlson, J.E.; De Moraes, C.M.; Mescher, M.C.; Schultz, J.C. Within-plant signalling via volatiles overcomes vascular constraints on systemic signalling and primes responses against herbivores. *Ecol. Lett.* **2007**, *10*, 490–498. [CrossRef] [PubMed]

170. Howe, G.A.; Jander, G. Plant immunity to insect herbivores. *Annu. Rev. Plant Biol.* **2008**, *59*, 41–66. [CrossRef] [PubMed]

171. Kessler, A.; Baldwin, I.T. Defensive function of herbivore-induced plant volatile emissions in nature. *Science* **2001**, *291*, 2141. [CrossRef]

172. Baldwin, I.T.; Kessler, A.; Halitschke, R. Volatile signaling in plant–plant herbivore interactions: What is real? *Curr. Opin. Plant* **2002**, *5*, 351–354. [CrossRef]

173. Howe, G.; Schaller, A. Direct Defenses in Plants and Their Induction by Wounding and Insect Herbivores. In *Induced Plant Resistance to Herbivory*; Schaller, A., Ed.; Springer: Dordrecht, The Netherlands, 2008; pp. 7–29.

174. Zimmermann, M.R.; Maischak, H.; Mithöfer, A.; Boland, W.; Felle, H.H. System potentials, a novel electrical long-distance apoplastic signal in plants, induced by wounding. *Plant Physiol.* **2009**, *149*, 1593–1600. [CrossRef]

175. Wortemann, R.; Herbette, S.; Barigah, T.S.; Fumanal, B.; Alia, R.; Ducousso, A.; Gomory, D.; Roeckel-Drevet, P.; Cochard, H. Genotypic variability and phenotypic plasticity of cavitation resistance in *Fagus sylvatica* L. across Europe. *Tree Physiol.* **2011**, *31*, 1175–1182. [CrossRef]

176. Arimura, G.-I.; Ozawa, R.; Maffei, M.E. Recent advances in plant early signaling in response to herbivory. *Int. J. Mol. Sci.* **2011**, *12*, 3723–3739. [CrossRef]

177. Mithöfer, A.; Boland, W. Plant defense against herbivores: Chemical aspects. *Annu. Rev. Plant Biol.* **2012**, *63*, 431–450. [CrossRef]

178. Ninkovic, V.; Rensing, M.; Dahlin, I.; Markovic, D. Who is my neighbor? Volatile cues in plant interactions. *Plant Signal. Behav.* **2019**, *14*, 1634993. [CrossRef]

179. Ninkovic, V.; Markovic, D.; Rensing, M. Plant volatiles as cues and signals in plant communication. *Plant Cell Environ.* **2020**, *44*, 1030–1043. [CrossRef] [PubMed]

180. Ueda, H.; Kikuta, Y.; Matsuda, K. Plant communication: Mediated by individual or blended VOCs? *Plant Signal. Behav.* **2012**, *7*, 222–226. [CrossRef] [PubMed]

181. Yoneya, K.; Takabayashi, J. Plant—Plant communication mediated by airborne signals: Ecological and plant physiological perspectives. *Plant Biotechnol.* **2014**, *31*, 409–416.

182. Karban, R.; Maron, J. The fitness consequences of interspecific eavesdropping between plants. *Ecology* **2002**, *83*, 1209–1213. [CrossRef]

183. Lambers, H.; Chapin, F.S.; Pons, T.L. Ecological Biochemistry: Allelopathy and defense against herbivores. In *Plant Physiological Ecology*; Lambers, H., Chapin, F.S., Pons, T.L., Eds.; Springer: New York, NY, USA, 2008; pp. 445–477.

184. Anthony, R.G.; Waldin, T.R.; Ray, J.A.; Bright, S.W.J.; Hussey, P.J. Herbicide resistance caused by spontaneous mutation of the cytoskeletal protein tubulin. *Nature* **1998**, *393*, 260–263. [CrossRef]

185. Nishida, N.; Tamotsu, S.; Nagata, N.; Saito, C.; Sakai, A. Allelopathic effects of volatile monoterpenoids produced by *Salvia leucophylla*: Inhibition of Cell Proliferation and DNA Synthesis in the root apical meristem of *Brassica campestris* seedlings. *J. Chem. Ecol.* **2005**, *31*, 1187–1203. [CrossRef]

186. Chaimovitsh, D.; Abu-Abied, M.; Belausov, E.; Rubin, B.; Dudai, N.; Sadot, E. Microtubules are an intracellular target of the plant terpene citral. *Plant J.* **2010**, *61*, 399–408. [CrossRef]

187. Kriegs, B.; Jansen, M.; Hahn, K.; Peisker, H.; Samajová, O.; Beck, M.; Braun, S.; Ulbrich, A.; Baluška, F.; Schulz, M. Cyclic monoterpene mediated modulations of *Arabidopsis thaliana* phenotype: Effects on the cytoskeleton and on the expression of selected genes. *Plant Signal. Behav.* **2010**, *5*, 832–838. [CrossRef]

188. Sarheed, M.M.; Rajabi, F.; Kunert, M.; Boland, W.; Wetters, S.; Miadowitz, K.; Kaźmierczak, A.; Sahi, V.P.; Nick, P. Cellular base of mint allelopathy: Menthone affects plant microtubules. *Front. Plant Sci.* **2020**, *11*, 1320. [CrossRef]

189. Mofikoya, A.O.; Bui, T.N.T.; Kivimäenpää, M.; Holopainen, J.K.; Himanen, S.J.; Blande, J.D. Foliar behaviour of biogenic semi-volatiles: Potential applications in sustainable pest management. *Arthropod-Plant Interact.* **2019**, *13*, 193–212. [CrossRef]

190. Bajalan, I.; Zand, M.; Rezaee, S. The study on allelopathic effects of *Mentha longifolia* on seed germination of velvet flower and two cultivars of wheat. *Int. Res. J. Appl. Basic Sci.* **2013**, *4*, 2539–2543.

191. Cheng, F.; Cheng, Z. Research Progress on the use of Plant Allelopathy in Agriculture and the Physiological and Ecological Mechanisms of Allelopathy. *Front. Plant Sci.* **2015**, *6*, 1020. [CrossRef] [PubMed]

192. Edyta, S.; Peter, R.; Alina, S.-S.; Beata, B.-K.; Katarzyna, M. Allelopathic effect of aqueous extracts from the leaves of peppermint (*Mentha piperita* L.) on selected physiological processes of common sunflower (*Helianthus annuus* L.). *Not. Bot. Horti. Agrobot. Cluj Napoca* **2015**, *43*, 335–342.

193. Martini, X.; Pelz-Stelinski, K.; Stelinski, L. Plant pathogen-induced volatiles attract parasitoids to increase parasitism of an insect vector. *Front. Ecol. Evol.* **2014**, *2*, 8. [CrossRef]

194. Chitarrini, G.; Soini, E.; Riccadonna, S.; Franceschi, P.; Zulini, L.; Masuero, D.; Vecchione, A.; Stefanini, M.; Di Gaspero, G.; Mattivi, F.; et al. Identification of biomarkers for defense response to *Plasmopara viticola* in a resistant grape variety. *Front. Plant Sci.* **2017**, *8*, 1524. [CrossRef]

195. Camacho-Coronel, X.; Molina-Torres, J.; Heil, M. Sequestration of exogenous volatiles by plant cuticular waxes as a mechanism of passive associational resistance: A proof of concept. *Front. Plant Sci.* **2020**, *11*, 121. [CrossRef]

196. Ciubotaru, R.M.; Franceschi, P.; Zulini, L.; Stefanini, M.; Škrab, D.; Rossarolla, M.D.; Robatscher, P.; Oberhuber, M.; Vrhovsek, U.; Chitarrini, G. Mono-Locus and pyramided resistant grapevine cultivars reveal early putative biomarkers upon artificial inoculation with *Plasmopara viticola*. *Front. Plant Sci.* **2021**, *12*, 693887. [CrossRef]
197. Ricciardi, V.; Marcianò, D.; Sargolzaei, M.; Maddalena, G.; Maghradze, D.; Tirelli, A.; Casati, P.; Bianco, P.A.; Failla, O.; Fracassetti, D.; et al. From plant resistance response to the discovery of antimicrobial compounds: The role of volatile organic compounds (VOCs) in grapevine downy mildew infection. *Plant Physiol. Biochem.* **2021**, *160*, 294–305. [CrossRef]
198. Sharifi, R.; Ryu, C.-M. Social networking in crop plants: Wired and wireless cross-plant communications. *Plant Cell Environ.* **2021**, *44*, 1095–1110. [CrossRef]
199. Himanen, S.J.; Blande, J.D.; Klemola, T.; Pulkkinen, J.; Heijari, J.; Holopainen, J.K. Birch (*Betula* spp.) leaves adsorb and re-release volatiles specific to neighbouring plants—A mechanism for associational herbivore resistance? *New Phytol.* **2010**, *186*, 722–732. [CrossRef]
200. Hansen, U.; Seufert, G. Temperature and light dependence of β-caryophyllene emission rates. *J. Geophys. Res. Atmos.* **2003**, *108*. [CrossRef]
201. Schmelz, E.A.; Engelberth, J.; Alborn, H.T.; O'Donnell, P.; Sammons, M.; Toshima, H.; Tumlinson, J.H. 3rd, Simultaneous analysis of phytohormones, phytotoxins, and volatile organic compounds in plants. *Proc. Natl. Acad. Sci. USA* **2003**, *100*, 10552–10557. [CrossRef] [PubMed]
202. Craig, S.C.; Daniel, J.C.; Raymond, M.W.; Ara, M.; Robert, R.H. A System and methodology for measuring volatile organic compounds produced by hydroponic lettuce in a controlled environment. *J. Am. Soc. Hortic. Sci.* **1996**, *121*, 483–487.
203. Zepp, R.G.; Callaghan, T.V.; Erickson, D.J. Effects of increased solar ultraviolet radiation on biogeochemical cycles. *Ambio* **1995**, *24*, 181–187.
204. Harley, P.; Guenther, A.; Zimmerman, P. Environmental controls over isoprene emission in deciduous oak canopies. *Tree Physiol.* **1997**, *17*, 705–714. [CrossRef]
205. Heiden, A.C.; Hoffmann, T.; Kahl, J.; Kley, D.; Klockow, D.; Langebartels, C.; Mehlhorn, H.; Sandermann Jr, H.; Schraudner, M.; Schuh, G.; et al. Emission of volatile organic compounds from ozone-exposed plants. *Ecol. Appl.* **1999**, *9*, 1160–1167. [CrossRef]
206. Brüggemann, N.; Schnitzler, J.P. Comparison of isoprene emission, intercellular isoprene concentration and photosynthetic performance in water-limited oak (*Quercus pubescens* Willd. and *Quercus robur* L.) saplings. *Plant Biol.* **2002**, *4*, 456–463. [CrossRef]
207. Sharkey, T.D.; Wiberley, A.E.; Donohue, A.R. Isoprene emission from plants: Why and how. *Ann. Bot.* **2008**, *101*, 5–18. [CrossRef]
208. Vickers, C.E.; Gershenzon, J.; Lerdau, M.T.; Loreto, F. A unified mechanism of action for volatile isoprenoids in plant abiotic stress. *Nat. Chem. Biol.* **2009**, *5*, 283–291. [CrossRef]
209. Loreto, F.; Velikova, V. Isoprene produced by leaves protects the photosynthetic apparatus against ozone damage, quenches ozone products, and reduces lipid peroxidation of cellular membranes. *Plant Physiol.* **2001**, *127*, 1781. [CrossRef]
210. Bertamini, M.; Faralli, M.; Varotto, C.; Grando, M.S.; Cappellin, L. Leaf monoterpene emission limits photosynthetic downregulation under heat stress in field-grown grapevine. *Plants* **2021**, *10*, 181. [CrossRef]
211. Mahajan, P.; Singh, H.P.; Kaur, S.; Batish, D.R.; Kohli, R.K. β-pinene moderates Cr(VI) phytotoxicity by quenching reactive oxygen species and altering antioxidant machinery in maize. *Environ. Sci. Pollut. Res.* **2019**, *26*, 456–463. [CrossRef] [PubMed]
212. Pollastri, S.; Baccelli, I.; Loreto, F. Isoprene: An antioxidant itself or a molecule with multiple regulatory functions in plants? *Antioxidants* **2021**, *10*, 684. [CrossRef] [PubMed]
213. Landi, M. Airborne signals and abiotic factors: The neglected side of the plant communication. *Commun. Integr. Biol.* **2020**, *13*, 67–73. [CrossRef] [PubMed]
214. Mauch-Mani, B.; Baccelli, I.; Luna, E.; Flors, V. Defense priming: An adaptive part of induced resistance. *Annu. Rev. Plant Biol.* **2017**, *68*, 485–512. [CrossRef] [PubMed]
215. Karban, R. Communication between sagebrush and wild tobacco in the field. *Biochem. Syst. Ecol.* **2001**, *29*, 995–1005. [CrossRef]
216. Zakir, A.; Khallaf, M.A.; Hansson, B.S.; Witzgall, P.; Anderson, P. Herbivore-induced changes in cotton modulates reproductive behavior in the moth *Spodoptera littoralis*. *Front. Ecol. Evol.* **2017**, *5*, 49. [CrossRef]
217. Conrath, U. Systemic acquired resistance. *Plant Signal. Behav.* **2006**, *1*, 179–184. [CrossRef]
218. Martinez-Medina, A.; Flors, V.; Heil, M.; Mauch-Mani, B.; Pieterse, C.M.J.; Pozo, M.J.; Ton, J.; van Dam, N.M.; Conrath, U. Recognizing plant defense priming. *Trends Plant Sci.* **2016**, *21*, 818–822. [CrossRef]
219. Preston, C.A.; Laue, G.; Baldwin, I.T. Plant–plant Signaling: Application of trans- or cis-methyl jasmonate equivalent to sagebrush releases does not elicit direct defenses in native tobacco. *J. Chem. Ecol.* **2004**, *30*, 2193–2214. [CrossRef]
220. Hagiwara, T.; Ishihara, M.I.; Takabayashi, J.; Hiura, T.; Shiojiri, K. Effective distance of volatile cues for plant–plant communication in beech. *Ecol. Evol.* **2021**, *11*, 12445–12452. [CrossRef] [PubMed]
221. van Hulten, M.; Pelser, M.; van Loon, L.C.; Pieterse, C.M.; Ton, J. Costs and benefits of priming for defense in *Arabidopsis*. *Proc. Natl. Acad. Sci. USA* **2006**, *103*, 5602–5607. [CrossRef] [PubMed]
222. Baldwin, I.T.; Halitschke, R.; Paschold, A.; von Dahl, C.C.; Preston, C.A. Volatile signaling in plant-plant interactions: 'Talking trees' in the Genomics era. *Science* **2006**, *311*, 812. [CrossRef] [PubMed]
223. Heil, M.; Ton, J. Long-distance signalling in plant defence. *Trends Plant Sci.* **2008**, *13*, 264–272. [CrossRef] [PubMed]
224. Katz, V.A.; Thulke, O.U.; Conrath, U. A benzothiadiazole primes parsley cells for augmented elicitation of defense responses. *Plant Physiol.* **1998**, *117*, 1333. [CrossRef] [PubMed]

225. Zimmerli, L.; Métraux, J.-P.; Mauch-Mani, B. β-Aminobutyric acid-induced protection of *Arabidopsis* against the necrotrophic fungus *Botrytis cinerea*. *Plant Physiol.* **2001**, *126*, 517. [CrossRef] [PubMed]

226. Balmer, A.; Pastor, V.; Glauser, G.; Mauch-Mani, B. Tricarboxylates induce defense priming against bacteria in *Arabidopsis thaliana*. *Front. Plant Sci.* **2018**, *9*, 1221. [CrossRef]

227. Balmer, A.; Pastor, V.; Gamir, J.; Flors, V.; Mauch-Mani, B. The 'prime-ome': Towards a holistic approach to priming. *Trends Plant Sci.* **2015**, *20*, 443–452. [CrossRef]

228. Brosset, A.; Blande, J.D. Volatile-mediated plant-plant interactions: VOCs as modulators of receiver plant defence, growth and reproduction. *J. Exp. Bot.* **2021**, *73*, 511–528. [CrossRef]

229. Luna, E.; Bruce, T.J.; Roberts, M.R.; Flors, V.; Ton, J. Next-generation systemic acquired resistance. *Plant Physiol.* **2012**, *158*, 844–853. [CrossRef]

230. Crisp, P.A.; Ganguly, D.; Eichten, S.R.; Borevitz, J.O.; Pogson, B.J. Reconsidering plant memory: Intersections between stress recovery, RNA turnover, and epigenetics. *Sci. Adv.* **2016**, *2*, e1501340. [CrossRef] [PubMed]

231. Brilli, F.; Loreto, F.; Baccelli, I. Exploiting plant volatile organic compounds (VOCs) in agriculture to improve sustainable defense strategies and productivity of crops. *Front. Plant Sci.* **2019**, *10*, 264. [CrossRef] [PubMed]

232. Dicke, M.; Agrawal, A.A.; Bruin, J. Plants talk, but are they deaf? *Trends Plant Sci.* **2003**, *8*, 403–405. [CrossRef]

233. Theocharis, A.; Bordiec, S.; Fernandez, O.; Paquis, S.; Dhondt-Cordelier, S.; Baillieul, F.; Clement, C.; Barka, E.A. *Burkholderia phytofirmans* PsJN primes *Vitis vinifera* L. and confers a better tolerance to low nonfreezing temperatures. *Mol. Plant Microbe Interact.* **2012**, *25*, 241–249. [CrossRef] [PubMed]

234. Holopainen, J.K.; Gershenzon, J. Multiple stress factors and the emission of plant VOCs. *Trends Plant Sci.* **2010**, *15*, 176–184. [CrossRef] [PubMed]

235. Pierik, R.; Ballaré, C.L.; Dicke, M. Ecology of plant volatiles: Taking a plant community perspective. *Plant Cell Environ.* **2014**, *37*, 1845–1853. [CrossRef]

236. Mittler, R. Abiotic stress, the field environment and stress combination. *Trends Plant Sci.* **2006**, *11*, 15–19. [CrossRef]

237. Kessler, A. The information landscape of plant constitutive and induced secondary metabolite production. *Curr. Opin. Insect Sci.* **2015**, *8*, 47–53. [CrossRef]

238. Holopainen, J.K.; Blande, J.D. Molecular plant volatile communication. In *Sensing in Nature*; López-Larrea, C., Ed.; Springer: New York, NY, USA, 2012; pp. 17–31.

239. Tholl, D.; Boland, W.; Hansel, A.; Loreto, F.; Röse, U.S.R.; Schnitzler, J.-P. Practical approaches to plant volatile analysis. *Plant J.* **2006**, *45*, 540–560. [CrossRef]

240. Beck, J.; Smith, L.; Baig, N. An overview of plant volatile metabolomics, sample treatment and reporting considerations with emphasis on mechanical damage and biological control of weeds. *Phytochem. Anal.* **2014**, *25*, 331–341. [CrossRef]

241. Kfoury, N.; Scott, E.; Orians, C.; Robbat, A., Jr. Direct contact sorptive extraction: A robust method for sampling plant volatiles in the field. *J. Agric. Food Chem.* **2017**, *65*, 8501–8509. [CrossRef] [PubMed]

242. Fung, A.G.; Yamaguchi, M.S.; McCartney, M.M.; Aksenov, A.A.; Pasamontes, A.; Davis, C.E. SPME-based mobile field device for active sampling of volatiles. *Microchem. J.* **2019**, *146*, 407–413. [CrossRef] [PubMed]

243. Louw, S. Recent trends in the chromatographic analysis of volatile flavor and fragrance compounds: Annual review 2020. *Anal. Sci. Adv.* **2021**, *2*, 157–170. [CrossRef]

244. Stewart-Jones, A.; Poppy, G.M. Comparison of glass vessels and plastic bags for enclosing living plant parts for headspace analysis. *J. Chem. Ecol.* **2006**, *32*, 845–864. [CrossRef]

245. Timkovsky, J.; Gankema, P.; Pierik, R.; Holzinger, R. A plant chamber system with downstream reaction chamber to study the effects of pollution on biogenic emissions. *Environ. Sci. Process. Impacts* **2014**, *16*, 2301–2312. [CrossRef]

246. Beck, J.J.; Smith, L.; Merrill, G.B. In situ volatile collection, analysis, and comparison of three *Centaurea* species and their relationship to biocontrol with herbivorous insects. *J. Agric. Food Chem.* **2008**, *56*, 2759–2764. [CrossRef] [PubMed]

247. Zhu, W.; Koziel, J.A.; Cai, L.; Wright, D.; Kuhrt, F. Testing odorants recovery from a novel metallized fluorinated ethylene propylene gas sampling bag. *J. Air Waste Manag. Assoc.* **2015**, *65*, 1434–1445. [CrossRef] [PubMed]

248. Gross, J.H. Direct analysis in real time—A critical review on DART-MS. *Anal. Bioanal. Chem.* **2014**, *406*, 63–80. [CrossRef] [PubMed]

249. Soo, J.-C.; Lee, E.G.; LeBouf, R.F.; Kashon, M.L.; Chisholm, W.; Harper, M. Evaluation of a portable gas chromatograph with photoionization detector under variations of VOC concentration, temperature, and relative humidity. *J. Occup. Environ. Hyg.* **2018**, *15*, 351–360. [CrossRef]

250. Wei-Hao Li, M.; Ghosh, A.; Venkatasubramanian, A.; Sharma, R.; Huang, X.; Fan, X. High-sensitivity micro-gas chromatograph–photoionization detector for trace vapor detection. *ACS Sens.* **2021**, *6*, 2348–2355. [CrossRef]

251. Arce, L.; Gallegos, J.; Garrido, R.; Medina, L.; Sielemann, S.; Wortelmann, T. Ion mobility spectrometry a versatile analytical tool for metabolomics applications in food science. *Curr. Metab.* **2014**, *2*, 264–271. [CrossRef]

252. Kallenbach, M.; Veit, D.; Eilers, E.J.; Schuman, M.C. Application of silicone tubing for robust, simple, high-throughput, and time-resolved analysis of plant volatiles in field experiments. *Bio Protoc.* **2015**, *5*, e1391. [CrossRef] [PubMed]

253. Ellis, A.M.; Mayhew, C.A. *Proton Transfer Reaction Mass Spectrometry: Principles and Applications*; John Wiley & Sons: Chichester, UK, 2014.

254. Jordan, A.; Haidacher, S.; Hanel, G.; Hartungen, E.; Märk, L.; Seehauser, H.; Schottkowsky, R.; Sulzer, P.; Märk, T.D. A high resolution and high sensitivity proton-transfer-reaction time-of-flight mass spectrometer (PTR-TOF-MS). *Int. J. Mass Spectrom.* **2009**, *286*, 122–128. [CrossRef]

255. Schwolow, S.; Gerhardt, N.; Rohn, S.; Weller, P. Data fusion of GC-IMS data and FT-MIR spectra for the authentication of olive oils and honeys—Is it worth to go the extra mile? *Anal. Bioanal. Chem.* **2019**, *411*, 6005–6019. [CrossRef]

256. Ranganathan, Y.; Borges, R.M. To transform or not to transform. *Plant Signal. Behav.* **2011**, *6*, 113–116. [CrossRef]

257. Houhou, R.; Bocklitz, T. Trends in artificial intelligence, machine learning, and chemometrics applied to chemical data. *Anal. Sci. Adv.* **2021**, *2*, 128–141. [CrossRef]

258. Marenco, R.A.; Siebke, K.; Farquhar, G.D.; Ball, M.C. Hydraulically based stomatal oscillations and stomatal patchiness in *Gossypium hirsutum*. *Funct. Plant Biol.* **2006**, *33*, 1103–1113. [CrossRef]

259. Peng, J.; van Loon, J.J.A.; Zheng, S.; Dicke, M. Herbivore-induced volatiles of cabbage (*Brassica oleracea*) prime defence responses in neighbouring intact plants. *Plant Biol.* **2011**, *13*, 276–284. [CrossRef]

260. Silva, P.; Gerós, H. Regulation by salt of vacuolar H+-ATPase and H+-pyrophosphatase activities and Na+/H+ exchange. *Plant Signal. Behav.* **2009**, *4*, 718–726. [CrossRef]

261. Haider, M.S.; Zhang, C.; Kurjogi, M.M.; Pervaiz, T.; Zheng, T.; Zhang, C.; Lide, C.; Shangguan, L.; Fang, J. Insights into grapevine defense response against drought as revealed by biochemical, physiological and RNA-Seq analysis. *Sci. Rep.* **2017**, *7*, 13134. [CrossRef]

262. Hamilton, A.J.; Lycett, G.W.; Grierson, D. Antisense gene that inhibits synthesis of the hormone ethylene in transgenic plants. *Nature* **1990**, *346*, 284–287. [CrossRef]

263. Oeller, P.W.; Lu, M.W.; Taylor, L.P.; Pike, D.A.; Theologis, A. Reversible inhibition of tomato fruit senescence by antisense RNA. *Science* **1991**, *254*, 437. [CrossRef] [PubMed]

264. Vancanneyt, G.; Sanz, C.; Farmaki, T.; Paneque, M.; Ortego, F.; Castañera, P.; Sánchez-Serrano, J.J. Hydroperoxide lyase depletion in transgenic potato plants leads to an increase in aphid performance. *Proc. Natl. Acad. Sci. USA* **2001**, *98*, 8139–8144. [CrossRef] [PubMed]

265. Halitschke, R.; Baldwin, I.T. Antisense LOX expression increases herbivore performance by decreasing defense responses and inhibiting growth-related transcriptional reorganization in *Nicotiana attenuata*. *Plant J.* **2003**, *36*, 794–807. [CrossRef]

266. Halitschke, R.; Ziegler, J.; Keinänen, M.; Baldwin, I.T. Silencing of hydroperoxide lyase and allene oxide synthase reveals substrate and defense signaling crosstalk in *Nicotiana attenuata*. *Plant J.* **2004**, *40*, 35–46. [CrossRef]

267. Paschold, A.; Halitschke, R.; Baldwin, I.T. Using 'mute' plants to translate volatile signals. *Plant J.* **2006**, *45*, 275–291. [CrossRef]

Perspective

Algal Ocelloids and Plant Ocelli

Felipe Yamashita and František Baluška *

Institute of Cellular and Molecular Botany, University of Bonn, 53115 Bonn, Germany
* Correspondence: baluska@uni-bonn.de

Abstract: Vision is essential for most organisms, and it is highly variable across kingdoms and domains of life. The most known and understood form is animal and human vision based on eyes. Besides the wide diversity of animal eyes, some animals such as cuttlefish and cephalopods enjoy so-called dermal or skin vision. The most simple and ancient organ of vision is the cell itself and this rudimentary vision evolved in cyanobacteria. More complex are so-called ocelloids of dinoflagellates which are composed of endocellular organelles, acting as lens- and cornea/retina-like components. Although plants have almost never been included into the recent discussions on organismal vision, their plant-specific ocelli had already been proposed by Gottlieb Haberlandt already in 1905. Here, we discuss plant ocelli and their roles in plant-specific vision, both in the shoots and roots of plants. In contrast to leaf epidermis ocelli, which are distributed throughout leaf surface, the root apex ocelli are located at the root apex transition zone and serve the light-guided root navigation. We propose that the plant ocelli evolved from the algal ocelloids, are part of complex plant sensory systems and guide cognition-based plant behavior.

Keywords: algae; cyanobacteria; eyes; eyespots; ocelloids; ocelli; plants; roots; shoots; vision

1. Introduction

Vision in animals is incredibly diverse and it evolved multiple times independently [1–3]. Despite a great diversity of visual organs, an eye can be defined as the existence of a cornea and/or lens which focuses the light towards a sensory region, such as eye retina or other light-sensitive structures and tissues, with photo-responsive proteins transforming the light signal first into electrical and then into chemical signals [4–6].

In 1905, Gottlieb Haberlandt proposed the plant ocelli concept for leaf epidermis in which the upper epidermal cells resemble convex or planoconvex lens, converging light rays on the light-sensitive subepidermal cells [7]. The Haberlandt plant ocelli theory is not surprising if we consider that various organisms such as bacteria, algae, and fungi (as discussed below) have cells with similar light-sensing properties. However, plant ocelli theory was almost forgotten and only recently revived [8,9]. Supporting this leaf epidermal ocelli scenario, leaf epidermis cells, with the exception of stomata guard cells, do not generate photosynthetic chloroplasts, although they have the best position with respect to the amount of light they receive.

This concept was recaptured some 70 years later when young seedlings of tropical vine *Monstera gigantea* were reported to localize and suitably support host trees using growth towards darkness termed *skototropism*—the directional movement of plant organ towards darkness [10]. Due to observations, and apart from other theories, Strong and Ray (1975) found skototropism to be the relevant mechanism in the finding of host trees by the Monstera vine. They provided evidence that shoot skototropism is an independent mechanism. Nevertheless, they assumed it to be a modification of negative phototropism. Additionally, they reported a negative effect of increasing distance and a positive effect of increasing host stem diameter on the shoot skototropism. Importantly, the larger a potential host tree is and the closer it is located to the vine seedling, the stronger the skototropic response will be [10].

Citation: Yamashita, F.; Baluška, F. Algal Ocelloids and Plant Ocelli. *Plants* **2023**, *12*, 61. https://doi.org/10.3390/plants12010061

Academic Editor: Kay Schneitz

Received: 16 November 2022
Revised: 15 December 2022
Accepted: 19 December 2022
Published: 22 December 2022

2. Chlamydomonas Algal Eyespot: Rhizoplast and Rootlet Connections

The green alga *Chlamydomonas reinhardtii* also has a subcellular eyespot apparatus. Algal eyespots are anchored at the Chlamydomonas cell periphery via so-called D4 bundles of microtubules, organized by the basal body (Figure 1). In addition, an important—but often neglected—feature of Chlamydomonas is the rhizoplast, which is a contractile centrin-based structure connecting basal bodies of flagella with the nuclear surface [11–13]. These so-called rhizoplasts or fibrous flagellar roots anchor nuclei to the flagellar or ciliar basal bodies [14–19]. The eyespot of Chlamydomonas is anchored to the D4 rootlet, extending from the peripheral flagellar basal bodies into the cell interior [20–22]. Intriguingly, similarly to the scenario with the ocelloids of the warnowiid dinoflagellates discussed below, these algal eyespots are also assembled from putatively symbiotic components. Besides the chloroplasts, there is cellular evidence suggesting that the nucleus–basal body–flagellum/cilium complex is of symbiotic origin, representing the guest cell of the host–guest cell consortium [23,24].

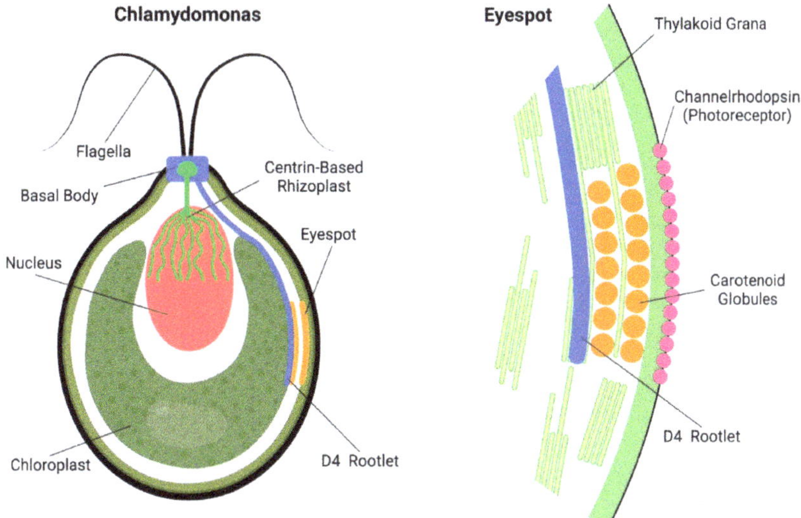

Figure 1. Algal Eyespot of Chlamydomonas. Chlamydomonas alga with two flagella associated with the basal bodies which intracellularly organize intracellular bundles of microtubules (known as rootlets) of which the D4 bundle anchors the eyespot. This eyespot is constructed from chloroplast thylakoid membranes and carotenoid globules, aligned under the plasma membrane which is enriched with photoreceptor channelrhodopsin. Besides the bundles of microtubules, the basal body also organizes the centrin-based contractile nucleo-basal body connector anchoring the nucleus. M4, M2 and D2 rootlets are not shown in this simplified scheme.

Chlamydomonas green algae have two vision responses. The first one is swimming in towards or away from light ray source, called phototaxis, depending on the total amount of reactive oxygen species (ROS) inside the cell [25,26]. The second is when they freeze for a few moments after receiving a strong light stimulus, followed by a backstroke, and then swimming normally in any direction. This second one is called photo-shock response: as the name implies, the algae stop their natural movement for seconds [27,28]. Under a microscope, it is easy to find the eyespots, as they are composed of orange carotenoid globules located under the plasma membrane enriched with photoreceptor proteins, channelrhodopsinsChR1 and ChR2 [29]. In green alga *Chlamydomonas reinhardtii*, the eyespot apparatus is composed of two layers of carotenoid globules (Figure 1) sandwiched between the thylakoid membranes of the chloroplast [28,30,31]. The eyespot apparatus is activated through light stimuli, and afterwards controls flagella to accomplish phototaxic behavior [30]. An important aspect is that the light-induced eyespot electric currents activate and control the flagellar currents via the electric action potential-like transmission [32–35]. Rapid calcium influxes and bioelectric currents integrate sensory events at the eyespot with control of flagella beating and phototaxis [27,32,33,36].

Another algae protist that evolved a light-sensitive apparatus adapted for unicellular vision is *Euglena gracilis*. It shows two basic types of photo-movements in response to light stimuli, known as photophobic and phototactic behaviors. Similarly, as in the eyespot of Chlamydomonas, *Euglena gracilis* carotenoids are important for photo-movements. The plastids do not develop into chloroplasts due to the lack of chlorophyll synthesis [37,38]. Recent studies have reported that mutants, deficient in carotenoid production, lose their phototactic responsiveness [38]. Carotenoids are obviously essential for light perception of the Euglena eyespot. Similarly, as in Chlamydomonas, the eyespot of Euglena is associated with the microtubules-based flagella [37,39,40]. However, *Euglena gracilis* obtained their plastids much later via the secondary endosymbiosis and are evolutionary distant, belonging to Archaezoans [41]. Thus, it is not surprising that Euglena and Chlamydomonas rely on different photoreceptors in their ocelli.

3. From Algal Ocelloids to Plant Ocelli

In 1967, David Francis described an eyespot in *Nematodinium armatum*, describing lenses capable of focusing light rays and concentrating them into a structure called a pigment cup. This structure is supposed to be a light-sensitive retinoid and may have a role in image formation [42]. In 2015, further unexpected support for the plant ocelli theory of Gottlieb Haberlandt was provided with the surprising discovery of eye-like ocelloids in warnowiid dinoflagellates [43,44]. These planktonic unicellular organisms use symbiotic organelles which act as eye-like ocelloids. A mitochondria-based layer generates a cornea-like surface across a lens structure, whereas the retinal body of ocelloids develops from a membrane network formed from plastids (Figure 2). To verify these microscopically based findings, the scientists sequenced the DNA of a warnowiid retinal body, which had a substantially greater percentage of DNA originating from plastids than comparable samples from the total cell [43]. Warnowiid dinoflagellates are the only unicellular microbial organism having camera-type eye-like organs for camera-type vision-like modus [4,42–45].

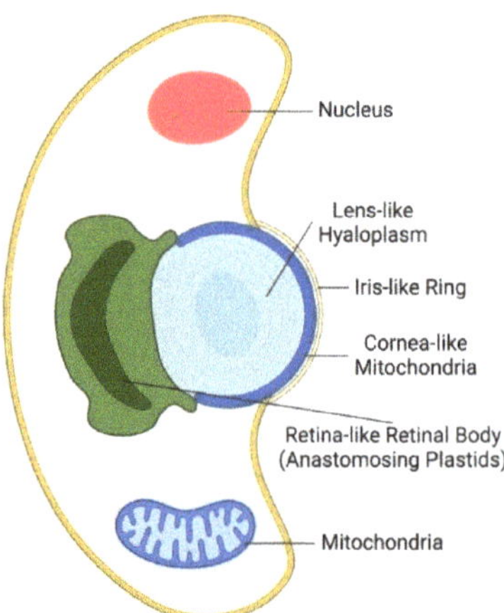

Figure 2. Algal Ocelloid of Dinoflagellates. Camera-like ocelloid of warnowiid dinoflagellates is composed of cornea-like mitochondrion enclosing hyaloplasm acting as lens and chloroplast-based retinal body. Similarly, as in the algal eyespot, the chloroplast plays the central role in the microbial vision. Adapted according [43].

4. Bacterial Vision: Cyanobacterium Synechocystis

The next surprising discovery followed one year later, when Schuergers et al. (2016) reported prokaryotic bacterial vision in cyanobacterium Synechocystis sp. PCC 6803 [46–49]. Here, the whole cell acts as a lens, focusing light on a small patch of the plasma membrane (Figure 3). A similar principle, in which the whole cell acts as a lens, was found also in eukaryotic volvocine algae [50]. Therefore, it should not be surprising if plant cells also rely on this feature via their ocelli. Importantly, biological evolution repeatedly uses all the successfully elaborated structures and processes which improve the organismal survival chances. Even the most complex organs of vision, such as animal and human eyes, represent the inherent part of the long evolutionary continuum. In the case multi-cellular volvocine algae, light-focusing roles of cells affect the adjacent cells in a manner which participates in morphological symmetries and colony behavior as relevant information [50]. In Synechocystis, light perception at the photosensitive patch of the plasma membrane electrically controls type IV pili-based motility apparatus [51] in such a manner that pili close to the light focal spot are inactivated, whereas pili on the opposite side of the cell (facing the light source) are active and allow movement towards the light source [46–49]. As cyanobacteria evolved more than three billion years ago, it is obvious that this ancient prokaryotic vision based on the type IV pili complex is a very successful solution to their environmental challenges [52,53].

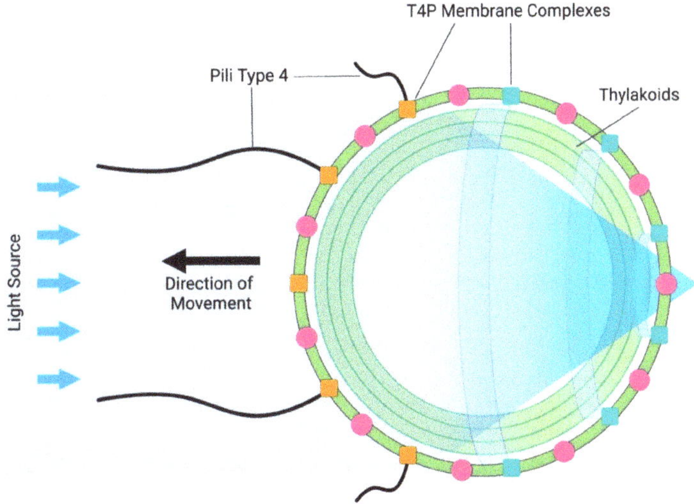

Figure 3. Bacterial Vision: Cyanobacterium Synechocystis. The whole cyanobacterial cell acts as a lens, focusing light beams on a small patch of the plasma membrane which controls the type-IV pili-based motility apparatus anchored in the plasma membrane via T4P complexes. Under the plasma membrane are thylakoid membranes. This model was adapted according to [49].

5. Plant Vision: *Boquila trifoliata*

Another example of an organism that can change its structures is the interesting plant *Boquila trifoliolata*, which can change its original three-lobed leaf shape into longitudinal leaves or any other shape, depending on the host plant next to its leaves. This is what the experiment by White & Yamashita (2021) illustrated [54]. The Boquila leaves were placed next to plastic leaves of non-living host plant, and the surprising result was that the Boquila mimicked the plastic plant leaves by changing leaf shape to a longitudinal shape, mimicking the plastic leaves of the non-living model plant. This experiment refutes two hypotheses proposed by other researchers. The first hypothesis was that horizontal gene transfer is mediated by the airborne microbes involved, thus allowing the Boquila to modify its leaves according to the leaves of the host plant. The second hypothesis was that the Boquila modified its leaves following some volatile chemical signals released by the host plant. As the plastic leaves of non-living host plants were able to induce mimicking response in the Boquila, the hypothesis of horizontal gene transfer and the hypothesis of volatile substances can be dismissed. The plastic leaves might release some volatile substance under sunlight exposure, but these are biologically not relevant. This is very strong support for the proposal that plant-specific vision based on leaf ocelli is behind the mimicking responses of Boquila plants. This would also explain that the Boquila leaves can actively identify their surrounding environment, and modify not only leaf sizes and forms, but also color, leaf vein networks and other anatomical patterns. Future experimental research is needed to understand how all this can be accomplished.

6. Root Apex Vision: Root Skototropism

Although all roots of plants growing out in the nature are underground in darkness, they express all photoreceptors at their root apices [55]. While a dim light is not stressful for roots, they try to escape from stronger lights, which represent a stress factor for roots [56–58]. In order to avoid the direct illumination of roots in young seedlings grown in laboratory conditions using transparent Petri dishes, we have proposed the use of partially darkened dishes which allow us to keep roots in darkness [59–61]. Alternatively, the D-Root system was established as an alternative method to maintain roots in the shaded environment [62–64]. Surprisingly, roots grew even faster when grown within the D-Root system and our analysis revealed that this was due to steep light–darkness gradient provided by the D-Root system, which roots evaluate as a potent growth stimulant [65,66]. The process of speeding up the root growth under the steep light–darkness gradient of the D-Root system is based on the TOR complex activity, as its specific inhibition blocked this light escape tropism of illuminated roots [66]. Interestingly, roots placed in the illuminated portion of the shaded Petri dishes could recognize the dark portions of dishes, even when placed up to 2 mm from the light/darkness border (Figure 4). This implies some kind of root apex vision in the root apex skototropism response. The root apex ocelli proposed for this root skototropism are based on the blue-light phot 1 photoreceptor [55]. In contrast to diffusely distributed leaf epidermis ocelli, the root apex ocelli are assembled locally [67,68] at the root apex transition zone [69,70]. This position is optimal for the root apex vision, guiding the root apex navigation towards darkness [71].

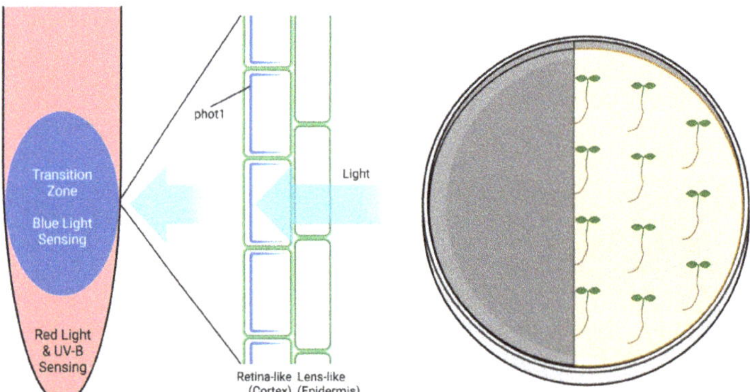

Figure 4. Root Apex Ocelli. Arabidopsis root apex expresses phot1 blue-light photoreceptor in cortex cells of the transition zone. The phot1 photoreceptors are arranged in the U-shape arrangements under the root epidermis cells which are devoid of phot 1 and are proposed to act as a lens cells, focusing the light on the underlying cortex cells. The root apex ocelli are proposed to allow root skototropism when roots grown within the illuminated portion of Petri dish can recognize the dark area and navigate the root growth towards it.

7. Conclusions and Perspectives

Vision via the whole organismal surface is known from some animals, such as cuttlefish and cephalopods [72,73]. Similarly, sea urchins and brittle stars have dispersed visual systems [74,75], all resembling the situation in plant leaf ocelli. Other lower animals have local eyes which resemble rather the root apex ocelli. Starfish have compound eyes at the arm tips [76,77]. Cnidarian medusae have eyes at the bases of their tentacles or on special sensory structures (rhopalia) which contain two lens-eyes flanked by two pairs of lens-less eyes [78]. Recent genetic studies have shown that the genes *Pax6*, *six1* and *six3* play key roles in the development of the eye in organisms from planaria to humans, arguing strongly for a monophyletic origin of the animal eye [79]. Nevertheless, there is no single regulatory gene in the formation of all animals. Diversity of vision in different animals must be based on gene expression as a tool and include the function of critical genes as mechanisms of the visual organ formation [79]. The hypothesis of phytochrome gene transfer from cyanobacteria, generating the first plastid in eukaryotes, paves the way for the presence of carotenoids in algae, which in turn are of extreme importance in eyespots [80]. Obviously, the leaf ocelli of plants conform well with algae and animal visual systems and represent obvious examples of convergent evolution. Root apex ocelli, based on the phot1 blue-light photoreceptor, represent another solution for the plant vision. It can be speculated that every cell with chloroplast has a cellular vision, resembling cells of cyanobacteria, algae, and plants. Albrecht-Buehler proposed 30 years ago that animal cells enjoy rudimentary vision [81–85] because they sense infrared wavelengths via their microtubules (Figure 5). This cellular vision is based on radial microtubules converging at their organizing centers (MTOCs), including centrosomes, basal bodies of cilia, and nuclear surfaces [86,87]. In future, it will be interesting to investigate the possible roles of microtubules in algal ocelloids and eyespots, as well as in plant leaves and root ocelli.

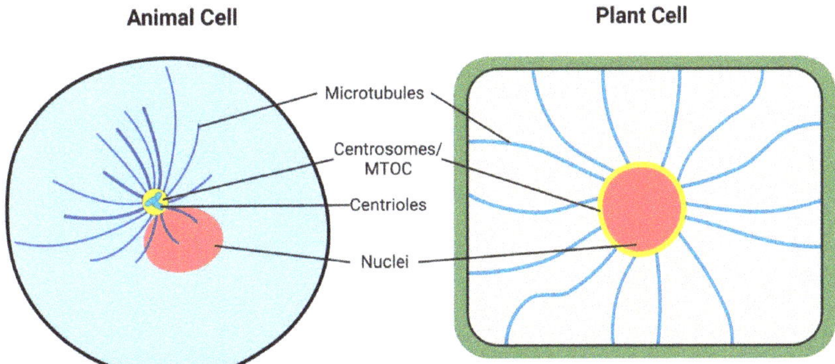

Figure 5. Microtubules-MTOC in Rudimentary Cell Vision of Eukaryotic Cells. Albrecht-Beuhler's rudimentary cellular vision is accomplished via microtubules conveying infrared wavelengths along microtubules towards the perinuclear centrosome of animal cells. In the plant cells, the centrosome is not corpuscular but is distributed diffusely along the whole nuclear surface.

In conclusion, it emerges that vision is an ancient sensory faculty which evolved some three billion years ago with the very first cyanobacteria. Evolution never discards successful innovations, and the algal and plant vision is based on that of chloroplasts too.

Author Contributions: Conceptualization, analysis, and writing, F.Y. and F.B.; Investigation and resources, F.Y. and F.B.; writing—review and editing, F.Y. and F.B.; supervision, F.B. All authors have read and agreed to the published version of the manuscript.

Funding: This research was funded by Stiftung Zukunft jetzt! (Munich, Germany).

Institutional Review Board Statement: Not applicable.

Informed Consent Statement: Not applicable.

Data Availability Statement: Not applicable.

Acknowledgments: The authors acknowledge Stiftung Zukunft jetzt! (Munich, Germany) for the scholarships to F.Y. All figures were created with BioRender.com.

Conflicts of Interest: The authors declare no conflict of interest.

References

1. Nilsson, D.-E.; Bok, M.J. Low-resolution vision—At the hub of eye evolution. *Integr. Comp. Biol.* **2017**, *57*, 1066–1070. [CrossRef] [PubMed]
2. Nilsson, D.-E. The diversity of eyes and vision. *Annu. Rev. Vis. Sci.* **2021**, *7*, 19–41. [CrossRef]
3. Nilsson, D.E. The evolution of visual roles—Ancient vision versus object vision. *Front. Neuroanat.* **2022**, *16*, 789375. [CrossRef]
4. Richards, T.A.; Gomes, S.L. How to build a microbial eye. *Nature* **2015**, *523*, 166–167. [CrossRef] [PubMed]
5. Hayakawa, S.; Takaku, Y.; Hwang, J.S.; Horiguchi, T.; Suga, H.; Gehring, W.; Ikeo, K.; Gojobori, T. Function and evolutionary origin of unicellular camera-type eye structure. *PLoS ONE* **2015**, *10*, e0118415. [CrossRef]
6. Galindo, L.J.; Milner, D.S.; Gomes, S.L.; Richards, T.A. A Light-sensing system in the common ancestor of the fungi. *Curr. Biol.* **2022**, *32*, 3146–3153.e3. [CrossRef] [PubMed]
7. Haberlandt, G. *Die Lichtsinnesorgane der Laubblätter*; Engelmann: Leipzig, Germany, 1905.
8. Baluška, F.; Mancuso, S. Vision in plants via plant-specific ocelli? *Trends Plant Sci.* **2016**, *21*, 727–730. [CrossRef] [PubMed]
9. Mancuso, S.; Baluška, F. Plant ocelli for visually guided plant behavior. *Trends Plant Sci.* **2017**, *22*, 5–6. [CrossRef]
10. Strong, D.R.; Ray, T.S. Host tree location behavior of a tropical vine (*Monstera gigantea*) by skototropism. *Science* **1975**, *190*, 804–806. [CrossRef]
11. Salisbury, J.L.; Baron, A.; Surek, B.; Melkonian, M. Striated Flagellar roots: Isolation and partial characterization of a calcium-modulated contractile organelle. *J. Cell Biol.* **1984**, *99*, 962–970. [CrossRef]
12. Salisbury, J.L.; Baron, A.T.; Sanders, M.A. The Centrin-based cytoskeleton of *Chlamydomonas reinhardtii*: Distribution in interphase and mitotic cells. *J. Cell Biol.* **1988**, *107*, 635–641. [CrossRef] [PubMed]
13. Wright, R.L.; Adler, S.A.; Spanier, J.G.; Jarvik, J.W. Nucleus-basal body connector in *Chlamydomonas*: Evidence for a role in basal body segregation and against essential roles in mitosis or in determining cell polarity. *Cell Motil. Cytoskelet.* **1989**, *14*, 516–526. [CrossRef] [PubMed]
14. Lechtreck, K.-F.; Melkonian, M. An update on fibrous flagellar roots in green algae. In *The Cytoskeleton of Flagellate and Ciliate Protists*; Springer: Vienna, Austria, 1991; pp. 38–44.
15. Salisbury, J.L. Roots. *J. Eukaryot. Microbiol.* **1998**, *45*, 28–32. [CrossRef] [PubMed]
16. Dutcher, S.K. Elucidation of basal body and centriole functions in *Chlamydomonas reinhardtii*. *Traffic* **2003**, *4*, 443–451. [CrossRef]
17. Geimer, S.; Melkonian, M. Centrin scaffold in *Chlamydomonas reinhardtii* revealed by immunoelectron microscopy. *Eukaryot. Cell* **2005**, *4*, 1253–1263. [CrossRef]
18. Mahen, R. The structure and function of centriolar rootlets. *J. Cell Sci.* **2021**, *134*. [CrossRef]
19. Owa, M.; Furuta, A.; Usukura, J.; Arisaka, F.; King, S.M.; Witman, G.B.; Kamiya, R.; Wakabayashi, K. Cooperative binding of the outer arm-docking complex underlies the regular arrangement of outer arm dynein in the axoneme. *Proc. Natl. Acad. Sci. USA* **2014**, *111*, 9461–9466. [CrossRef]
20. Mittelmeier, T.M.; Thompson, M.D.; Öztürk, E.; Dieckmann, C.L. Independent localization of plasma membrane and chloroplast components during eyespot assembly. *Eukaryot. Cell* **2013**, *12*, 1258–1270. [CrossRef]
21. Mittelmeier, T.M.; Boyd, J.S.; Lamb, M.R.; Dieckmann, C.L. Asymmetric properties of the *Chlamydomonas reinhardtii* cytoskeleton direct rhodopsin photoreceptor localization. *J. Cell Biol.* **2011**, *193*, 741–753. [CrossRef]
22. Boyd, J.S.; Gray, M.M.; Thompson, M.D.; Horst, C.J.; Dieckmann, C.L. the daughter four-membered microtubule rootlet determines anterior-posterior positioning of the eyespot in *Chlamydomonas reinhardtii*. *Cytoskeleton* **2011**, *68*, 459–469. [CrossRef]
23. Baluška, F.; Lyons, S. Energide–cell body as smallest unit of eukaryotic life. *Ann. Bot.* **2018**, *122*, 741–745. [CrossRef] [PubMed]
24. Baluška, F.; Lyons, S. Archaeal origins of eukaryotic cell and nucleus. *Biosystems* **2021**, *203*, 104375. [CrossRef] [PubMed]
25. Morishita, J.; Tokutsu, R.; Minagawa, J.; Hisabori, T.; Wakabayashi, K. Characterization of *Chlamydomonas reinhardtii* mutants that exhibit strong positive phototaxis. *Plants* **2021**, *10*, 1483. [CrossRef] [PubMed]
26. Wakabayashi, K.; Misawa, Y.; Mochiji, S.; Kamiya, R. Reduction-oxidation poise regulates the sign of phototaxis in *Chlamydomonas reinhardtii*. *Proc. Natl. Acad. Sci. USA* **2011**, *108*, 11280–11284. [CrossRef] [PubMed]
27. Wakabayashi, K.; Isu, A.; Ueki, N. Channelrhodopsin-dependent photo-behavioral responses in the unicellular green alga *Chlamydomonas reinhardtii*. *Optogenetics* **2021**, *1293*, 21–33.
28. Schmidt, M.; Geßner, G.; Luff, M.; Heiland, I.; Wagner, V.; Kaminski, M.; Geimer, S.; Eitzinger, N.; Reißenweber, T.; Voytsekh, O.; et al. Proteomic analysis of the eyespot of *Chlamydomonas reinhardtii* provides novel insights into its components and tactic movements. *Plant Cell* **2006**, *18*, 1908–1930. [CrossRef] [PubMed]
29. Nagel, G.; Ollig, D.; Fuhrmann, M.; Kateriya, S.; Musti, A.M.; Bamberg, E.; Hegemann, P. Channelrhodopsin-1: A light-gated proton channel in green algae. *Science* **2002**, *296*, 2395–2398. [CrossRef]

30. Ueki, N.; Ide, T.; Mochiji, S.; Kobayashi, Y.; Tokutsu, R.; Ohnishi, N.; Yamaguchi, K.; Shigenobu, S.; Tanaka, K.; Minagawa, J.; et al. Eyespot-dependent determination of the phototactic sign in *Chlamydomonas reinhardtii*. *Proc. Natl. Acad. Sci. USA* **2016**, *113*, 5299–5304. [CrossRef]

31. Kreimer, G. The green algal eyespot apparatus: A primordial visual system and more? *Curr. Genet.* **2009**, *55*, 19–43. [CrossRef]

32. Hegemann, P. Algal sensory photoreceptors. *Annu. Rev. Plant Biol.* **2008**, *59*, 167–189. [CrossRef]

33. Hegemann, P. Vision in microalgae. *Planta* **1997**, *203*, 265–274. [CrossRef]

34. Holland, E.M.; Harz, H.; Uhl, R.; Hegemann, P. Control of phobic behavioral responses by rhodopsin-induced photocurrents in *Chlamydomonas*. *Biophys. J.* **1997**, *73*, 1395–1401. [CrossRef] [PubMed]

35. Sineshchekov, O.A.; Spudich, J.L. Sensory rhodopsin signaling in green flagellate algae. In *Handbook of Photosensory Receptors*; Wiley-VCH: Weinheim, Germany, 2005; pp. 25–42. [CrossRef]

36. Berthold, P.; Tsunoda, S.P.; Ernst, O.P.; Mages, W.; Gradmann, D.; Hegemann, P. Channelrhodopsin-1 initiates phototaxis and photophobic responses in *Chlamydomonas* by immediate light-induced depolarization. *Plant Cell* **2008**, *20*, 1665–1677. [CrossRef] [PubMed]

37. Tamaki, S.; Tanno, Y.; Kato, S.; Ozasa, K.; Wakazaki, M.; Sato, M.; Toyooka, K.; Maoka, T.; Ishikawa, T.; Maeda, M.; et al. Carotenoid accumulation in the eyespot apparatus required for phototaxis is independent of chloroplast development in *Euglena gracilis*. *Plant Sci.* **2020**, *298*, 110564. [CrossRef]

38. Kato, S.; Ozasa, K.; Maeda, M.; Tanno, Y.; Tamaki, S.; Higuchi-Takeuchi, M.; Numata, K.; Kodama, Y.; Sato, M.; Toyooka, K.; et al. Carotenoids in the eyespot apparatus are required for triggering phototaxis in *Euglena gracilis*. *Plant J.* **2020**, *101*, 1091–1102. [CrossRef]

39. Hyams, J.S. The Euglena paraflagellar rod: Structure, relationship to other flagellar components and preliminary biochemical characterization. *J. Cell Sci.* **1982**, *55*, 199–210. [CrossRef]

40. Ozasa, K.; Kang, H.; Song, S.; Tamaki, S.; Shinomura, T.; Maeda, M. Regeneration of the eyespot and flagellum in *Euglena gracilis* during cell division. *Plants* **2021**, *10*, 2004. [CrossRef]

41. Burki, F. The eukaryotic tree of life from a global phylogenomic perspective. *Cold Spring Harb. Perspect. Biol.* **2014**, *6*, a016147. [CrossRef]

42. Francis, D. On the eyespot of the dinoflagellate, *Nematodinium*. *J. Exp. Biol.* **1967**, *47*, 495–501. [CrossRef]

43. Gavelis, G.S.; Hayakawa, S.; White III, R.A.; Gojobori, T.; Suttle, C.A.; Keeling, P.J.; Leander, B.S. Eye-like ocelloids are built from different endosymbiotically acquired components. *Nature* **2015**, *523*, 204–207. [CrossRef]

44. Nilsson, D.-E.; Marshall, J. Lens eyes in protists. *Curr. Biol.* **2020**, *30*, R458–R459. [CrossRef] [PubMed]

45. Colley, N.J.; Nilsson, D.-E. Photoreception in phytoplankton. *Integr. Comp. Biol.* **2016**, *56*, 764–775. [CrossRef] [PubMed]

46. Schuergers, N.; Lenn, T.; Kampmann, R.; Meissner, M.V.; Esteves, T.; Temerinac-Ott, M.; Korvink, J.G.; Lowe, A.R.; Mullineaux, C.W.; Wilde, A. Cyanobacteria use micro-optics to sense light direction. *eLife* **2016**, *5*, e12620. [CrossRef] [PubMed]

47. Schuergers, N.; Mullineaux, C.W.; Wilde, A. Cyanobacteria in motion. *Curr. Opin. Plant Biol.* **2017**, *37*, 109–115. [CrossRef]

48. Dieckmann, C.; Mittelmeier, T. Life in focus. *eLife* **2016**, *5*, e14169. [CrossRef]

49. Nilsson, D.-E.; Colley, N.J. Comparative Vision: Can bacteria really see? *Curr. Biol.* **2016**, *26*, R369–R371. [CrossRef]

50. Kessler, J.O.; Nedelcu, A.M.; Solari, C.A.; Shelton, D.E. Cells acting as lenses: A possible role for light in the evolution of morphological asymmetry in multicellular volvocine algae. In *Evolutionary Transitions to Multicellular Life*; Springer: Berlin/Heidelberg, Germany, 2015; pp. 225–243.

51. Harwood, T.V.; Zuniga, E.G.; Kweon, H.; Risser, D.D. The cyanobacterial taxis protein HmpF regulates type IV pilus activity in response to light. *Proc. Natl. Acad. Sci. USA* **2021**, *118*, e2023988118. [CrossRef]

52. Timsit, Y.; Lescot, M.; Valiadi, M.; Not, F. Bioluminescence and photoreception in unicellular organisms: Light-signalling in a bio-communication perspective. *Int. J. Mol. Sci.* **2021**, *22*, 11311. [CrossRef]

53. Allaf, M.M.; Peerhossaini, H. Cyanobacteria: Model microorganisms and beyond. *Microorganisms* **2022**, *10*, 696. [CrossRef]

54. White, J.; Yamashita, F. *Boquila trifoliolata* mimics leaves of an artificial plastic host plant. *Plant Signal. Behav.* **2022**, *17*, 1977530. [CrossRef]

55. Mo, M.; Yokawa, K.; Wan, Y.; Baluška, F. How and why do root apices sense light under the soil surface? *Front. Plant Sci.* **2015**, *6*, 775. [CrossRef] [PubMed]

56. Yokawa, K.; Fasano, R.; Kagenishi, T.; Baluška, F. Light as stress factor to plant roots—Case of root halotropism. *Front. Plant Sci.* **2014**, *5*, 718. [CrossRef]

57. Burbach, C.; Markus, K.; Zhang, Y.; Schlicht, M.; Baluška, F. Photophobic behavior of maize roots. *Plant Signal. Behav.* **2012**, *7*, 874–878. [CrossRef]

58. Yokawa, K.; Kagenishi, T.; Kawano, T.; Mancuso, S.; Baluška, F. Illumination of *Arabidopsis* roots induces immediate burst of ROS production. *Plant Signal. Behav.* **2011**, *6*, 1460–1464. [CrossRef] [PubMed]

59. Xu, W.; Ding, G.; Yokawa, K.; Baluška, F.; Li, Q.-F.; Liu, Y.; Shi, W.; Liang, J.; Zhang, J. An improved agar-plate method for studying root growth and response of *Arabidopsis thaliana*. *Sci. Rep.* **2013**, *3*, 1273. [CrossRef] [PubMed]

60. Yokawa, K.; Kagenishi, T.; Baluška, F. Root photomorphogenesis in laboratory-maintained *Arabidopsis* seedlings. *Trends Plant Sci.* **2013**, *18*, 117–119. [CrossRef] [PubMed]

61. Novák, J.; Černý, M.; Pavlů, J.; Zemánková, J.; Skalák, J.; Plačková, L.; Brzobohatý, B. Roles of proteome dynamics and cytokinin signaling in root to hypocotyl ratio changes induced by shading roots of *Arabidopsis* seedlings. *Plant Cell Physiol.* **2015**, *56*, 1006–1018. [CrossRef] [PubMed]

62. Silva-Navas, J.; Moreno-Risueno, M.A.; Manzano, C.; Pallero-Baena, M.; Navarro-Neila, S.; Téllez-Robledo, B.; Garcia-Mina, J.M.; Baigorri, R.; Gallego, F.J.; del Pozo, J.C. D-Root: A system for cultivating plants with the roots in darkness or under different light conditions. *Plant J.* **2015**, *84*, 244–255. [CrossRef]

63. Lacek, J.; García-González, J.; Weckwerth, W.; Retzer, K. Lessons learned from the studies of roots shaded from direct root illumination. *Int. J. Mol. Sci.* **2021**, *22*, 12784. [CrossRef]

64. Miotto, Y.E.; da Costa, C.T.; Offringa, R.; Kleine-Vehn, J.; Maraschin, F.D.S. Effects of light intensity on root development in a D-root growth system. *Front. Plant Sci.* **2021**, *12*, 778382. [CrossRef] [PubMed]

65. Qu, Y.; Liu, S.; Bao, W.; Xue, X.; Ma, Z.; Yokawa, K.; Baluška, F.; Wan, Y. Expression of root genes in *Arabidopsis* seedlings grown by standard and improved growing methods. *Int. J. Mol. Sci.* **2017**, *18*, 951. [CrossRef] [PubMed]

66. Yan, X.; Yamashita, F.; Njimona, I.; Baluška, F. Root and hypocotyl growth of *Arabidopsis* seedlings grown under different light conditions and influence of TOR kinase inhibitor AZD. *Int. J. Biotechnol. Mol. Biol. Res.* **2022**, *12*, 22–30. [CrossRef]

67. Wan, Y.-L.; Eisinger, W.; Ehrhardt, D.; Kubitscheck, U.; Baluska, F.; Briggs, W. The subcellular localization and blue-light-induced movement of Phototropin 1-GFP in etiolated seedlings of *Arabidopsis thaliana*. *Mol. Plant* **2008**, *1*, 103–117. [CrossRef] [PubMed]

68. Wan, Y.; Jasik, J.; Wang, L.; Hao, H.; Volkmann, D.; Menzel, D.; Mancuso, S.; Baluška, F.; Lin, J. the signal transducer NPH3 integrates the Phototropin1 photosensor with PIN2-based polar auxin transport in *Arabidopsis* root phototropism. *Plant Cell* **2012**, *24*, 551–565. [CrossRef]

69. Baluška, F.; Mancuso, S.; Volkmann, D.; Barlow, P.W. Root Apex transition zone: A signalling–response nexus in the root. *Trends Plant Sci.* **2010**, *15*, 402–408. [CrossRef]

70. Baluška, F.; Mancuso, S. Root apex transition zone as oscillatory zone. *Front. Plant Sci.* **2013**, *4*, 354. [CrossRef]

71. Baluška, F.; Yamashita, F.; Mancuso, S. Root apex cognition: From neuronal molecules to root-fungal networks. In *Rhizobiology: Molecular Physiology of Plant Roots*; Springer: Cham, Switzerland, 2021; pp. 1–24.

72. Mäthger, L.M.; Roberts, S.B.; Hanlon, R.T. Evidence for distributed light sensing in the skin of cuttlefish, *Sepia officinalis*. *Biol. Lett.* **2010**, *6*, 600–603. [CrossRef]

73. Kingston, A.C.N.; Kuzirian, A.M.; Hanlon, R.T.; Cronin, T.W. Visual phototransduction components in cephalopod chromatophores suggest dermal photoreception. *J. Exp. Biol.* **2015**, *218*, 1596–1602. [CrossRef]

74. Yerramilli, D.; Johnsen, S. Spatial vision in the purple sea urchin *Strongylocentrotus purpuratus* (Echinoidea). *J. Exp. Biol.* **2010**, *213*, 249–255. [CrossRef]

75. Sumner-Rooney, L.; Rahman, I.A.; Sigwart, J.D.; Ullrich-Lüter, E. Whole-body photoreceptor networks are independent of 'lenses' in brittle stars. *Proc. R. Soc. B Biol. Sci.* **2018**, *285*, 20172590. [CrossRef]

76. Garm, A.; Nilsson, D.-E. Visual navigation in starfish: First evidence for the use of vision and eyes in starfish. *Proc. R. Soc. B Biol. Sci.* **2014**, *281*, 20133011. [CrossRef] [PubMed]

77. Korsvig-Nielsen, C.; Hall, M.; Motti, C.; Garm, A. Eyes and negative phototaxis in juvenile crown-of-thorns starfish, *Acanthaster* Species Complex. *Biol. Open* **2019**, *8*, bio041814. [CrossRef] [PubMed]

78. Nilsson, D.E.; Gislén, L.; Coates, M.M.; Skogh, C.; Garm, A. Advanced optics in a jellyfish eye. *Nature* **2005**, *435*, 201–205. [CrossRef]

79. Gehring, W.J. New perspectives on eye development and the evolution of eyes and photoreceptors. *J. Hered.* **2005**, *96*, 171–184. [CrossRef]

80. Duanmu, D.; Bachy, C.; Sudek, S.; Wong, C.-H.; Jiménez, V.; Rockwell, N.C.; Martin, S.S.; Ngan, C.Y.; Reistetter, E.N.; van Baren, M.J.; et al. Marine algae and land plants share conserved phytochrome signaling systems. *Proc. Natl. Acad. Sci. USA* **2014**, *111*, 15827–15832. [CrossRef] [PubMed]

81. Albrecht-Buehler, G. In defense of "nonmolecular" cell biology. *Int. Rev. Cytol.* **1990**, *120*, 191–241. [CrossRef] [PubMed]

82. Albrecht-Buehler, G. Rudimentary form of cellular "vision". *Proc. Natl. Acad. Sci. USA* **1992**, *89*, 8288–8292. [CrossRef]

83. Albrecht-Buehler, G. Cellular infrared detector appears to be contained in the centrosome. *Cell Motil. Cytoskelet.* **1994**, *27*, 262–271. [CrossRef]

84. Albrecht-Buehler, G. Changes of cell behavior by near-infrared signals. *Cell Motil. Cytoskelet.* **1995**, *32*, 299–304. [CrossRef]

85. Albrecht-Buehler, G. Altered drug resistance of microtubules in cells exposed to infrared light pulses: Are microtubules the "nerves" of cells? *Cell Motil. Cytoskelet.* **1998**, *40*, 183–192. [CrossRef]

86. Baluška, F.; Volkmann, D.; Barlow, P.W. Nuclear components with microtubule-organizing properties in multicellular eukaryotes: Functional and evolutionary considerations. *Int. Rev. Cytol.* **1997**, *175*, 91–135. [CrossRef]

87. Mazia, D. The chromosome cycle and the centrosome cycle in the mitotic cycle. *Int. Rev. Cytol.* **1987**, *100*, 49–92. [CrossRef] [PubMed]

Perspective

Decision Making in Plants: A Rooted Perspective

Jonny Lee [1,2], Miguel Segundo-Ortin [1,2,*] and Paco Calvo [1,2]

[1] Minimal Intelligence Laboratory (MINT Lab), University of Murcia, 30100 Murcia, Spain; leejonathan.cw@gmail.com (J.L.); fjcalvo@um.es (P.C.)
[2] Department of Philosophy, University of Murcia, 30100 Murcia, Spain
[*] Correspondence: miguel.segundo@um.es

Abstract: This article discusses the possibility of plant decision making. We contend that recent work on bacteria provides a pertinent perspective for thinking about whether plants make choices. Specifically, the analogy between certain patterns of plant behaviour and apparent decision making in bacteria provides principled grounds for attributing decision making to the former. Though decision making is our focus, the discussion has implications for the wider issue of whether and why plants (and non-neural organisms more generally) are appropriate targets for cognitive abilities. Moreover, decision making is especially relevant to the issue of plant intelligence as it is commonly taken to be characteristic of cognition.

Keywords: decision making; plant behaviour; bacteria; intelligence

1. Introduction

At the centre of debates over plant intelligence lies the question of whether plants possess cognitive abilities, such as learning, memory, numerosity, anticipation, and so on [1–4]. This paper focuses on plant decision making [5] and connects it with the widespread discussion of decision making in non-neural organisms. Generally speaking, an organism is said to make a decision whenever (i) it selects between alternative courses of action, and (ii) this selection is not random but is based on an evaluation of the alternatives in light of some collected information [6]. We contend that recent work on bacteria provides a pertinent perspective for thinking about whether plants make choices. Specifically, the analogy between certain patterns of plant behaviour and apparent decision making in bacteria provides principled grounds for attributing decision making to the former. Though decision making is our focus, the discussion has implications for the wider issue of whether and why plants (and non-neural organisms more generally) are appropriate targets for cognitive science. Moreover, whilst we avoid defending any position on the wider implications for plant intelligence, we note that decision making is commonly taken to be characteristic of cognition (e.g., [7], but see [8]) and is therefore pertinent to debates about plant intelligence.

We begin by introducing the notion of decision making and outlining recent work on bacteria (Section 2). We then turn to prima facie evidence for decision making in plants before discussing one reason to think that the analogy between single-celled organisms and plants does not hold, namely, because plants do not genuinely select between behaviours (Section 3). We close by forecasting the importance of future research (Section 4).

2. Decision Making in Bacteria (and Beyond)

As already mentioned above, decision making involves selecting between several possible options for behaviour based on information about the organism and/or its environment (e.g., see [9–11]). A perennial problem with assessing whether some atypical taxa (such as plants) exhibit a cognitive phenomenon (such as decision making) is defining the ability in question. Nevertheless, we take this generic characterisation to be sufficiently

Citation: Lee, J.; Segundo-Ortin, M.; Calvo, P. Decision Making in Plants: A Rooted Perspective. *Plants* 2023, 12, 1799. https://doi.org/10.3390/plants12091799

Academic Editor: Vittorio Rossi

Received: 10 March 2023
Revised: 19 April 2023
Accepted: 24 April 2023
Published: 27 April 2023

ecumenical as a starting point. The more liberally minded may insist that decision making need not be 'behavioural' but also expressible via physiological or cognitive changes (e.g., [6]). Although we do not preclude a broad definition of behaviour that encompasses physiological/cognitive changes, we must note that the notion of 'behaviour' is itself highly contested (see Section 3.2).

Before examining whether plants undertake decision making, it will be fruitful to turn first to established research on bacteria, insofar as this will furnish us with a clear-cut phylogenetic entry point as we transition from bacterial unicellularity into the acquisition of plant multicellularity, and from prokaryotic into eukaryotic forms of life. The first unicellular eukaryote is thought to have resulted from bacterial genome fusion and synergistic interactions between, probably, cyanobacteria and proteobacteria ancestors [12]. Subsequently, according to phylogenetic reconstruction, two bacterial endosymbiotic events resulted in the origins of the precursors of mitochondria and chloroplasts [13]. First, the uptake of an alpha-purple bacterium marked the origin of the mitochondria in the common ancestor of plants and animals, and at a later stage, the uptake of a photosynthetic cyanobacterium paved the way for chloroplasts, this time, exclusively in the plant lineage. Plants, therefore, presented an evolutionary innovation, whereas the rest of the eukaryotic life forms (up to and including humans) preserved their ancestral cellular organization [14]. One way or another, it is highly unlikely that a previously evolved adaptive trait is jettisoned at a later stage [15].

Following the principle of evolutionary conservatism, it is worth noting that the evolutionary origins of eukaryote neurobiology run very deep in the tree of life with many neural-based aspects of cognition already present in bacteria, serving to channel their cellular processes of survival (e.g., neural network-like signal transduction in bacteria) [16]. In a similar vein, the number of structural and functional similarities between neurons and plant cells being researched keeps growing [17]. Several proteins known to mediate neurotransmission synaptically in animals have been found in bacteria, throwing light upon the phylogenetic development of neurotransmitters; glutamate and gamma-aminobutyric acid (GABA) are among the chemicals that function, not as mere metabolites, but rather as plant signalling molecules ('biomediators', in plant physiological parlance to distinguish them from animal neurotransmitters). In addition, actin and other cellular motors are also found in plants [18].

It is increasingly common to claim that bacteria are capable of elementary forms of decision making. Among the supporting evidence is the discovery of 'control mechanisms' underlying locomotion. These are distributed, 'heterarchichally structured' mechanisms for obtaining information about the organism's internal and external conditions that facilitate the evaluation of alternative behaviours and the selection between them [19]. The efficacy of control mechanisms for producing adaptive behaviour is exemplified by locomotive chemotaxis in *E. coli*. In brief, these bacteria are faced with selecting between directions for locomotion, relying on their flagella (the hair-like structure protruding from the cell body) attached to a motor for moving around, and travelling up or down gradients of different substances that attract or repel them. The motor rotates either clockwise—which moves the organism forward—or counterclockwise—which causes the organism to tumble and turn to face another direction. These behaviours are not triggered randomly or as a simple reaction to perturbation. Rather, they are the result of 'control mechanisms' that gather information and, equally important, evaluate that information to govern 'production mechanisms' (those responsible for the behavioural output) [4]. In particular, *E. coli*, as well as many other bacteria, use a two-component regulatory system (TCS) [20], functionally similar to the nervous system of animals, which serves the role of a memory and inner connection between sensors and effectors. Courtesy of this system, *E. coli* can take sequential measurements of the substance concentration whose net result is a systematic form of chemotaxis [21]. These control mechanisms, however minimal, are adequate for adaptively determining between different possible actions. It is for this reason that many theorists attribute a form of decision making to bacteria.

As *E. coli* demonstrate, the primary appeal of attributing decision making to bacteria is their ability to switch between behaviours based on the receipt and evaluation of information, which resembles decision making in more paradigmatic cases (corresponding to our initial characterisation, above). Furthermore, describing such behaviour selection in bacteria as a form of decision making suggests a generic, non-idiosyncratic (non-taxa specific) notion. This is attractive because it implies that more-or-less similar abilities (i.e., 'decision-making abilities') may be identified and compared across different branches on the tree of life (see Section 3.2 below for related discussion).

As Bechtel and Bich explicate, decision making is 'an activity that all organisms as autonomous systems must perform to keep themselves viable [...] [g]iven the variable nature of the environment and the continual degradation of the organism' [22] (p. 1). In keeping with the bacteria case, the production of flexible behaviour required to survive in a dynamic environment requires organisms to regulate processes of production using mechanisms of control that measure environmental variables and evaluate the resulting information regarding certain standards (or 'norms') of viability. However, control mechanisms are not always hierarchal (i.e., mechanisms organised into successively higher-level control mechanisms) but typically heterarchical. In effect, control mechanisms can function with (more-or-less) independence in the absence of a centralised controller. In short, the case of bacteria demonstrates how selecting between different possible behaviours based on the receipt and evaluation of information according to certain norms of viability is possible without a centralised 'executive' mechanism. Notice that whilst an approach such as that advocated by Bechtel and Bich permits decision making to be widespread—allowing even single-celled organisms to make choices—it does not trivialize the concept, e.g., allowing every biological process to count as decision making. Rather, decision making involves identifiable (if highly distributed) mechanisms of control that measure and evaluate environmental variables.

A first-pass objection to the idea of decision making in bacteria is the assumption that the ability depends on the authority of an executive mechanism. Such a view likely results from modelling decision making on deliberative, conscious choices in humans, where familiar decisions at least seem to be determined by a centralised controller.

However, it is debatable whether the assumption holds in most forms of decision making. For instance, the medicinal leech (*Hirudo verbena*) selects between swimming and crawling but does not depend on a centralised neural mechanism, but rather on the emergent effect of 21 independent ganglia located between its 'head and tail brains' [23] (p. 3). Similarly, extensive work on domesticated cats, for example, has demonstrated that decision-making mechanisms in neural organisms with brains are distributed across cortical and subcortical structures. The neural circuitry responsible for decision making in these cases is critically modulated by a range of often broadly diffused chemical signals carrying information about the state of the environment and organism [19] (p. 1061). Brains, so the evidence shows, do not obviate heterarchical organisation, at the very least. In fact, some human behaviour may emerge from the coordinated activity of heterarchical control mechanisms as well (for extended discussion, see [22,23]).

In summary, even neural organisms rely on decentralised mechanisms and non-neural components when making decisions. One could, of course, still insist that only deliberative decision making of the sort familiar to human introspection is bona fide decision making, hence any similarities between processes in bacteria (or leeches) and human decision making remain superficial when it comes to determining cognitive abilities. We note that this position leads to an excessively restrictive notion of decision making that would exclude even paradigmatic cases of non-conscious decision making in humans which are standardly accepted by cognitive science (e.g., see [10]; see Section 4 below for related discussion).

A related worry stemming from a 'cognitivist' approach is that any genuine cognitive ability must be underwritten by a representational process [24,25]. Hence, for non-neural organisms to make genuine choices in the same (cognitive) sense, it is necessary for them to trade in representations. One might argue that this is the case [26]. However, it is worth noting that cognitivism is no longer the default assumption in the field, and many would reject its conception of cognition nowadays. We cannot delve into these murky issues here. However, notice that even if the elementary forms of decision making surveyed in this paper are not considered bona fide cognition, then the ramifications for understanding the role and distribution of decision-making abilities in the tree of life remain ambiguous; if not all 'decision making' is *truly* cognitive, perhaps *true* cognition is less vital than first thought.

3. Making Our Minds up about Plant Decision Making

Like bacteria and all other organisms, plants face myriad challenges to survival in an unpredictable world. To meet these challenges, plants must continually adapt to their dynamic surroundings by growing flexibly, deploying a range of defence mechanisms, and managing the uptake and distribution of nutrients. Given that plant physiology, like all physiology, incurs energetic costs, plants must constantly prioritise where to grow, which defence mechanisms to trigger, and what resources to favour. On the face of it, it is reasonable to conclude that plants must make choices too. In Section 3.1 we dig deeper into the idea of plant decision making. In Section 3.2 we discuss reasons one might remain sceptical.

3.1. A Potted Introduction to Plant Decision Making

Evidence for plant decision making can be found above and below ground [27]. Well-known above-ground examples are the dodder plant (*Cuscuta pentagona*) [28] and the tropical vine *Monstera gigantea* [29]. Given the choice to parasite a tomato plant (*Lycopersicon esculentum*) or a wheat seedling, the dodder plant will grow toward the former, rejecting the lower quality and less appealing wheat host. However, if wheat is the one and only option available in the vicinity, dodder will grow towards it, although more slowly and growing fewer tendrils [15]. In the case of *Monstera*, young seedlings can tell light and dark patches apart, growing toward the former, as dark patches correspond to the base of the trunks of potential hosts [29]. As the host is reached, *Monstera* seedlings will switch their skototropic, dark-oriented behaviour for a phototropic pattern of upward climbing.

Because these examples have been discussed at length, our focus in this section will be on the less well-known root growth (for similar discussions of decision making at the shoot level see [28,30–34]. Take, for instance, the so-called 'binary decision making' of maize roots [35]. When maize (*Zea mays* L.) roots reach the fork of a Y-maze (a growth space with the shape of an inverted Y), they can grow down one arm or the other. Unsurprisingly, in the absence of volatiles roots exhibit no preference, using only gravitational direction to determine growth. However, when a gradient of volatiles is introduced, roots are repelled or attracted, as inferred from their differential patterns of growth towards or against particular chemical gradients. If exposed to, say, diethyl ether or ethylene in one arm, roots will grow towards it; by contrast, exposition to methyl jasmonate in one arm will trigger an escape tropism, similar to the type of photophobic, avoidance behaviour [36] or halotropic (salt-stress) responses [37] observed in roots. More striking, root growth appears dependent on the combination of environmental conditions such as chemical volatiles, indicating 'that the different combinations of types/concentrations of diverse volatiles affect the root decision making' [35].

The sensitivity of root growth to combinations of environmental conditions instead of single factors has also been found, for example, in the preference of *Calamgrostis canadensis* for light plus warm soil over other combinations [5]. Forced choices between hydrotropism and root gravitropism for differing moisture gradients under the gravity pull have also been reported [38]. Note, in addition, that increased growth in one part of a plant's root network is frequently accompanied by decreased growth in another, indicating that plants coordinate root growth across the whole organism [39,40]. This implies that plants engage in a sort of trade-off evaluation, where the growth of some structures is prioritized over others in relation to the current needs.

Consider this other example. When grown alone, the roots of *Abutilon theophrasti* will distribute broadly and uniformly regardless of whether the nutrient distribution is heterogenous or homogenous [41]. When a competitor is introduced and nutrient distribution is homogenous, roots grow more selectively, avoiding contact (and thus, competition) with neighbouring roots. However, when another exemplar is introduced and nutrient distribution is heterogenous, roots exhibit reduced selectivity, and an increased tendency to grow in areas shared with neighbouring roots. This shows that growth patterns are dependent on integrating information about nutrients and neighbours. More generally, root growth patterns seem to rely on the detection and integration of myriad signals carrying resource and non-resource information [42]. Further work indicates that some plants discriminately distribute more resources to parts of roots in patches of soil with increasing levels of nutrients over those in areas with higher absolute but non-increasing levels of nutrients, meaning the plant root growth is sensitive to temporal change as well [43,44].

Finally, pea plants switch between risk-prone or risk-averse root growth depending on context. Dener et al. [45] grew split-root pea plants in such a way that their root tips could grow into separate pots in two conditions, sharing equal mean nutrient irrigation; in one condition, the pots contained constant levels whilst the other contained fluctuating concentrations. The study supported the conclusion that pea plants preferred soil with variable distribution in the context where mean nutrient levels were sufficiently low but constant distribution where mean nutrient levels were enough to meet their metabolic needs. The authors took this to demonstrate risk sensitivity, switching between risk-prone and risk-averse growth as a function of resource availability, congruent with predictions from risk sensitivity theory (for further discussion on the 'rationality' of root growth patterns, see [46]).

This small sample of the empirical literature suggests that when confronted with a dynamic and heterogeneous environment, plants adaptively select between growth patterns based on information about their environment. In other words, plants seem to choose where to grow in a way that suggests a sort of normative evaluation.

Compared with bacteria, the mechanisms for such apparent decision making in plants are less certain (in part because their physiology is more complex, with processes spanning across the cellular level—say, touch receptors—and the levels of both organs and organism— say, sensitive cells and sensitive hairs, respectively [27]) and harder to generalise (because their physiology varies more across species). However, a sketch is possible: plants achieve behaviours such as selective root growth in response to the environment by exploiting receptors sensitive to a range of stimuli (akin to animals), distributed internal electrical and chemical signalling systems for information integration (akin to single-celled organisms and animals in some cases), and mechanisms for organism-level behaviour, often through phenotypic changes via gene expression (e.g., [47]). This contrasts with the view that plant behaviour is purely genetically determined by natural selection or epigenetically determined by the environment (e.g., [48]).

In summary, though many details are still lacking, plants appear capable of organism-level decision making through distributed mechanisms, such as bacteria. We say 'appear' because one may harbour lingering doubts as to whether the analogy between plants and bacteria holds because only bacteria select between genuine behaviours. We deal with this objection in the following section.

3.2. Growing Pains

With the aid of a microscope, one can appreciate the buzz of bacterial activity. However, gazing at a potted cactus or strip of grass, plants can appear tediously immobile. Compared with bacteria, it is harder to think of plants as behaving, and one might insist that, unlike the former, plants do not selectively move by integrating information. In this section, we offer an answer to both concerns. First, we argue that it is not clear that movement is required for behaviour. Second, we contend that plants, like bacteria, do select between movements, albeit (a) at a slower time scale and (b) primarily via phenotypic plasticity (e.g., patterns of growth), rather than locomotion. Taking into account the evidence surveyed above, we hold that the analogy between bacteria and plants is strengthened: both select between movement-based behaviours (*mutatis mutandis*) based on the evaluation and integration of information via distributed (non-centralised) mechanisms. Thus, if one grants decision making to bacteria, one ought to grant decision making to plants.

'Behaviour' is a notoriously vague concept, with disparate definitions found across disciplines. In responding to this ambiguity, Levitis et al. [49] propose a discipline-neutral definition based on a meta-study of responses across biology: 'behaviour is the internally coordinated responses (actions or inactions) of whole living organisms (individuals or groups) to internal and/or external stimuli, excluding responses more easily understood as developmental changes' (p. 103). This non-idiosyncratic definition comfortably encompasses plants alongside bacteria. Notice, however, that the definition does not depend on movement; if plants do not move, they are not thereby excluded from behaving. Rather, what matters is whether organisms internally coordinate actions and our examples above suggest that plants do. Thus, taking such a characterisation for granted, there is no reason to deny decision making to plants on the basis that they do not move [50].

However, even if one insists on a more restrictive definition of behaviour that required movement (e.g., [51]), we see no reason to exclude plants [5,52] either. The idea that plants move, via idiosyncratic means, stretches at least as far back as Darwin (for example, see 'The Power of Movement in Plants'; [53]). Darwin appreciated that plants are constantly in motion (for a book-length tribute to the pioneering work of Darwin, see [15]). Of course, plants do not locomote. Instead, plants primarily achieve motion via directional growth responses to the environment (such as phototropism and gravitropism), as well as non-directional movements that are typically regulated by turgor pressure or electrical stimulation (such as thigmonasty and thermonasty). Some plant movement is incredibly fast; *Mimosa pudica* folds its leaves in response to touch in around 5 s, whilst Venus flytraps (*Dionaea muscipula*) close their traps around 100 ms (neither are growth-based movements). However, most plant movement is growth-based and slow compared with animal movement, and imperceptible to the human eye. This likely goes some way to account for our tendency to think of plants as stationary. The stark reality of plant motion is laid bear with timelapse photography which allows plant motion to be perceptible at our timescale. Timelapse photography does for our appreciation of plants what microscopes do for our appreciation of bacteria.

Plants thus move slowly and largely by growth but, following Darwin, they do move. Thus, even if decision making requires selecting between movements, then plants are not excluded from decision making. The analogy between bacteria and plants is saved. To be clear, the claim is not that all plant movement counts as behaviour (or decision making for that matter) any more than all animal movement does. Knee-jerk reactions are excluded, for example. Rather, we are claiming that there are more ways to move than locomotion.

To see this more clearly, consider the well-studied example of *Physarum polycephalum* (aka 'slime mould'). *P. polycephalum* is a unicellular protist which has received much attention for the complex behaviour it shows during its multinucleate plasmodial phase. At this stage, slime mould consists of a network of tubules which carry protoplasm throughout the entire organism courtesy of a series of oscillators that pulse, expanding and contracting the tubules, depending on external circumstances and the state of the nearby oscillators. When the organism detects an attractant, pulses nearest to the attractant increase, causing the organism to grow towards it. The opposite occurs when the organism detects a repellent: activity of the oscillators decreases, reducing the flow of protoplasm in this area.

Not unlike plants, *P. polycephalum* has been tested in multiple protocols adapted from human and animal decision-making studies [6]. These experiments have shown that slime mould compares the relative properties of multiple options in making choices [54] in that it can discriminate high-calorie over low-calorie food, and that it can make sophisticated trade-offs when access to some nutrient source involves exposure to danger [55,56]. More strikingly, it has been reported that slime mould is susceptible to some biases previously observed in human and non-human animals [57]. Overall, these studies reinforce the view that brainless organisms can sample and integrate information from different internal and external parameters in order to make adaptive decisions. As Smith-Ferguson and Beekman explicate, '[t]he coupling of neighbouring oscillators means information can be encoded or "entrained" into oscillation frequencies and transferred to parts of the plasmodium which are too far to detect the chemical cues. Hence, the physiology of the organism—its fluid dynamics—allows it to transfer information throughout the organism without the need for a nervous system' [58] (p. 467). Locomotion is not here considered a necessary condition for behaviour and decision making.

In summary, the relevant (functional) analogy holds between bacteria, plants, and other organisms such as protists. If we grant idiosyncratic forms of behaviour selection in different organisms, it becomes easier to accept decision making in plants. In other words, if we (i) accept minimal decision-making abilities in taxa such as prokaryotes and protists alongside (ii) movement via growth, the argument for extending decision-making abilities to plants is strengthened. Alternatively, pressure is placed on the sceptics of plant decision making to either deny decision making in bacteria (and protists) or demonstrate some non-arbitrary difference between the former and plant behaviour.

4. Future Research

Research on bacteria suggests that prokaryotes may serve as 'experimental organisms' for studying decision making more broadly (up to the level of non-conscious human decision making), with an emphasis placed on the fact that discovering the ability in question in simpler organisms assists in revealing the core characteristics of the mechanisms underlying that phenomenon. For example, Huang et al. [19] argue that by identifying mechanisms for decision making in these (relatively) simple cases, we may gain insight into the mechanisms for decision making in more prototypical cases, as in humans and other animals (p. 1064). As we have seen, this lesson extends beyond prokaryotes to include other 'minimal' decision makers (see the example of slime moulds, which are eukaryotic), with the potential to include plants. It goes without saying that the specific mechanisms will vary by necessity. In the aforementioned illustration of root growth behaviour, different volatiles may serve to modulate cellular membrane properties at the root apex, which in turn would explain the differential distribution of the plant hormone auxin that results in the positive or negative tropism exhibited [35]. Yet at a higher level of description, membrane properties will serve to identify common threads, as plant–animal comparative electrophysiology reveals [18]. The response to anaesthesia by both animals and plants, whereby the integrity of the plasma membrane is compromised with the alteration of key membrane properties [59,60] provides a clear-cut illustration of this.

A comparison of traits across different taxa may also offer insight into the evolutionary history of decision making. As Petrillo and Rosati [61] write 'the broad lesson is that evolutionary explanations for a given species' pattern of decision-making need to account for how that strategy plays out for specific species in their specific ecological context' (p. 780). Using the example of diverging preferences in decision making about the temporal and spatial distribution of rewards in cotton-top tamarins and common marmosets, the authors go on to note that '[e]mpirical evidence from comparative studies suggests that some differences in species decision-making strategies map onto differences in these species' wild ecology' (p. 781). Whilst De Petrillo and Rosati are concerned with comparative animal cognition, we can see how their comparative method might apply, on a greater scale, across the tree of life.

Promising insights from studying decision making in experimental organisms, such as bacteria and plants, for our understanding of decision making in more prototypical cases, such as humans and other animals, itself provides justification for attributing genuine decision-making abilities to the experimental organisms. If studying abilities in experimental organisms that resemble decision making in prototypical cases, such that research in the former leads to discoveries in the latter, then we should consider recognising that the experimental organisms possess that ability. Or more pragmatically, by treating organisms such as bacteria and plants as capable of making choices, we gain insight into less contested cases of decision making in other organisms. Ultimately, one may fear that any refusal to rubber-stamp the decision-making credentials of bacteria or plants reflects a mere semantic (but potentially unhelpful) preference if bacteria and plant processes do resemble paradigmatic decision making to the extent that the former guides discoveries about the latter (for related discussion see [62]).

The search for decision making in plants may further expand our use of non-neural taxa for the identification of key components in decision making across the tree of life. In addition to engaging with the broader philosophical debate around the extension of psychological predicates, future work should further detail the control mechanisms for plant decision making and the potential of plants as experimental organisms, whilst also still exploring how plants make choices by idiosyncratic, plant-specific means.

5. Conclusions

We should take seriously the possibility that plants make choices. This paper presented recent research that evidences decision making in bacteria, thus supporting the broader notion that decision making does not require a centralised system for processing information. However, one might think there is a breakdown in the analogy between plants and bacteria because only the latter select between an array of genuine behaviours; in particular, plants do not move. We argued that we ought to accept that plants behave in the same sense as bacteria (*mutatis mutandis*) because plants do move, albeit at a slower timescale than most animal movements and primarily via growth. If we accept decision making in bacteria, and we accept that plants select between movements in response to their environment, then we have firm grounds to accept that plants make decisions.

Author Contributions: Conceptualization, J.L., M.S.-O. and P.C.; investigation, M.S.-O. and P.C.; writing—original draft preparation, J.L.; writing—review and editing, J.L., M.S.-O. and P.C. All authors have read and agreed to the published version of the manuscript.

Funding: J.L. was supported by a Juan de la Cierva Fellowship from Ministerio de Ciencia e Innovación del Gobierno de España (Award # FJC2019-041071-I); M.S.-O. was supported by a Ramón y Cajal Fellowship from Ministerio de Ciencia e Innovación del Gobierno de España (Award # RYC2021-031242-I).

Data Availability Statement: The data is contained within the manuscript.

Conflicts of Interest: The authors declare no conflict of interest.

References

1. Segundo-Ortin, M.; Calvo, P. Are plants cognitive? A reply to Adams. *Stud. Hist. Philos. Sci. Part A* **2018**, *73*, 64–71. [CrossRef] [PubMed]
2. Segundo-Ortin, M.; Calvo, P. Consciousness and cognition in plants. *WIREs Cogn. Sci.* **2021**, *13*. [CrossRef]
3. Gagliano, M. Inside the Vegetal Mind: On the Cognitive Abilities of Plants. In *Memory and Learning in Plants*; Baluska, F., Gagliano, M., Witzany, G., Eds.; Springer: Berlin/Heidelberg, Germany, 2018; pp. 215–220.
4. Castiello, U. (Re)claiming plants in comparative psychology. *J. Comp. Psychol.* **2021**, *135*, 127–141. [CrossRef] [PubMed]
5. Trewavas, A.J. *Plant Behaviour and Intelligence*, 1st ed.; Oxford University Press: Oxford, UK, 2014.
6. Reid, C.R.; Garnier, S.; Beekman, M.; Latty, T. Information integration and multiattribute decision making in non-neuronal organisms. *Anim. Behav.* **2015**, *100*, 44–50. [CrossRef]
7. Shettleworth, S.J. *Cognition, Evolution, and Behavior*, 2nd ed.; Oxford University Press: Oxford, UK; New York, NY, USA, 2010.
8. Keijzer, F. Making Decisions does not Suffice for Minimal Cognition. *Adapt. Behav.* **2003**, *11*, 266–269. [CrossRef]
9. Wang, Y.; Ruhe, G. The Cognitive Process of Decision Making. *Int. J. Cogn. Inform. Nat. Intell.* **2007**, *1*, 73–85. [CrossRef]
10. Stevens, J.R. Mechanisms for Decisions about the Future. In *Animal Thinking: Contemporary Issues in Comparative Cognition*; Menzel, R., Fisher, J., Eds.; The MIT Press: Cambridge, MA, USA, 2011; pp. 93–104.

11. Budaev, S.; Jørgensen, C.; Mangel, M.; Eliassen, S.; Giske, J. Decision-Making From the Animal Perspective: Bridging Ecology and Subjective Cognition. *Front. Ecol. Evol.* **2019**, *7*, 164. [CrossRef]

12. Simonson, A.B.; Servin, J.A.; Skophammer, R.G.; Herbold, C.W.; Rivera, M.C.; Lake, J.A. Decoding the genomic tree of life. *Proc. Natl. Acad. Sci. USA* **2005**, *102*, 6608–6613. [CrossRef] [PubMed]

13. Meyerowitz, E.M. Plants Compared to Animals: The Broadest Comparative Study of Development. *Science* **2002**, *295*, 1482–1485. [CrossRef] [PubMed]

14. Bouteau, F.; Grésillon, E.; Chartier, D.; Arbelet-Bonnin, D.; Kawano, T.; Baluška, F.; Mancuso, S.; Calvo, P.; Laurenti, P. Our sisters the plants? notes from phylogenetics and botany on plant kinship blindness. *Plant Signal. Behav.* **2021**, *16*, 2004769. [CrossRef]

15. Calvo, P.; Lawrence, N. *Planta Sapiens: Unmasking Plant Intelligence*; The Bridge Street Press: London, UK, 2022.

16. Baluška, F.; Mancuso, S. Deep evolutionary origins of neurobiology: Turning the essence of 'neural' upside-down. *Commun. Integr. Biol.* **2009**, *2*, 60–65. [CrossRef]

17. Baluška, F. Recent surprising similarities between plant cells and neurons. *Plant Signal. Behav.* **2010**, *5*, 87–89. [CrossRef] [PubMed]

18. Lee, J.; Calvo, P. The Potential of Plant Action Potentials. 2022. Available online: http://philsci-archive.pitt.edu/21287/ (accessed on 31 October 2022).

19. Huang, L.T.-L.; Bich, L.; Bechtel, W. Model Organisms for Studying Decision-Making: A Phylogenetically Expanded Perspective. *Philos. Sci.* **2021**, *88*, 1055–1066. [CrossRef]

20. Véscovi, E.G.; Sciara, M.I.; Castelli, M.E. Two component systems in the spatial program of bacteria. *Curr. Opin. Microbiol.* **2010**, *13*, 210–218. [CrossRef] [PubMed]

21. Garzón, P.C.; Keijzer, F. Plants: Adaptive behavior, root-brains, and minimal cognition. *Adapt. Behav.* **2011**, *19*, 155–171. [CrossRef]

22. Bechtel, W.; Bich, L. Grounding cognition: Heterarchical control mechanisms in biology. *Philos. Trans. R. Soc. B Biol. Sci.* **2021**, *376*, 20190751. [CrossRef] [PubMed]

23. Huang, L.T.-L.; Bechtel, W. A Phylogenetic Perspective on Distributed Decision-Making Mechanisms. Available online: http://mechanism.ucsd.edu/research/Huang-2020-A%20phylogenetic%20perspective%20on%20distr.pdf (accessed on 10 March 2023).

24. Adams, F.R.; Garrison, R. The Mark of the Cognitive. *Minds Mach.* **2023**, *23*, 339–352. [CrossRef]

25. Colaço, D. Why studying plant cognition is valuable, even if plants aren't cognitive. *Synthese* **2022**, *200*, 453. [CrossRef]

26. Garzón, F.C. The Quest for Cognition in Plant Neurobiology. *Plant Signal. Behav.* **2007**, *2*, 208–211. [CrossRef] [PubMed]

27. Barlow, P.W. Reflections on 'plant neurobiology'. *Biosystems* **2008**, *92*, 132–147. [CrossRef]

28. Runyon, J.B. Volatile Chemical Cues Guide Host Location and Host Selection by Parasitic Plants. *Science* **2006**, *313*, 1964–1967. [CrossRef] [PubMed]

29. Strong, D.R.; Ray, T.S. Host Tree Location Behavior of a Tropical Vine (*Monstera gigantea*) by Skototropism. *Science* **1975**, *190*, 804–806. [CrossRef]

30. Darwin, C. *On the Movements and Habits of Climbing Plants*; Cambridge University Press: Cambridge, UK, 1875.

31. Saito, K. A study on diameter-dependent support selection of the tendrils of Cayratia japonica. *Sci. Rep.* **2022**, *12*, 4461. [CrossRef]

32. Gianoli, E. The behavioural ecology of climbing plants. *AoB PLANTS* **2015**, *7*, plv013. [CrossRef] [PubMed]

33. Kelly, C.K. Resource choice in Cuscuta europaea. *Proc. Natl. Acad. Sci. USA* **1992**, *89*, 12194–12197. [CrossRef] [PubMed]

34. Gruntman, M.; Groß, D.; Májeková, M.; Tielbörger, K. Decision-making in plants under competition. *Nat. Commun.* **2017**, *8*, 2235. [CrossRef]

35. Yokawa, K.; Derrien-Maze, N.; Mancuso, S.; Baluška, F. Binary Decisions in Maize Root Behavior: Y-Maze System as Tool for Unconventional Computation in Plants. *Int. J. Unconv. Comput.* **2014**, *10*, 381–390.

36. Burbach, C.; Markus, K.; Zhang, Y.; Schlicht, M.; Baluška, F. Photophobic behavior of maize roots. *Plant Signal. Behav.* **2012**, *7*, 874–878. [CrossRef] [PubMed]

37. Galvan-Ampudia, C.S.; Julkowska, M.M.; Darwish, E.; Gandullo, J.; Korver, R.A.; Brunoud, G.; Haring, M.A.; Munnik, T.; Vernoux, T.; Testerink, C. Halotropism Is a Response of Plant Roots to Avoid a Saline Environment. *Curr. Biol.* **2013**, *23*, 2044–2050. [CrossRef] [PubMed]

38. Takahashi, N.; Yamazaki, Y.; Kobayashi, A.; Higashitani, A.; Takahashi, H. Hydrotropism interacts with gravitropism by degrading amyloplasts in seedling roots of Arabidopsis and radish. *Plant Physiol.* **2003**, *132*, 805–810. [CrossRef] [PubMed]

39. Hodge, A. Root decisions. *Plant Cell Environ.* **2009**, *32*, 628–640. [CrossRef] [PubMed]

40. Firn, R. Plant Intelligence: An Alternative Point of View. *Ann. Bot.* **2004**, *93*, 345–351. [CrossRef] [PubMed]

41. Cahill, J.F.; McNickle, G.G.; Haag, J.J.; Lamb, E.G.; Nyanumba, S.M.; Clair, C.C.S. Plants Integrate Information About Nutrients and Neighbors. *Science* **2010**, *328*, 1657. [CrossRef] [PubMed]

42. Novoplansky, A. What plant roots know? *Semin. Cell Dev. Biol.* **2019**, *92*, 126–133. [CrossRef]

43. Shemesh, H.; Arbiv, A.; Gersani, M.; Ovadia, O.; Novoplansky, A. The Effects of Nutrient Dynamics on Root Patch Choice. *PLoS ONE* **2010**, *5*, e10824. [CrossRef]

44. Novoplansky, A. Future Perception in Plants. In *Anticipation across Disciplines*; Nadin, M., Ed.; Springer: Berlin/Heidelberg, Germany, 2015; pp. 57–70.

45. Dener, E.; Kacelnik, A.; Shemesh, H. Pea Plants Show Risk Sensitivity. *Curr. Biol.* **2016**, *26*, 1763–1767. [CrossRef]

46. Schmid, B. Decision-Making: Are Plants More Rational than Animals? *Curr. Biol.* **2016**, *26*, R675–R678. [CrossRef] [PubMed]

47. Karban, R.; Orrock, J.L. A judgment and decision-making model for plant behavior. *Ecology* **2018**, *99*, 1909–1919. [CrossRef]

48. Taiz, L.; Alkon, D.; Draguhn, A.; Murphy, A.; Blatt, M.; Thiel, G.; Robinson, D.G. Reply to Trewavas et al. and Calvo and Trewavas. *Trends Plant Sci.* **2020**, *25*, 218–220. [CrossRef] [PubMed]

49. Levitis, D.A.; Lidicker, W.Z.; Freund, G. Behavioural biologists do not agree on what constitutes behaviour. *Anim. Behav.* **2009**, *78*, 103–110. [CrossRef] [PubMed]

50. Cvrčková, F.; Žárský, V.; Markoš, A. Plant Studies May Lead Us to Rethink the Concept of Behavior. *Front. Psychol.* **2016**, *7*, 622. [CrossRef] [PubMed]

51. Tinbergen, N. *The Study of Instinct*; Oxford University Press: New York, NY, USA, 1951.

52. Trewavas, A. The foundations of plant intelligence. *Interface Focus* **2017**, *7*, 20160098. [CrossRef] [PubMed]

53. Darwin, C. *The Power of Movement in Plants: Darwin's 1880 Book on Phototropism in Plants*; D. Appelton and Company: New York, NY, USA, 2022.

54. Reid, C.R.; MacDonald, H.; Mann, R.P.; Marshall, J.A.R.; Latty, T.; Garnier, S. Decision-making without a brain: How an amoeboid organism solves the two-armed bandit. *J. R. Soc. Interface* **2016**, *13*, 20160030. [CrossRef]

55. Beekman, M.; Latty, T. Brainless but Multi-Headed: Decision Making by the Acellular Slime Mould Physarum polycephalum. *J. Mol. Biol.* **2015**, *427*, 3734–3743. [CrossRef]

56. Latty, T.; Beekman, M. Food quality and the risk of light exposure affect patch-choice decisions in the slime mold Physarum polycephalum. *Ecology* **2010**, *91*, 22–27. [CrossRef]

57. Latty, T.; Beekman, M. Speed–accuracy trade-offs during foraging decisions in the acellular slime mould *Physarum polycephalum. Proc. R. Soc. B Boil. Sci.* **2010**, *278*, 539–545. [CrossRef]

58. Smith-Ferguson, J.; Beekman, M. Who needs a brain? Slime moulds, behavioural ecology and minimal cognition. *Adapt. Behav.* **2019**, *28*, 465–478. [CrossRef]

59. Jakšová, J.; Rác, M.; Bokor, B.; Petřík, I.; Novák, O.; Reichelt, M.; Mithöfer, A.; Pavlovič, A. Anaesthetic diethyl ether impairs long-distance electrical and jasmonate signaling in Arabidopsis thaliana. *Plant Physiol. Biochem.* **2021**, *169*, 311–321. [CrossRef]

60. Pavlovič, A.; Jakšová, J.; Kučerová, Z.; Špundová, M.; Rác, M.; Roudnický, P.; Mithöfer, A. Diethyl ether anesthesia induces transient cytosolic [Ca2+] increase, heat shock proteins, and heat stress tolerance of photosystem II in Arabidopsis. *Front. Plant Sci.* **2022**, *13*, 995001. [CrossRef]

61. De Petrillo, F.; Rosati, A.G. Decision naking in animals: Rational choices and adaptive strategies. In *The Cambridge Handbook of Animal Cognition*, 1st ed.; Kaufman, A.B., Call, J., Kaufman, J.C., Eds.; Cambridge University Press: Cambridge, UK, 2021; pp. 770–791.

62. Calvo, P.; Trewavas, A. Physiology and the (Neuro)biology of Plant Behavior: A Farewell to Arms. *Trends Plant Sci.* **2020**, *25*, 214–216. [CrossRef] [PubMed]

plants

Article

Changes in the Amount and Distribution of Soil Nutrients and Neighbours Have Differential Impacts on Root and Shoot Architecture in Wheat (*Triticum aestivum*)

Habba F. Mahal [1], Tianna Barber-Cross [1], Charlotte Brown [2], Dean Spaner [3] and James F. Cahill, Jr. [1,*]

[1] Department of Biological Sciences, University of Alberta, Edmonton, AB T6G 2E9, Canada; habbamahal@gmail.com (H.F.M.)

[2] Département de Biologie, Université de Sherbrooke, Sherbrooke, QC J1K 2R1, Canada

[3] Department of Agricultural, Food and Nutritional Science, University of Alberta, Edmonton, AB T6G 2P5, Canada

* Correspondence: james.cahill@ualberta.ca

Abstract: Plants exhibit differential behaviours through changes in biomass development and distribution in response to environmental cues, which may impact crops uniquely. We conducted a mesocosm experiment in pots to determine the root and shoot behavioural responses of wheat, *T. aestivum*. Plants were grown in homogeneous or heterogeneous and heavily or lightly fertilized soil, and alone or with a neighbour of the same or different genetic identity (cultivars: CDC Titanium, Carberry, Glenn, Go Early, and Lillian). Contrary to predictions, wheat did not alter relative reproductive effort in the presence of neighbours, more nutrients, or homogenous soil. Above and below ground, the plants' tendency to use potentially shared space exhibited high levels of plasticity. Above ground, they generally avoided shared, central aerial space when grown with neighbours. Unexpectedly, nutrient amount and distribution also impacted shoots; plants that grew in fertile or homogenous environments increased shared space use. Below ground, plants grown with related neighbours indicated no difference in neighbour avoidance. Those in homogenous soil produced relatively even roots, and plants in heterogeneous treatments produced more roots in nutrient patches. Additionally, less fertile soil resulted in pot-level decreases in root foraging precision. Our findings illustrate that explicit coordination between above- and belowground biomass in wheat may not exist.

Keywords: *Triticum aestivum*; intraspecific competition; kin recognition; nutrient addition; reproductive effort; fitness; aboveground; crown shyness; belowground; root foraging

Citation: Mahal, H.F.; Barber-Cross, T.; Brown, C.; Spaner, D.; Cahill, J.F., Jr. Changes in the Amount and Distribution of Soil Nutrients and Neighbours Have Differential Impacts on Root and Shoot Architecture in Wheat (*Triticum aestivum*). *Plants* **2023**, *12*, 2527. https://doi.org/10.3390/plants12132527

Academic Editors: Frantisek Baluska and Gustavo Maia Souza

Received: 15 May 2023
Revised: 23 June 2023
Accepted: 24 June 2023
Published: 2 July 2023

1. Introduction

Resource capture and plant–plant social interactions are inherently spatially explicit, driven by specific plant organs placed in specific locations at specific points in time. Further, the location of specific organs can influence immediate and future resource capture [1–3]. Thus, there has likely been powerful selection pressures on plants to exhibit high levels of plasticity in organ placement in response to local conditions, a phenomenon falling within the larger domain of plant behavioural ecology [4–7]. Accordingly, current research suggests that plants use environmental information to inform behavioural responses that impact their fitness. For example, plants may react to neighbours' indirect effects, including shading and resource depletion [6]. The behavioural reactions of plants to spatially explicit features of landscapes involve localized movement and changes in overall development and growth [8–10]. Commonly described behaviours include alterations in biomass development and spatial distribution, including differential root production and stem/leaf orientation shifts in response to local conditions [11–14]. Most studies, however, investigate these behaviours in isolation rather than taking an integrated approach exploring multiple behavioural responses within an individual or population. Thus, whether plants integrate information about stimuli and exhibit overall coordination in these above- or belowground responses, or whether there are differences in shoot and root responses to the same stimuli, remains unknown.

A plant's responses above and below ground due to other organisms and their placement in the environment may result in altered resource capture [1,15–18]. These responses have cascading effects on neighbouring plants since competition is a prevalent ecological and evolutionary driver [17]. For example, plants may exhibit crown shyness [19–22], altering shoot placement to reduce leaf overlap in response to neighbouring plants within a potentially shared canopy space. Such a behavioural response alters net photosynthesis [23,24], potentially impacting plant growth and fitness. Behavioural responses to neighbour shoots are not limited to forested systems, and there could be fitness benefits in any system with moderate to dense plant communities. For example, individuals in dense habitats such as crop fields could be victims of density-dependent mortality, but they may prevent this by shifting their crowns strategically [24,25]. Within root systems, plants may move root biomass depending on nutrient availability, soil patchiness, and the presence and identity of neighbouring plants [26–28]. Although these behaviours are well supported by ecological and behavioural theory, it is also well established that in animal species analogous behaviours are context-dependent [29–32]. However, this contextual information is lacking for most plant species, particularly non-model species.

Evolutionary theory suggests that individuals may alter competition with neighbours under certain conditions as a function of their degree of relatedness [33,34]. There is mixed support for kin-dependent behavioural responses in plants [35,36], and these kin-dependent effects may be locally contingent upon the relative costs or benefits of different behavioural decisions. Kin selection theory ties together inclusive fitness and altruism between close relatives in animal and plant species, which is highly important for crop plants, typically genetic monocultures. Many plant species have developed mechanisms to detect and determine both the neighbours' presence and identity, including changes in light quality or shading [37], touch [38,39], and volatile organic compounds [22,40,41]. Plants then alter aboveground biomass to facilitate or hinder neighbour development [22]. Accordingly, with plant reproduction, studies have shown that some species might have higher reproductive success when grown with siblings than with unrelated plants [7,21,42], some have decreased seed yields [43–47], and some exhibit no difference in reproduction [45]. Similarly, many wild and domesticated species can distinguish between neighbours below ground, indicating kin versus non-kin [22,48] and self-recognition versus non-self-recognition [49–51]. Studies looking at root-mediated kin recognition show shifts in biomass allocation [52–55], lateral root density [55], root branching intensity [55], specific root length [56], and resource uptake rate [20,57]. These species may engage in a tragedy of the commons, allocating more resources to root development to maximize nutrient capture at the expense of their competitors [49,58–61]. This is highlighted in studies focusing on root foraging precision. Plants exhibit plasticity in root development and spatial distribution when dealing with the interplay of nutrient density or neighbour presence [18,22,62–64], typically producing more roots in high-nutrient, high-yield soil patches [49,59,65–67].

However, whether plants coordinate above- and belowground behavioural responses to nutrients and neighbours or whether root and shoot systems respond individually to local conditions remains unknown. Studies on plant behaviours in patchy environments have been concerned with changes in a single plant's above- or belowground biomass distribution, leaving significant literature gaps. Accordingly, direct connections between changes below ground and the aboveground repercussions remain unclear. Finally, no studies have examined the combined effects of intraspecific kin recognition, soil fertility, and nutrient placement both above and below ground. Our study explored the coordination between shoot and root behavioural responses in wheat (*Triticum aestivum* L.). We used wheat because of its global importance as a food crop, and though many studies have investigated the morphological traits and yield of different cultivars [68–70], few have focused on behaviour. Furthermore, though there is evidence that wheat alters root and shoot proliferation in the presence of neighbours [61,71] and resource patches [72–74], no studies have explored the effects of both factors simultaneously, making it an ideal choice.

We conducted a mesocosm experiment on five different wheat varieties where we manipulated soil nutrient amounts, soil heterogeneity, and neighbour genotypic identity. We hypothesized that *Triticum aestivum* L. plants, in the presence of a related neighbour, would have increased reproductive effort and reproductive biomass due to kin-related behavioural adjustments. Furthermore, when looking at kin effects above ground, we hypothesized that plants grown with neighbours would exhibit increased neighbour avoidance and crown shyness, decreasing the use of the shared aerial space between the plants. Consequently, plants would decrease crown shyness and aboveground avoidance in highly nutritious or homogenous soil as the cost of using shared aerial space is likely lessened due to potentially increased aboveground biomass and overall photosynthetic capabilities. Finally, in studying kin effects below ground, we predicted plants would exhibit increased neighbour avoidance in the presence of related plants, decreasing their foraging precision in nutrient patches and the directly shared soil space to mitigate competition. Accordingly, the value of the nutrient patch is likely greater at low soil fertility levels, resulting in increased root distribution in the nutrient patches due to increased root foraging efforts by the plant.

2. Results

2.1. Fitness Effects

We found that *Triticum aestivum* L. plants generally had high levels of reproductive effort (Figure 1). However, it was slightly higher when plants were grown alone than with a neighbour, regardless of the neighbour's identity, soil fertility, or nutrient distribution (df1, df2 = 2, 583, F = 2.758, p = 0.0643, Table 1). Additionally, contrary to our hypothesis, plants grown in lower nutrient conditions illustrated higher levels of reproductive effort than those grown in higher nutrient conditions, irrespective of nutrient distribution (df1, df2 = 1, 583, F = 7.243, p =0.0073, Table 1). However, there was no difference in the reproductive effort between plants grown in homogeneous or patchy soil (df1, df2 = 1, 583, F = 0.66, p = 0.4169, Table 1). For the overall interaction between nutrient level and soil homogeneity, there were no significant differences between plants grown alone or with neighbours (df1, df2 = 2, 583, F = 0.271, p = 0.7628, Table 1).

Table 1. The effects of neighbour identity, soil fertility, and nutrient distribution on shoot and root asymmetry, root precision, and reproductive effort. The results are from 4 beta regressions, and bolded values represent significant relationships (alpha = 0.05).

Model	F-Value (df1, df2)	p-Value
Reproductive effort		
neighbour	2.518 (2, 580)	0.0815
fertility	4.494 (1, 580)	**0.0344**
distribution	1.005 (1, 580)	0.3166
neighbour:fertility	1.938 (2, 580)	0.1449
neighbour:distribution	0.979 (2, 580)	0.3763
fertility:distribution	0.100 (1, 580)	0.7524
neighbour:fertility:distribution	0.331 (2, 580)	0.7185
Shoot asymmetry		
neighbour	1.522 (2, 583)	0.219
fertility	7.243 (1, 583)	**0.0073**
distribution	0.66 (1, 583)	0.4169
neighbour:fertility	2.758 (2, 583)	0.0643
neighbour:distribution	0.617 (2, 583)	0.5397
fertility:distribution	0.071 (1, 583)	0.7905
neighbour:fertility:distribution	0.271 (2, 583)	0.7628

Table 1. *Cont.*

Model	F-Value (df1, df2)	*p*-Value
Root asymmetry		
neighbour	0.152 (2, 582)	0.8589
fertility	1.622 (1, 582)	0.2034
distribution	0.538 (1, 582)	0.4634
neighbour:fertility	0.044 (2, 582)	0.9572
neighbour:distribution	1.062 (2, 582)	0.3463
fertility:distribution	11.975 (1, 582)	**0.0006**
neighbour:fertility:distribution	0.878 (2, 582)	0.4163
Root precision		
neighbour	0.634 (2, 582)	0.5311
fertility	10.739 (1, 582)	**0.0011**
distribution	330.789 (1, 582)	**<0.0001**
neighbour:fertility	0.233 (2, 582)	0.7926
neighbour:distribution	0.696 (2, 582)	0.4991
fertility:distribution	0.615 (1, 582)	0.4333
neighbour:fertility:distribution	0.577 (2, 582)	0.562

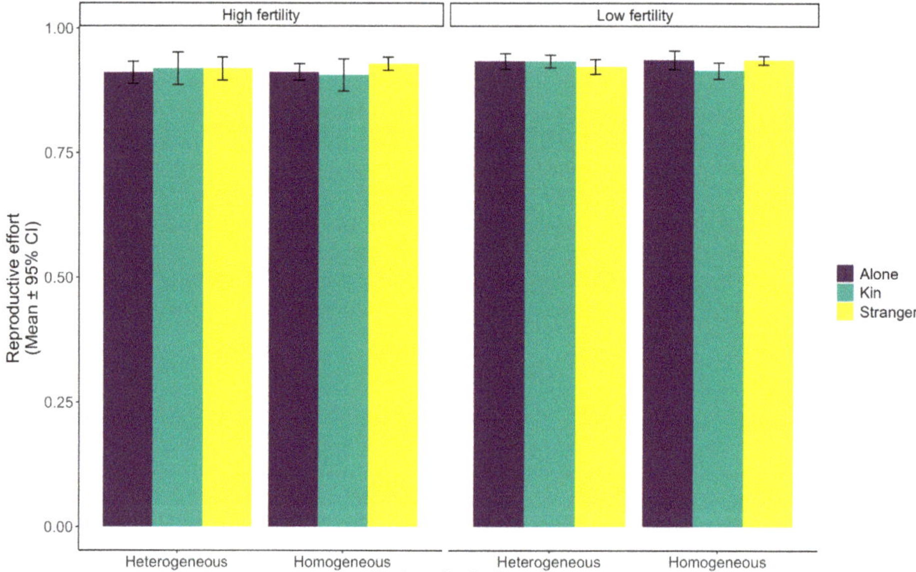

Figure 1. Wheat (*Triticum aestivum*) reproductive effort (mean ± 95% confidence interval) in response to identity, nutrient distribution, and fertility. The reproductive effort was measured as the amount of reproductive biomass the plant grew divided by the plant's total aboveground biomass.

2.2. Aboveground Biomass Distribution

We did not observe a significant overall effect of neighbours (df1, df2 = 2, 583, F = 1.522, *p* = 0.219, Table 1, Figure 2). However, plants grown with neighbours exhibited slightly increased shoot asymmetry under high-fertility soil conditions, regardless of neighbour identity (df1, df2 = 2, 583, F = 2.758, *p* = 0.0643, Table 1). Similarly, under low-fertility conditions, shoot asymmetry was slightly lower when plants were grown alone than with neighbours, irrespective of whether the neighbour was kin or stranger (df1, df2 = 2, 583, F = 2.758, *p* = 0.0643, Table 1). Confirming our hypotheses, plants that grew in fertile environments showed significantly higher levels of growth towards the neighbour and central shared area than away (df1, df2 = 1, 583, F = 7.243, *p* = 0.0073, Table 1). When looking at the interactions between the fixed factors, there was no overall impact of neighbour presence, neighbour identity, and soil treatment (df1, df2 = 2, 583, F = 0.271, p = 0.7628, Table 1).

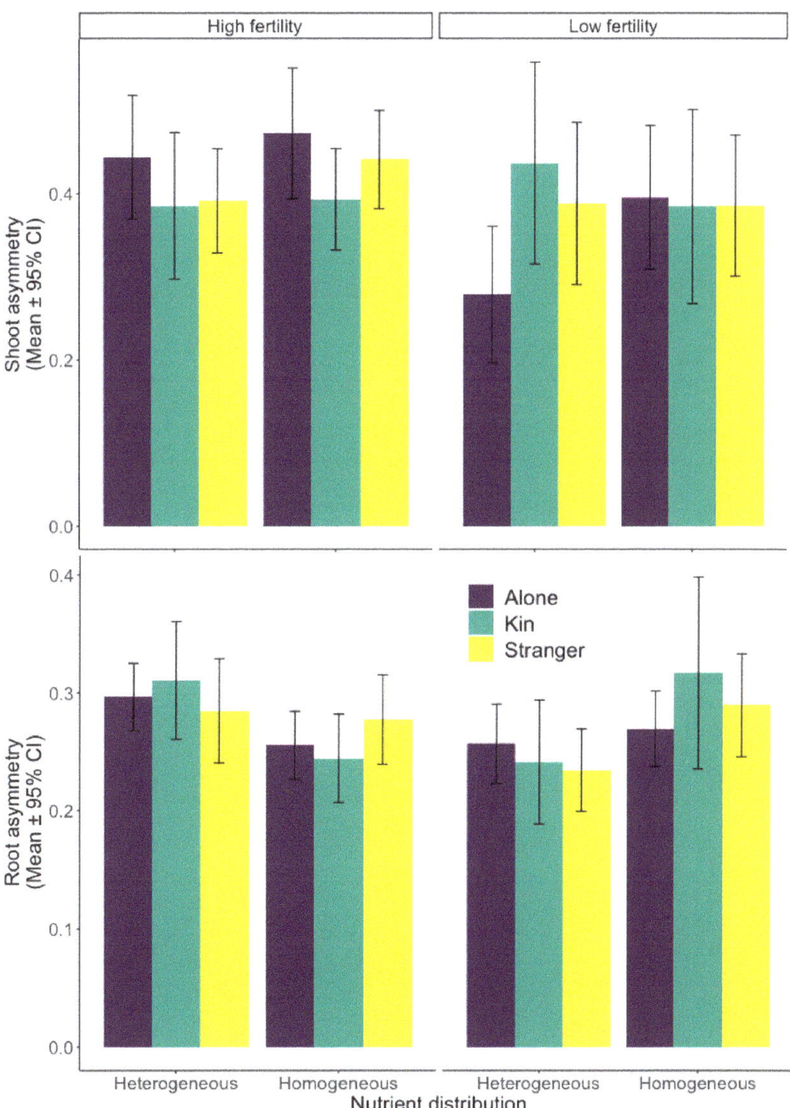

Figure 2. Wheat (*Triticum aestivum*) shoot and root asymmetry (mean ± 95% confidence interval) in response to identity, nutrient distribution, and fertility. Shoot asymmetry is the proportion of each plant's shoot biomass grown toward the centre of the pot and a neighbour divided by the plant's overall shoot biomass grown. The root asymmetry is the proportion of roots harvested in the central shared soil space divided by the total roots harvested in the central shared space and the soil between each plant and the pot edge.

2.3. Belowground Biomass Distribution

We found that root asymmetry was highly contingent on the nutrient level and soil heterogeneity but not neighbour presence or identity, contrary to our hypothesis (df1, df2 = 2, 582, F = 0.152, p = 0.8589, Table 1, Figure 2). However, we did find a significant effect of soil heterogeneity on root asymmetry, depending on the nutrient treatment. Root asymmetry was higher in high-nutrient pots with heterogeneous soil than in low-nutrient, heterogeneous soil (df1, df2 = 582, t = 3.301, p = 0.001, Table 2). When overall soil fertility was low, root asymmetry was higher in homogeneous soil than in heterogeneous soil (df1, df2 = 582, t = −2.931, p = 0.0035, Table 2). However, when soil fertility was high, there was no discernible difference in root asymmetry between homogeneous and heterogeneous soil treatments (df1, df2 = 582, t = 1.952, p = 0.0514, Table 2).

Table 2. Planned contrasts with neighbour identity and nutrient treatments for impacts on reproductive effort, shoot and root asymmetry, and root precision. For neighbour contrasts, results are averaged over the levels of nutrient treatments, while for nutrient treatments, results are averaged over the levels of kin. Estimates and standard errors represent log odds ratios, and bolded values represent significant relationships (alpha = 0.05).

Model	Estimate	S.E.	df	t Ratio	p-Value
Reproductive effort					
Alone—Neighbour	0.115	0.0573	580	2.003	**0.0456**
Kin—Stranger	−0.115	0.0811	580	−1.413	0.1583
Heterogeneous High fertility—Homogeneous High fertility	0.0796	0.0835	580	0.954	0.3407
Heterogeneous High fertility—Heterogeneous Low fertility	−0.11	0.0875	580	−1.257	0.2094
Homogeneous High fertility—Homogeneous Low fertility	−0.1481	0.0839	580	−1.765	**0.0781**
Heterogeneous Low fertility—Homogeneous Low fertility	0.0415	0.0873	580	0.475	0.6349
Shoot asymmetry					
Alone—Neighbour	−0.1828	0.105	583	−1.733	0.0837
Kin—Stranger	0.0329	0.154	583	0.214	0.8309
Heterogeneous High fertility—Homogeneous High fertility	−0.1221	0.16	583	−0.763	0.4455
Heterogeneous High fertility—Heterogeneous Low fertility	0.2756	0.162	583	1.703	0.0891
Homogeneous High fertility—Homogeneous Low fertility	0.3358	0.159	583	2.113	**0.035**
Heterogeneous Low fertility—Homogeneous Low fertility	−0.0619	0.16	583	−0.386	0.6995
Root asymmetry					
Alone—Neighbour	−0.0166	0.0538	582	−0.309	0.7571
Kin—Stranger	0.0401	0.0777	582	0.515	0.6064
Heterogeneous High fertility—Homogeneous High fertility	0.157	0.0803	582	1.952	0.0514
Heterogeneous High fertility—Heterogeneous Low fertility	0.272	0.0824	582	3.301	**0.001**
Homogeneous High fertility—Homogeneous Low fertility	−0.126	0.0801	582	−1.57	0.1169
Heterogeneous Low fertility—Homogeneous Low fertility	−0.241	0.0822	582	−2.931	**0.0035**
Root precision					
Alone—Neighbour	−0.0397	0.0664	582	−0.599	0.5497
Kin—Stranger	0.1027	0.096	582	1.069	0.2853
Heterogeneous High fertility—Homogeneous High fertility	1.372	0.1022	582	13.427	**<0.0001**
Heterogeneous High fertility—Heterogeneous Low fertility	0.288	0.1078	582	2.675	**0.0077**
Homogeneous High fertility—Homogeneous Low fertility	0.177	0.0924	582	1.917	0.0557
Heterogeneous Low fertility—Homogeneous Low fertility	1.261	0.1005	582	12.55	**<0.0001**

There was no significant effect of neighbour presence and soil treatment interaction on root precision (df1, df2 = 2, 582, F = 0.1577, p = 0.562, Table 1, Figure 3). However, we did find increased patch use when the soil was highly nutritious, regardless of nutrient distribution, indicating a lack of interaction (df1, df2 = 1, 582, F = 10.739, p = 0.0011, Table 1). Similarly, root biomass in the patches was higher in heterogeneous soil, irrespective of soil fertility (df1, df2 = 1, 582, F = 330.789, $p \leq$ 0.0001, Table 1).

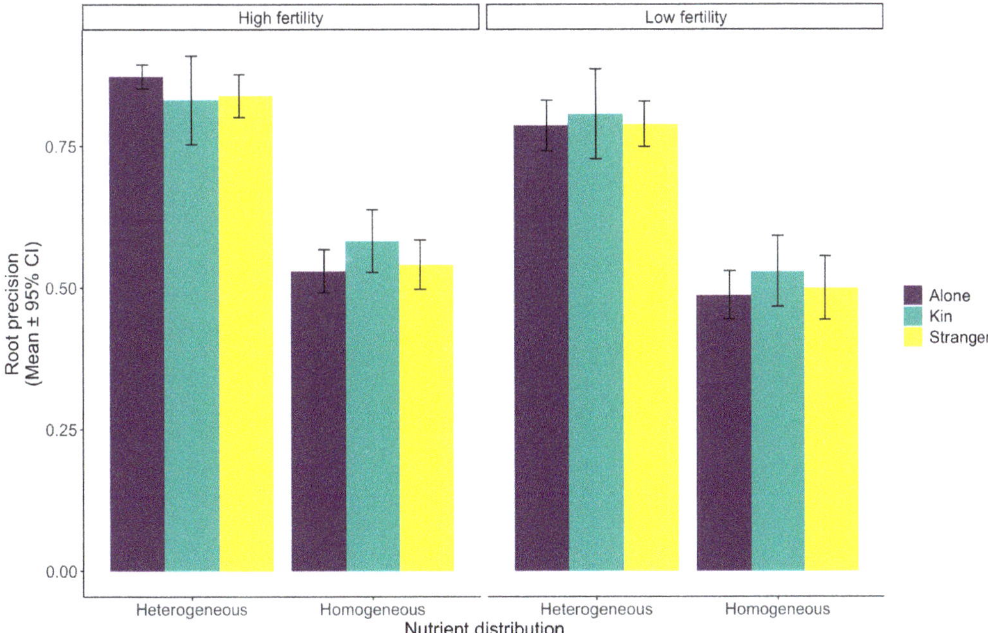

Figure 3. Wheat (*Triticum aestivum*) root precision (mean ± 95% confidence interval) in response to identity, nutrient distribution, and fertility. Precision was measured as the proportion of biomass harvested where the nutrient patch was in the pot divided by the total biomass harvested from the patch and an area equidistant to the plants without a nutrient patch present. This created a pot-level measure.

3. Discussion

Our results indicate stark differences in wheat's above- and belowground behavioural responses to neighbours and nutrients. The reproductive effort had only slightly significant kin effects, with reproductive effort marginally higher for plants grown alone in low-nutrient conditions or with neighbours in high-nutrient soil (Figure 1). When looking at the overall aboveground biomass distribution changes, it is apparent that soil fertility, nutrient level, and neighbour presence have a significant impact (Figure 2). Finally, root distribution was affected by the nutrient level, soil heterogeneity, and neighbour presence, with changes in shared soil areas and patch use (Figures 3 and 4).

3.1. Reproductive Effort

For reproductive effort, we found that none of the soil structure elements or social interactions we investigated—including soil fertility, nutrient homogeneity, neighbour presence, and neighbour identity—had a drastic impact (Figure 1). Li et al. found that yield was stable across various plant densities when studying maize [47,75,76]. However, output linearly declined when the plant density was above an optimum level set by the species [75]. Over time, the extensive breeding of crop plants may have caused the lack of distinct impact of soil structure and social interactions on the reproductive effort. Crop plants are often bred for traits emphasizing group fitness over individual performance [18,77–80]. Hence, past selection for inclusive fitness may have favoured more cooperative plant genotypes with features such as shorter stems, erect leaves, and restrained roots [81]. Overall fitness is improved when the negative consequences of competition between kin, including clones, full-siblings, and half-siblings, are minimized [43]. So, the ability of crop species to recognize kin may increase yield by reducing competitive effects [48,81–83].

Accordingly, in a study on rice, the data showed that cultivars with kin recognition in mixed cultures increased grain yields, but interestingly, not all cultivars possessed this ability [83]. Furthermore, numerous studies have shown that cooperation based on kin may decrease the prevalence of competitive traits [84–86]. Cooperation also allowed for optimized above- and belowground resource capture [35] and subsequent increases in overall fitness [7,87,88].

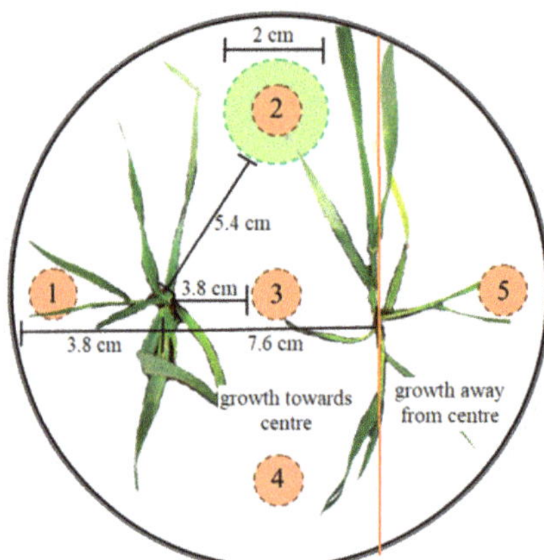

Figure 4. Planting design of wheat, *Triticum aestivum* L., in heterogenous and homogenous soil treatments. The distances between the plants (7.6 cm), the plants and pot edge (3.8 cm), and the plants and centre of the pot (3.8 cm) are displayed. Control treatments had a single plant placed in one of these two locations. The nutrient patch, depicted by the green circle, was placed equidistant from both plants, and the diameter of the patch (2 cm) and its distance from the plants (5.4 cm) are illustrated. At harvest, shoot biomass was separated into growth towards or away from the pot's centre, as shown by the orange line. We also took five root cores, illustrated by the numbered brown circles.

However, Kiers and Denison refute the notion of crop plants' emphasis on group fitness, stating that high genetic relatedness, particularly siblings or clones, does not necessarily select for cooperation [81]. Furthermore, single-genotype crop fields would not necessarily provide more significant reproductive effort or yield [82,89]. For example, *Lupinus angustifolius* plants produce significantly more flowers and seeds when grown with unrelated neighbours than with siblings [44]. These differences in kinship effects could result from specific biotic or abiotic environments [45,90]. Our study did not show this above or below ground, indicating that the effect of the relatedness between neighbours is difficult to predict, especially regarding crop species. We lack conclusive evidence of the impacts of kin recognition.

3.2. Aboveground Biomass Distribution

Above ground, our results show that the wheat plants increased growth towards the centre of the pot when a neighbouring plant was absent (Figure 2). Some studies have suggested that plants may over-proliferate shoots when grown with a neighbour [49]. Many plant species interpret environmental cues such as shading, volatile organic compounds, touch, and root exudates to determine whether a neighbour is present [22]. Over-proliferation would prove wasteful, however, if the neighbour responds similarly. They would also create a tragedy of the commons, collectively exhausting the resources [1,91]. In our study, the unconstrained growth due to a lack of neighbours enables solitary plants to organize their development solely concerning resource availability, modelling Ideal Free Distribution (IFD) [22].

Our study did not show any distinct impact of neighbour identity on the distribution of aboveground biomass (Figure 2). Wheat may consider all the neighbours as related rather than segregating them into stranger and kin categories if the 'stranger' neighbours are not genetically distant enough. This could be attributed to similar genetic backgrounds and close genetic relatedness within the crop [82,83]. In our study, the soil's nutrient level significantly impacted shoot asymmetry, with the plants increasing neighbour avoidance under low-nutrient conditions (Figure 2). The plants may follow IFD under low-nutrient conditions, minimizing overlap of shared aerial space when belowground resources are already limited but abandoning IFD when belowground resources are plentiful enough for the plants to compete aggressively. The decrease in shading at low nutrient levels could be due to decreased chlorophyll requirements when plants must conserve nutrients due to limited availability. A greenhouse pot experiment on wheat showed that decreasing fertilizer reduced chlorophyll content in the leaves [92]. With a greater need to place roots below ground due to an enhanced necessity for foraging and structural integrity considering the competition, plants would be far less free-handed with allocating resources towards aboveground competition. In some plant species, increasing soil fertility under homogeneous conditions decreased aboveground size-asymmetric competition [93]. Alternatively, increasing nutrient levels in the soil can also inadvertently increase size-asymmetric competition by prompting shoot growth, effectively altering the competition to above ground from below ground [94–96], which is more akin to what we observed in our study (Figure 2). Hence, it is apparent from our research that aboveground architecture changes in direct response to alterations in belowground environmental conditions.

3.3. Belowground Biomass Distribution and Patch Use

Our study of belowground plant behaviour indicates that soil heterogeneity and nutrient level affected root distribution and patch use (Figures 3 and 4). We saw even root distribution in homogeneous soil treatments, while heterogeneous treatments elicited significant increases in patch use (Figure 4). In the heterogeneous treatments, the plants decreased root growth in all the core locations without a nutrient patch, likely reallocating resources to growth proliferation within the patches, especially in highly nutritive soil. Hence, we observed a cascading effect throughout the pot where the plants use environmental information to optimize root foraging behaviour [66,97–101]. A trade-off exists between maximizing resource intake from the high-quality soil patch and prolonged exploration.

Our study, however, does not indicate any direct effects of neighbour presence or identity on belowground biomass placement (Figure 3). It is likely that wheat either cannot recognize kin or has been bred to disregard familial connections when placing roots below ground. The lack of over-proliferation in the presence of neighbours, especially strangers, has been seen in prior studies [5,102–105]. However, we did observe the effects of the interaction between nutrient heterogeneity in the soil and neighbour presence. The plants allocated more roots opposite the nutrient patch when they were not facing competition for resources in a limited space with a proximate neighbour (Figure 4). When dealing with a predominately low-nutrient environment with a singular high-nutrient patch equidistant to a neighbour, the plant will expend energy in maximizing resource capture, engaging in a tragedy of the commons. When the plants are free of these restraints and are alone in homogenous soil, they are free to explore the entirety of the soil.

Interestingly, the interaction of nutrient level and soil heterogeneity impacted root growth between the plants and the edge of the pots; when the soil was highly nutritious and homogeneous, more root biomass was placed in these locations (Figure 3). The plants more readily utilized the soil they had first access to when not facing direct competition for a patch equally accessible to both plants, and the soil was fertile. This foraging strategy would reduce the need to extensively search the rest of the pot for a potential nutrient patch or higher-quality soil. Plants experience a trade-off between exploration of the environment and exploitation of resources [106,107], which is comparable to animals as they move and forage across landscapes [108,109]. Thus, they may invest more energy in pre-empting resources within a high-nutrient patch, ensuring direct access by increasing preliminary root growth [49,100].

3.4. Future Directions

Crop systems are drastically different from most natural biological systems, precisely due to their organization in closely planted monoculture fields to promote maximum yield from minimum space on the landscape. The dynamics of plant–plant interactions in these synthetically constructed plant communities have been prone to drastic change through intentional and unintentional artificial selection, especially in cereal crops such as wheat [18,110]. A new avenue of exploration could involve looking at past crop selection and the extent to which plants select for or against kin. Future studies could apply an evolutionary lens by comparing wild ancestors and domesticated cultivars.

Additionally, most research on kin/non-kin effects on spatial distribution and fitness has been genetically limited. Most studies look at half- to full-siblings versus strangers without quantification or scale of genetic relatedness. Further testing would be needed to fill this gap in the literature and determine the genetic distance required for a kin/stranger effect within wheat. These studies would be especially pertinent since kin selection can lead to better group fitness outcomes directly tied to increases in grain production [57,81,83,111]. These findings may have important implications for ecosystem functioning and agriculture. We must understand the underlying mechanisms better to apply this knowledge and enhance crop performance effectively.

4. Materials and Methods

4.1. Study Species

We chose five cultivars of Canada Western Red Spring (CWRS) wheat: Lillian, Glenn, Carberry, C.D.C. Titanium, and Go Early. Lillian is a solid-stemmed cultivar with high protein content and resistance to wheat stem sawfly and stripe rust [69,112]. Glenn also has high protein content, with additional resistance to Fusarium head blight, leaf rust, and stem rust [69,113]. Carberry is high-yielding with leaf rust resistance [69,70,114,115], while CDC Titanium is midge-tolerant with resistance to Fusarium head blight and leaf and stripe rusts [69,116]. Finally, as the name suggests, Go Early matures early and produces a high yield with common bunt and rust resistance [69,117]. We attained seeds for all cultivars from the Spaner Research Lab at the University of Alberta in Edmonton, AB.

4.2. Experimental Design Overview

In brief, we created experimental mesocosms in round 15.2 cm plastic pots with a volume of 1.75 L filled with low-nutrient soil [118–121] (Figure 4). We planted one or two of the same or different wheat varieties in each pot. Each pot received 1 of 4 soil nutrient treatments (high vs. low fertility × heterogeneous vs. homogeneous distribution), resulting in 90 treatment combinations. Pots were organized into 5 replicate blocks, each containing 1 or 3 replicates of each treatment combination (3 replicates if plants were grown individually; 1 replicate if 2 plants were in each pot). In total, there were 600 experimental pots [(15 pairwise combinations × 4 soil treatments × 5 blocks) + (5 cultivars × 3 replicates × 4 soil treatments × 5 blocks)], all grown on the roof of the University of Alberta Biotron in a randomized block design.

4.3. Soil Treatments, Neighbour Treatments, and Plant Growth

The soil in all pots consisted of a low-nutrient soil mix (3:1 sand-to-topsoil mixture (Canar Rock Products Ltd., Edmonton, Canada)), which has been used in other root foraging experiments within the lab group [14,122]. We added 2 levels (0.551 g/L and 4.403 g/L) of slow-release 14:14:14 NPK fertilizer to create low- and high-soil-fertility treatments. For homogenous treatments, the fertilizer was mixed evenly throughout the soil. The fertilizer was placed in a 1 cm patch spanning the entire pot depth in the heterogeneous treatments. When present, nutrient patches were placed equidistant from both plants

(Figure 4). The locations of the soil patches in the single and neighbour treatments were the same. Additionally, the total nutrients added per pot were identical between the homogenous and heterogenous treatments within a soil fertility treatment, either high or low.

Three seeds of a given genotype were planted at each location, and seedlings were then thinned to one individual per location within three days, retaining the largest individual. Treatments comprising two plants had each plant placed halfway between the edge and centre of the pot (Figure 4). We planted the seeds in the same pattern for the plants grown alone, with one of the planting locations in the pot remaining empty (Figure 4). The plants received natural sunlight, and we watered them throughout their growth.

4.4. Harvest

The plants grew for an average of 43 days, from 7 June to 20 July 2018. We harvested the plants after the seeds had developed to capture maximum reproductive growth. However, to prevent plants from outgrowing the pots below ground and altering root distribution, we harvested the plants before the seeds were fully ripened. To harvest, we clipped each plant at the soil surface. We separated the biomass into two categories: growth towards or away from the centre of the pot relative to the initial planting location (Figure 4). We used this separation to quantify aboveground shifts from a neighbour's presence. These 2 categories were split into reproductive and non-reproductive biomass for each plant, dried at 65 °C for a minimum of 48 h, and weighed.

According to our research questions, we took five root cores per pot in specified locations using steel root corers (Figure 4). First, to determine patch use, we harvested roots from where the nutrient patch was placed in heterogeneous treatments (Figure 4, core 2). To assess preferential patch use, we also harvested root biomass in the nutrient-empty space directly opposite the patches (Figure 4, core 4). Then, to investigate the use of shared soil, we took soil cores in the central shared soil space (Figure 4, core 3) as well as the spaces each plant had singular access to, between the plants and the pot edges (Figure 4, cores 1 and 5). Finally, we collected all remaining root fragments in the pots. Each root core was gently washed over a 1 mm sieve, removing the soil. After extracting the root fragments using tweezers, we dried them in a drying oven at 65 °C for 48 h prior to weighing them. However, due to a lack of visual differentiation between the roots of the wheat plants, especially those with kin neighbours of the same genotype, cultivar determination and separation were impossible for the cores or root fragments.

4.5. Statistical Analysis

To analyse our data, we ran a priori contrasts between the nutrient level and nutrient homogeneity and the overall effect of neighbouring plants, allowing us to compare differences between plants grown with a neighbour or without. We also ran a priori contrasts looking at the effects of kin versus strangers, disregarding single plants. We created beta regression models using the glmmTMB package [123] in R [124] to assess biomass allocation, fitness metrics, and root proliferation, with blocks as random effects. In each of the four beta regression models, the individual effects of neighbours, soil fertility, and nutrient distribution were modelled. Additionally, the two-way interactions between neighbour identity and fertility, neighbour identity and distribution, and soil fertility and nutrient distribution were modelled. Finally, the effects of the three-way interaction between neighbour identity, soil fertility, and nutrient distribution were also analysed. We also ran planned contrasts with neighbour identity and nutrient treatments to determine their impacts on reproductive effort, shoot and root asymmetry, and root precision. For neighbour contrasts, results were averaged over the levels of nutrient treatments, while for nutrient treatments, results were averaged over the levels of kin.

We calculated reproductive effort by taking the proportion of biomass containing reproductive structures and dividing it by the aboveground biomass for each plant. Aboveground biomass asymmetry was measured as the proportion of shoots, including both reproductive and non-reproductive parts, of each plant that grew towards the shared space in the centre of the pot compared with the biomass in shared space that grew away from the centre. Similarly, to study root asymmetry, we measured the proportion of root biomass in the shared soil space (Figure 4, core 3) and compared it with the overall root biomass found in both shared and independent space (Figure 4, cores 1, 3, and 5). To prevent dividing by 0 when there was no biomass growing away from the centre, we added 0.005 g to all biomass values. Finally, to examine root foraging precision, we took the proportion of roots harvested in the patch location (Figure 4, core 2) and divided it by the root biomass harvested in the patch and an equidistant location without a nutrient patch within the pot (Figure 4, cores 2 and 4), creating a pot-level measure. For the belowground measures, we utilized a pot-level measure since visual differentiation between fragmented roots from two neighbouring wheat plants was impossible, preventing investigation of individual space use.

Author Contributions: Conceptualization and methodology, H.F.M., T.B.-C. and J.F.C.J.; software and formal analysis, H.F.M. and C.B.; investigation, data curation, writing—original draft preparation, and visualization, H.F.M.; validation, project administration, and writing—review and editing, H.F.M. and J.F.C.J.; resources and supervision, J.F.C.J. and D.S.; funding acquisition, H.F.M., J.F.C.J. and D.S. All authors have read and agreed to the published version of the manuscript.

Funding: This study was primarily funded by NSERC Discovery Grants awarded to J.F.C.J. and D.S. The study also used grants from the Alberta Crop Industry Development Fund, Alberta Wheat Commission, Saskatchewan Wheat Development Commission, Agriculture and Agri-Food Canada, Western Grains Research Foundation Endowment Fund, and Core Program Check-off funds awarded to D.S., and an Alberta Graduate Excellence Award awarded to H.F.M.

Data Availability Statement: The data presented in this study are available upon reasonable request from the corresponding author.

Acknowledgments: We thank V. Wagner, C. Brown, M. Dettlaff, A. Batbaatar, J. Grenke, M. Ljubotina, T. Bao, A. Filazzola, K. Salimbayeva, Y. Fan, E. Holden, T. Barber-Cross, K. Oppon, I. Peetoom Heida, T. Blenkinsopp, B. Nagtegaal, A. Mahal, M. Mahal, S. Ahmed, J. Wild, Q. Stanczak, and R. Glover for their assistance throughout this study.

Conflicts of Interest: The authors declare no conflict of interest.

References

1. Novoplansky, A. Picking Battles Wisely: Plant Behaviour under Competition. *Plant Cell Environ.* **2009**, *32*, 726–741. [CrossRef]
2. Ballaré, C.L.; Sánchez, R.A.; Scopel, A.L.; Casal, J.J.; Ghersa, C.M. Early Detection of Neighbour Plants by Phytochrome Perception of Spectral Changes in Reflected Sunlight. *Plant Cell Environ.* **1987**, *10*, 551–557.
3. Novoplansky, A.; Cohen, D.; Sachs, T. How Portulaca Seedlings Avoid Their Neighbors. *Oecologia* **1990**, *82*, 490–493. [CrossRef] [PubMed]
4. Péret, B.; Desnos, T.; Jost, R.; Kanno, S.; Berkowitz, O.; Nussaume, L. Root Architecture Responses: In Search of Phosphate. *Plant Physiol.* **2014**, *166*, 1713–1723. [CrossRef] [PubMed]
5. Nord, E.A.; Zhang, C.; Lynch, J.P. Root Responses to Neighbouring Plants in Common Bean Are Mediated by Nutrient Concentration Rather than Self/Non-Self Recognition. *Funct. Plant Biol.* **2011**, *38*, 941–952. [CrossRef] [PubMed]
6. Pierik, R.; Mommer, L.; Voesenek, L.A. Molecular Mechanisms of Plant Competition: Neighbour Detection and Response Strategies. *Funct. Ecol.* **2013**, *27*, 841–853. [CrossRef]
7. Donohue, K. The Influence of Neighbor Relatedness on Multilevel Selection in the Great Lakes Sea Rocket. *Am. Nat.* **2003**, *162*, 77–92. [CrossRef]
8. Wang, P.; Heijmans, M.M.P.D.; Mommer, L.; Van Ruijven, J.; Maximov, T.C.; Berendse, F. Belowground Plant Biomass Allocation in Tundra Ecosystems and Its Relationship with Temperature. *Environ. Res. Lett.* **2016**, *11*, 055003. [CrossRef]
9. Gruntman, M.; Groß, D.; Májeková, M.; Tielbörger, K. Decision-Making in Plants under Competition. *Nat. Commun.* **2017**, *8*, 2235. [CrossRef]
10. Semchenko, M.; Zobel, K.; Hutchings, M.J. To Compete or Not to Compete: An Experimental Study of Interactions between Plant Species with Contrasting Root Behaviour. *Evol. Ecol.* **2010**, *24*, 1433–1445. [CrossRef]
11. Hodge, A. The Plastic Plant: Root Responses to Heterogeneous Supplies of Nutrients. *New Phytol.* **2004**, *162*, 9–24. [CrossRef]
12. Semchenko, M.; Hutchings, M.J.; John, E.A. Challenging the Tragedy of the Commons in Root Competition: Confounding Effects of Neighbour Presence and Substrate Volume. *J. Ecol.* **2007**, *95*, 252–260. [CrossRef]

13. Vicherová, E.; Glinwood, R.; Hájek, T.; Šmilauer, P.; Ninkovic, V. Bryophytes Can Recognize Their Neighbours through Volatile Organic Compounds. *Sci. Rep.* **2020**, *10*, 7045. [CrossRef] [PubMed]
14. Karst, J.D.; Belter, P.R.; Bennett, J.A.; Cahill, J.F. Context Dependence in Foraging Behaviour of Achillea Millefolium. *Oecologia* **2012**, *170*, 925–933. [CrossRef] [PubMed]
15. Schmid, C.; Bauer, S.; Bartelheimer, M. Should I Stay or Should I Go? Roots Segregate in Response to Competition Intensity. *Plant Soil* **2015**, *391*, 283–291. [CrossRef]
16. Goldberg, D.E.; Werner, P.A. Equivalence of Competitors in Plant Communities: A Null Hypothesis and a Field Experimental Approach. *Am. J. Bot.* **1983**, *70*, 1098–1104. [CrossRef]
17. Aschehoug, E.T.; Brooker, R.; Atwater, D.Z.; Maron, J.L.; Callaway, R.M. The Mechanisms and Consequences of Interspecific Competition among Plants. *Annu. Rev. Ecol. Evol. Syst.* **2016**, *47*, 263–281. [CrossRef]
18. Anten, N.P.R.; Chen, B.J.W. Detect Thy Family: Mechanisms, Ecology and Agricultural Aspects of Kin Recognition in Plants. *Plant Cell Environ.* **2021**, *44*, 1059–1071. [CrossRef]
19. Li, L.; Tilman, D.; Lambers, H.; Zhang, F.-S. Plant Diversity and Overyielding: Insights from Belowground Facilitation of Intercropping in Agriculture. *New Phytol.* **2014**, *203*, 63–69. [CrossRef]
20. Takigahira, H.; Yamawo, A. Competitive Responses Based on Kin-Discrimination Underlie Variations in Leaf Functional Traits in Japanese Beech (*Fagus Crenata*) Seedlings. *Evol. Ecol.* **2019**, *33*, 521–531. [CrossRef]
21. Yamawo, A.; Mukai, H. Outcome of Interspecific Competition Depends on Genotype of Conspecific Neighbours. *Oecologia* **2020**, *193*, 415–423. [CrossRef] [PubMed]
22. Bilas, R.D.; Bretman, A.; Bennett, T. Friends, Neighbours and Enemies: An Overview of the Communal and Social Biology of Plants. *Plant Cell Environ.* **2021**, *44*, 997–1013. [CrossRef]
23. Goudie, J.W.; Polsson, K.R.; Ott, P.K. An Empirical Model of Crown Shyness for Lodgepole Pine (*Pinus Contorta* Var. Latifolia [Engl.] Critch.) in British Columbia. *Ecol. Manag.* **2009**, *257*, 321–331. [CrossRef]
24. Uria-Diez, J.; Pommerening, A. Crown Plasticity in Scots Pine (*Pinus sylvestris* L.) as a Strategy of Adaptation to Competition and Environmental Factors. *Ecol. Model.* **2017**, *356*, 117–126. [CrossRef]
25. Getzin, S.; Wiegand, K. Asymmetric Tree Growth at the Stand Level: Random Crown Patterns and the Response to Slope. *Ecol. Manag.* **2007**, *242*, 165–174. [CrossRef]
26. Hamilton, W.D. The Genetical Evolution of Social Behaviour. I. *J. Theor. Biol.* **1964**, *7*, 1–16. [CrossRef]
27. Hamilton, W.D. The Genetical Evolution of Social Behavior. II. *J. Theor. Biol.* **1964**, *7*, 17–52. [CrossRef] [PubMed]
28. Rankin, D.J.; Bargum, K.; Kokko, H. The Tragedy of the Commons in Evolutionary Biology. *Trends Ecol. Evol.* **2007**, *22*, 643–651. [CrossRef]
29. Gamboa, G.J. Kin Recognition in Eusocial Wasps George. *Ann. Zool Fenn.* **2004**, *41*, 789–808.
30. Krause, E.T.; Krüger, O.; Kohlmeier, P.; Caspers, B.A. Olfactory Kin Recognition in a Songbird. *Biol. Lett.* **2012**, *8*, 327–329. [CrossRef]
31. Sharp, S.P.; McGowan, A.; Wood, M.J.; Hatchwell, B.J. Learned Kin Recognition Cues in a Social Bird. *Nature* **2005**, *434*, 1127–1130. [CrossRef] [PubMed]
32. Lihoreau, M.; Rivault, C. Kin Recognition via Cuticular Hydrocarbons Shapes Cockroach Social Life. *Behav. Ecol.* **2009**, *20*, 46–53. [CrossRef]
33. Ehlers, B.K.; Bilde, T. Inclusive Fitness, Asymmetric Competition and Kin Selection in Plants. *Oikos* **2019**, *128*, 765–774. [CrossRef]
34. Broz, A.K.; Broeckling, C.D.; De-la-Peña, C.; Lewis, M.R.; Greene, E.; Callaway, R.M.; Sumner, L.W.; Vivanco, J.M. Plant Neighbor Identity Influences Plant Biochemistry and Physiology Related to Defense. *BMC Plant Biol.* **2010**, *10*, 115. [CrossRef] [PubMed]
35. Lepik, A.; Abakumova, M.; Zobel, K.; Semchenko, M. Kin Recognition Is Density-Dependent and Uncommon among Temperate Grassland Plants. *Funct. Ecol.* **2012**, *26*, 1214–1220. [CrossRef]
36. File, A.L.; Murphy, G.P.; Dudley, S.A. Fitness Consequences of Plants Growing with Siblings: Reconciling Kin Selection, Niche Partitioning and Competitive Ability. *Proc. R. Soc. B Biol. Sci.* **2012**, *279*, 209–218. [CrossRef]
37. Roig-Villanova, I.; Martínez-García, J.F. Plant Responses to Vegetation Proximity: A Whole Life Avoiding Shade. *Front. Plant Sci.* **2016**, *7*, 236. [CrossRef]
38. Anten, N.P.R.; Alcalá-Herrera, R.; Schieving, F.; Onoda, Y. Wind and Mechanical Stimuli Differentially Affect Leaf Traits in Plantago Major. *New Phytol.* **2010**, *188*, 554–564. [CrossRef]
39. Markovic, D.; Nikolic, N.; Glinwood, R.; Seisenbaeva, G.; Ninkovic, V. Plant Responses to Brief Touching: A Mechanism for Early Neighbour Detection? *PLoS ONE* **2016**, *11*, e0165742. [CrossRef]
40. Heil, M.; Karban, R. Explaining Evolution of Plant Communication by Airborne Signals. *Trends Ecol. Evol.* **2010**, *25*, 137–144. [CrossRef]
41. Baldwin, I.T. Plant Volatiles. *Curr. Biol.* **2010**, *20*, R392–R397. [CrossRef] [PubMed]
42. Dudley, S.A. Plant Cooperation. *AoB Plants* **2015**, *7*, plv113. [CrossRef] [PubMed]
43. Cheplick, G.P.; Kane, K.H. Genetic Relatedness and Competition in Triplasis Purpurea (Poaceae): Resource Paritioning or Kin Selection? *Int. J. Plant Sci.* **2004**, *165*, 623–630. [CrossRef]
44. Milla, R.; Forero, D.M.; Escudero, A.; Iriondo, J.M. Growing with Siblings: A Common Ground for Cooperation or for Fiercer Competition among Plants? *Proc. R. Soc. B* **2009**, *276*, 2531–2540. [CrossRef] [PubMed]

45. Masclaux, F.; Hammond, R.L.; Meunier, J.; Gouhier-Darimont, C.; Keller, L.; Reymond, P. Competitive Ability Not Kinship Affects Growth of Arabidopsis Thaliana Accessions. *New Phytol.* **2010**, *185*, 322–331. [CrossRef] [PubMed]
46. Postma, J.A.; Hecht, V.L.; Hikosaka, K.; Nord, E.A.; Pons, T.L.; Poorter, H. Dividing the Pie: A Quantitative Review on Plant Density Responses. *Plant Cell Environ.* **2021**, *44*, 1072–1094. [CrossRef]
47. Li, J.; Xie, R.Z.; Wang, K.R.; Ming, B.; Guo, Y.Q.; Zhang, G.Q.; Li, S.K.; Zhang, G.; Xie, R.; Wang, K.; et al. Variations in Maize Dry Matter, Harvest Index, and Grain Yield with Plant Density. *Crop Econ. Prod. Manag.* **2015**, *107*, 829–834. [CrossRef]
48. Chen, B.J.W.; During, H.J.; Anten, N.P.R. Detect Thy Neighbor: Identity Recognition at the Root Level in Plants. *Plant Sci.* **2012**, *195*, 157–167. [CrossRef]
49. Gersani, M.; Brown, J.; O'Brien, E.; Maina, G.; Abramsky, Z. Tragedy of the Commons as a Result of Root Competition. *J. Ecol.* **2001**, *89*, 660–669. [CrossRef]
50. Holzapfel, C.; Alpert, P. Root Cooperation in a Clonal Plant: Connected Strawberries Segregate Roots. *Oecologia* **2003**, *134*, 72–77. [CrossRef]
51. Gruntman, M.; Novoplansky, A. Physiologically Mediated Self Non-Self Discrimination in Roots. *Proc. Natl. Acad. Sci. USA* **2004**, *101*, 3863–3867. [CrossRef] [PubMed]
52. Dudley, S.; File, A. Kin Recognition in an Annual Plant. *Biol. Lett.* **2007**, *3*, 435–438. [CrossRef] [PubMed]
53. Murphy, G.P.; Dudley, S.A. Kin Recognition: Competition and Cooperation in Impatiens (Balsaminaceae). *Am. J. Bot.* **2009**, *96*, 1990–1996. [CrossRef] [PubMed]
54. Chen, B.J.W.; During, H.J.; Vermeulen, P.J.; de Kroon, H.; Poorter, H.; Anten, N.P.R. Corrections for Rooting Volume and Plant Size Reveal Negative Effects of Neighbour Presence on Root Allocation in Pea. *Funct. Ecol.* **2015**, *29*, 1383–1391. [CrossRef]
55. Palmer, A.G.; Ali, M.; Yang, S.; Parchami, N.; Bento, T.; Mazzella, A.; Oni, M.; Riley, M.C.; Schneider, K.; Massa, N. Kin Recognition Is a Nutrient-Dependent Inducible Phenomenon. *Plant Signal Behav.* **2016**, *11*, e1224045. [CrossRef] [PubMed]
56. Semchenko, M.; Saar, S.; Lepik, A. Plant Root Exudates Mediate Neighbour Recognition and Trigger Complex Behavioural Changes. *New Phytol.* **2014**, *204*, 631–637. [CrossRef]
57. Zhang, L.; Liu, Q.; Tian, Y.; Xu, X.; Ouyang, H. Kin Selection or Resource Partitioning for Growing with Siblings: Implications from Measurements of Nitrogen Uptake. *Plant Soil* **2016**, *398*, 79–86. [CrossRef]
58. O'Brien, E.E.; Gersani, M.; Brown, J.S. Root Proliferation and Seed Yield in Response to Spatial Heterogeneity of Below-Ground Competition. *New Phytol.* **2005**, *168*, 401–412. [CrossRef]
59. Semchenko, M.; John, E.; Hutchings, M. Effects of Physical Connection and Genetic Identity of Neighbouring Ramets on Root-Placement Patterns in Two Clonal Species. *New Phytol.* **2007**, *176*, 644–654. [CrossRef]
60. Maina, G.; Brown, J.; Gersani, M. Intra-Plant versus Inter-Plant Root Competition in Beans: Avoidance, Resource Matching or Tragedy of the Commons. *Plant Ecol.* **2002**, *160*, 235–247. [CrossRef]
61. Zhu, Y.-H.; Weiner, J.; Li, F.-M. Root Proliferation in Response to Neighbouring Roots in Wheat (*Triticum aestivum*). *Basic Appl. Ecol.* **2019**, *39*, 10–14. [CrossRef]
62. Brady, D.J.; Gregory, P.J.; Fillery, I.R.P. The Contribution of Different Regions of the Seminal Roots of Wheat to Uptake of Nitrate from Soil. *Plant Soil* **1993**, *155*, 155–158. [CrossRef]
63. McNickle, G.G.; Cahill Jr, J.F. Plant Root Growth and the Marginal Value Theorem. *Proc. Natl. Acad. Sci. USA* **2009**, *106*, 4747–4751. [CrossRef] [PubMed]
64. VanVuuren, M.M.I.; Robinson, D.; Griffiths, B.S. Nutrient Inflow and Root Proliferation during the Exploitation of a Temporally and Spatially Discrete Source of Nitrogen in Soil. *Plant Soil* **1996**, *178*, 185–192. [CrossRef]
65. Cahill, J.F.; McNickle, G.G. The Behavioral Ecology of Nutrient Foraging by Plants. *Annu. Rev. Ecol. Evol. Syst.* **2011**, *42*, 289–311. [CrossRef]
66. Fransen, B.; de Kroon, H.; de Kovel, C.G.F.; van den Bosch, F. Disentangling the Effects of Root Foraging and Inherent Growth Rate on Plant Biomass Accumulation in Heterogeneous Environments: A Modelling Study. *Ann. Bot.* **1999**, *84*, 305–311. [CrossRef]
67. Wijesinghe, D.K.; John, E.A.; Beurskens, S.; Hutchings, M.J. Root System Size and Precision in Nutrient Foraging: Responses to Spatial Pattern of Nutrient Supply in Six Herbaceous Species 973 Root System Size and Precision in Nutrient Foraging. *J. Ecol.* **2001**, *89*, 972–983. [CrossRef]
68. Chen, H.; Nguyen, K.; Iqbal, M.; Beres, B.L.; Hucl, P.J.; Spaner, D. The Performance of Spring Wheat Cultivar Mixtures under Conventional and Organic Management in Western Canada. *Agrosyst. Geosci. Environ.* **2019**, *3*, e20003. [CrossRef]
69. Randhawa, H.S.; Asif, M.; Pozniak, C.; Clarke, J.M.; Graf, R.J.; Fox, S.L.; Humphreys, D.G.; Knox, R.E.; Depauw, R.M.; Singh, A.K.; et al. Application of Molecular Markers to Wheat Breeding in Canada. *Plant Breed.* **2013**, *132*, 458–471. [CrossRef]
70. Chen, H.; Bemister, D.H.; Iqbal, M.; Strelkov, S.E.; Spaner, D.M. Mapping Genomic Regions Controlling Agronomic Traits in Spring Wheat under Conventional and Organic Managements. *Crop Sci.* **2020**, *60*, 2038–2052. [CrossRef]
71. Fréville, H.; Roumet, P.; Rode, N.O.; Rocher, A.; Latreille, M.; Muller, M.-H.; David, J. Preferential Helping to Relatives: A Potential Mechanism Responsible for Lower Yield of Crop Variety Mixtures? *Evol. Appl.* **2019**, *12*, 1837–1849. [CrossRef] [PubMed]
72. Hackett, C. A Method of Applying Nutrients Locally to Roots under Controlled Conditions, and Some Morphological Effects of Locally Applied Nitrate on the Branching of Wheat Roots. *Aust. J. Biol. Sci.* **1972**, *3*, 1169–1180. [CrossRef]
73. Bhatt, G.M.; Derera, N.F. Genotype x Environment Interactions for, Heritabilities of, and Correlations among Traits in Wheat. *Euphytica* **1975**, *24*, 597–604. [CrossRef]

74. Zhu, L.; Zhang, D. Donald's Ideotype and Growth Redundancy: A Pot Experimental Test Using an Old and a Modern Spring Wheat Cultivar. *PLoS ONE* **2013**, *8*, 70006. [CrossRef]
75. Tollenaar, M. Is Low Plant Density a Stress in Maize? *Maydica* **1992**, *37*, 305–311.
76. Echarte, L.; Andrade, F.H. Harvest Index Stability of Argentinean Maize Hybrids Released between 1965 and 1993. *Field Crops Res.* **2003**, *82*, 1–12. [CrossRef]
77. Donald, C.M. The Breeding of Crop Ideotypes. *Euphytica* **1968**, *17*, 385–403. [CrossRef]
78. Denison, R.F. Past Evolutionary Tradeoffs Represent Opportunities for Crop Genetic Improvement and Increased Human Lifespan. *Evol. Appl.* **2010**, *4*, 216–224. [CrossRef]
79. Weiner, J.; Andersen, S.B.; Wille, K.-M.; Griepentrog, H.W.; Olsen, J.M. Evolutionary Agroecology: The Potential for Cooperative, High Density, Weed-Suppressing Cereals. *Evol. Appl.* **2010**, *3*, 473–479. [CrossRef]
80. Anten, N.P.R.; Vermeulen, P.J. Tragedies and Crops: Understanding Natural Selection to Improve Cropping Systems. *Trends Ecol. Evol.* **2016**, *31*, 429–439. [CrossRef]
81. Kiers, E.T.; Denison, R.F. Inclusive Fitness in Agriculture. *Philos. Trans. R. Soc. B* **2014**, *369*, 20130367. [CrossRef] [PubMed]
82. Murphy, G.P.; van Acker, R.; Rajcan, I.; Swanton, C.J. Identity Recognition in Response to Different Levels of Genetic Relatedness in Commercial Soya Bean. *R. Soc. Open Sci.* **2017**, *4*, 160879. [CrossRef] [PubMed]
83. Yang, X.-F.; Li, L.-L.; Xu, Y.; Kong, C.-H. Kin Recognition in Rice (*Oryza sativa*) Lines. *New Phytol.* **2018**, *220*, 567–578. [CrossRef] [PubMed]
84. Cahill, J.F.; McNickle, G.G.; Haag, J.J.; Lamb, E.G.; Nyanumba, S.M.; Clair, C.C.S. Plants Integrate Information about Nutrients and Neighbors. *Science* **2010**, *328*, 1657. [CrossRef] [PubMed]
85. Bhatt, M.V.; Khandelwal, A.; Dudley, S.A. Kin Recognition, Not Competitive Interactions, Predicts Root Allocation in Young Cakile Edentula Seedling Pairs. *New Phytol.* **2011**, *189*, 1135–1142. [CrossRef] [PubMed]
86. Fang, S.; Clark, R.T.; Zhenge, Y.; Iyer-Pascuzzia, A.S.; Weitz, J.S.; Kochiand, L.V.; Edelsbrunnere, H.; Liaob, H.; Benfey, P.N. Genotypic Recognition and Spatial Responses by Rice Roots. *Proc. Natl. Acad. Sci. USA* **2013**, *110*, 2670–2675. [CrossRef]
87. Biernaskie, J.M. Evidence for Competition and Cooperation among Climbing Plants. *Proc. R. Soc. B Biol. Sci.* **2011**, *278*, 1989–1996. [CrossRef]
88. Torices, R.; Gómez, J.M.; Pannell, J.R. Kin Discrimination Allows Plants to Modify Investment towards Pollinator Attraction. *Nat. Commun.* **2018**, *9*, 2018. [CrossRef]
89. Taylor, P.D. Altruism in Viscous Populations—An Inclusive Fitness Model. *Evol. Ecol.* **1992**, *6*, 352–356. [CrossRef]
90. Goodnight, C.J. The Influence of Environmental Variation on Group and Individual Selection in a Cress. *Evolution* **1985**, *39*, 545–558. [CrossRef]
91. Smyeka, J.; Herben, T. Phylogenetic Patterns of Tragedy of Commons in Intraspecific Root Competition. *Plant Soil* **2017**, *417*, 87–97. [CrossRef]
92. Brown, C.; Oppon, K.J.; Cahill, J.F. Species-Specific Size Vulnerabilities in a Competitive Arena: Nutrient Heterogeneity and Soil Fertility Alter Plant Competitive Size Asymmetries. *Funct. Ecol.* **2019**, *33*, 1491–1503. [CrossRef]
93. Shah, S.H.; Houborg, R.; McCabe, M.F. Response of Chlorophyll, Carotenoid and SPAD-502 Measurement to Salinity and Nutrient Stress in Wheat (*Triticum aestivum* L.). *Agronomy* **2017**, *7*, 61. [CrossRef]
94. Cahill, J.F. Ferilizations Effects on Interactions between Above- and Belowground Competition in an Old Field. *Ecology* **1999**, *80*, 466–480. [CrossRef]
95. Hautier, Y.; Niklaus, P.A.; Hector, A. Competition for Light Causes Plant Biodiversity Loss after Eutrophication. *Science* **2009**, *324*, 636–638. [CrossRef] [PubMed]
96. Lamb, E.G.; Kembel, S.W.; Cahill, J.F. Shoot, but Not Root, Competition Reduces Community Diversity in Experimental Mesocosms. *J. Ecol.* **2009**, *97*, 155–163. [CrossRef]
97. Mahall, B.; Callaway, R. Root Communication among Desert Shrubs. *Proc. Natl. Acad. Sci. USA* **1991**, *88*, 874–896. [CrossRef] [PubMed]
98. Schenk, H.J. Root Competition: Beyond Resource Depletion. *J. Ecol.* **2006**, *94*, 725–739. [CrossRef]
99. Belter, P.R.; Cahill, J.F. Disentangling Root System Responses to Neighbours: Identification of Novel Root Behavioural Strategies. *AoB Plants* **2015**, *7*, plv059. [CrossRef]
100. Ljubotina, M.K.; Cahill, J.F. Effects of Neighbour Location and Nutrient Distributions on Root Foraging Behaviour of the Common Sunflower. *Proc. R. Soc. B Biol. Sci.* **2019**, *286*, 20190955. [CrossRef]
101. Jennings, P.R.; DeJesus, J. Studies on Competition in Rice I. Competition in Mixtures of Varieties. *Evolution* **1968**, *22*, 119–124. [CrossRef]
102. Lankinen, A. Root Competition Influences Pollen Competitive Ability in Viola Tricolor: Effects of Presence of a Competitor beyond Resource Availability? *J. Ecol.* **2008**, *96*, 756–765. [CrossRef]
103. Markham, J.; Halwas, S. Effect of Neighbor Presence and Soil Volume on the Growth of Andropogon Gerardii Vitman. *Plant Ecol. Divers* **2011**, *4*, 265–268. [CrossRef]
104. Meier, I.C.; Angert, A.; Falik, O.; Shelef, O.; Rachmilevitch, S. Increased Root Oxygen Uptake in Pea Plants Responding to Non-Self Neighbor. *Planta* **2013**, *238*, 577–586. [CrossRef]
105. McNickle, G.G.; Brown, J.S. An Ideal Free Distribution Explains the Root Production of Plants That Do Not Engage in a Tragedy of the Commons Game. *J. Ecol.* **2014**, *102*, 963–971. [CrossRef]

106. Semchenko, M.; Zobel, K.; Heinemeyer, A.; Hutchings, M.J. Foraging for Space and Avoidance of Physical Obstructions by Plant Roots: A Comparative Study of Grasses from Contrasting Habitats. *New Phytol.* **2008**, *179*, 1162–1170. [CrossRef] [PubMed]
107. Peng, Y.H.; Niklas, K.J.; Sun, S.C. Do Plants Explore Habitats before Exploiting Them? An Explicit Test Using Two Stoloniferous Herbs. *Chin. Sci. Bull.* **2012**, *57*, 2425–2432. [CrossRef]
108. Stephens, D.W. On Economically Tracking a Variable Environment. *Theor. Popul. Biol.* **1987**, *32*, 15–25. [CrossRef]
109. Nimmo, D.G.; Avitabile, S.; Banks, S.C.; Bliege Bird, R.; Callister, K.; Clarke, M.F.; Dickman, C.R.; Doherty, T.S.; Driscoll, D.A.; Greenville, A.C.; et al. Animal Movements in Fire-Prone Landscapes. *Biol. Rev.* **2019**, *94*, 981–998. [CrossRef]
110. Khush, G.S. Green Revolution: Preparing for the 21st Century. *Genome* **1999**, *42*, 646–655. [CrossRef]
111. Murphy, G.P.; Swanton, C.J.; van Acker, R.C.; Dudley, S.A. Kin Recognition, Multilevel Selection and Altruism in Crop Sustainability. *J. Ecol.* **2017**, *105*, 930–934. [CrossRef]
112. DePauw, R.; Townley-Smith, T.; Humphreys, G.; Knox, R.; Clarke, F.; Clarke, J. Lillian Hard Red Spring Wheat. *Can. J. Plant Sci.* **2005**, *85*, 397–401. [CrossRef]
113. Mergoum, M.; Frohberg, R.; Stack, R.; Olson, T.; Friesen, T.; Rasmussen, J. Registration of Glenn Wheat. *Crop Sci.* **2006**, *46*, 473–475. [CrossRef]
114. DePauw, R.; Knox, R.; McCaig, T.; Clarke, F.; Clarke, J. Carberry Hard Red Spring Wheat. *Can. J. Plant Sci.* **2011**, *91*, 529–534. [CrossRef]
115. Chen, H.; Moakhar, N.P.; Iqba, M.; Pozniak, C.; Hucl, P.; Spaner, D. Genetic Variation for Flowering Time and Height Reducing Genes and Important Traits in Western Canadian Spring Wheat. *Euphytica* **2016**, *208*, 377–390. [CrossRef]
116. Hucl, P. CDC Titanium Canadian Food Inspection Agency. Available online: http://www.inspection.gc.ca/english/plaveg/pbrpov/%0Acropreport/whe/app00009612e.shtml (accessed on 23 February 2023).
117. Spaner, D. Go Early. *Plant Var. J.* **2017**, *104*. Available online: https://inspection.canada.ca/english/plaveg/pbrpov/cropreport/whe/app00009713e.shtml (accessed on 12 February 2023).
118. Tumbleson, M.E.; Kommedahl, T. Reproductive Potential of Cyperus Esculentus by Tubers. *Weeds* **1961**, *9*, 646–653. [CrossRef]
119. Youngner, V.B. Growth of U-3 Bermudagrass Under Various Day and Night Temperatures and Light Intensities. *Agron J.* **1959**, *51*, 557–559. [CrossRef]
120. Miyasaka, S.C.; Habte, M.; Friday, J.B.; Johnson, E.V. Manual on Arbuscular Mycorrhizal Fungus Production and Inoculation Techniques. *Soil Crop Manag.* **2003**, 1–4.
121. Gerdemann, J.W. Vesicular-Arbuscular Mycorrhizae Formed on Maize and Tuliptree by Endogone Fasciculata. *Mycologia* **1965**, *57*, 562–575. [CrossRef]
122. Martínková, J.; Klimeš, A.; Klimešová, J. No Evidence for Nutrient Foraging in Root-Sprouting Clonal Plants. *Basic Appl. Ecol.* **2018**, *28*, 27–36. [CrossRef]
123. Brooks, M.E.; Kristensen, K.; Van Benthem, K.J.; Magnusson, A.; Berg, C.W.; Nielsen, A.; Skaug, H.J.; Machler, M.; Bolker, B.M. GlmmTMB Balances Speed and Flexibility Among Packages for Zero-Inflated Generalized Linear Mixed Modeling. *R J.* **2017**, *9*, 378–400. [CrossRef]
124. R Core Team. *R: A Language and Environment for Statistical Computing*; R Foundation for Statistical Computing: Vienna, Austria, 2019.

Article

The Electrome of a Parasitic Plant in a Putative State of Attention Increases the Energy of Low Band Frequency Waves: A Comparative Study with Neural Systems

André Geremia Parise [1,*], Thiago Francisco de Carvalho Oliveira [2], Marc-Williams Debono [3,*] and Gustavo Maia Souza [2]

[1] School of Biological Sciences, University of Reading, Reading RG6 6AH, UK
[2] Laboratory of Plant Cognition and Electrophysiology (LACEV), Department of Botany, Institute of Biology, Federal University of Pelotas, Capão do Leão 96160-000, RS, Brazil; fthicar@gmail.com (T.F.d.C.O.)
[3] PSA Research Group, 91120 Palaiseau, France
[*] Correspondence: andregparise@gmail.com (A.G.P.); psa-rg@plasticites-sciences-arts.org (M.-W.D.)

Abstract: Selective attention is an important cognitive phenomenon that allows organisms to flexibly engage with certain environmental cues or activities while ignoring others, permitting optimal behaviour. It has been proposed that selective attention can be present in many different animal species and, more recently, in plants. The phenomenon of attention in plants would be reflected in its electrophysiological activity, possibly being observable through electrophytographic (EPG) techniques. Former EPG time series obtained from the parasitic plant *Cuscuta racemosa* in a putative state of attention towards two different potential hosts, the suitable bean (*Phaseolus vulgaris*) and the unsuitable wheat (*Triticum aestivum*), were revisited. Here, we investigated the potential existence of different band frequencies (including low, delta, theta, mu, alpha, beta, and gamma waves) using a protocol adapted from neuroscientific research. Average band power (ABP) was used to analyse the energy distribution of each band frequency in the EPG signals, and time dispersion analysis of features (TDAF) was used to explore the variations in the energy of each band. Our findings indicated that most band waves were centred in the lower frequencies. We also observed that *C. racemosa* invested more energy in these low-frequency waves when suitable hosts were present. However, we also noted peaks of energy investment in all the band frequencies, which may be linked to extremely low oscillatory electrical signals in the entire tissue. Overall, the presence of suitable hosts induced a higher energy power, which supports the hypothesis of attention in plants. We further discuss and compare our results with generic neural systems.

Keywords: selective attention; low waves; gamma waves; parasitic plants; EPG activity; plant cognition; plant-plant interaction; plant electrophysiology; average band power

Citation: Parise, A.G.; Oliveira, T.F.d.C.; Debono, M.-W.; Souza, G.M. The Electrome of a Parasitic Plant in a Putative State of Attention Increases the Energy of Low Band Frequency Waves: A Comparative Study with Neural Systems. *Plants* **2023**, *12*, 2005. https://doi.org/10.3390/plants12102005

Academic Editor: Frantisek Baluska

Received: 30 March 2023
Revised: 10 May 2023
Accepted: 11 May 2023
Published: 16 May 2023

1. Introduction

To survive in a world where the parameters change constantly, organisms must keep track of these variations. This requires an active engagement with the cues that can be perceived by the sensory organs or surfaces and that will cause rearrangements in the internal structure of the organism. Ultimately, this leads to adjustments in the behaviour of the organism in relation to the perceived cues to keep its homeostasis within an acceptable range [1–3]. However, there are limits to both the capacity to sustain those engagements, and the behaviours that are possible to entail. To balance this out, attention is required.

Attention is the cognitive phenomenon that mediates the interaction between the internal states of an organism and the features of the environment with which the organism will actively engage [1,4,5]. Despite the focus on external stimuli, attention can be directed to internal stimuli as well (e.g., attention towards an organ or limb that is aching, working memory, or introspective thoughts in humans) [6]. Attention is a dynamic process that is constantly altered as both internal and external states change. In humans, a whole taxonomy of attention types has been recognised [5,6], but basal forms of attention should be observed in any organism that must make trade-offs between its needs and the many possibilities of engagements (or affordances, in Gibsonian parlance) the environment offers [7,8]. In this case, the most basal form of attention, *selective attention*, is probably shared by many different taxa [8,9]. Here, we discuss the possibility of plants being attentive.

Despite the structural simplicity of their bodies, plants have very complex physiologies, and many channels for engaging with the world. Plants have a panoply of senses that constantly monitor environmental variables [10,11]. All these senses are distributed by virtually all their bodies, and not concentrated in sense organs as often happens with animals. Consequently, plants are subjected to what could be an overwhelming number of stimuli, and responding reliably to those relevant for immediate and future survival is of the utmost importance. How, then, could plants perceive, process, and use all the information the environment potentially offers?

Part of the issue is solved by the modular constitution of plants, where the modules act semi-independently upon the stimuli that are relevant to them in their contexts, without requiring much interaction with the other modules. However, sometimes, plants face challenges, or perform actions that require the coordination of many or even all of the modules. Examples include resisting to abiotic stresses such as drought or excessive temperatures, growing away from potential competitors, growing or orientating leaves towards light sources even when light is absent [12–16], and movements towards structural supports in the case of climbing plants [17–19]. These behaviours necessarily require focusing on the most relevant stimuli to the task in hand and ignoring the stimuli irrelevant or not related to the task. In other words, they require attention [20].

In the case of plants, attention is likely observable using electrophytograms (EPGs) in the form of spontaneous bioelectrical variations of electrical potential. These potential variations usually operate in the microvolt range and are continuously produced by presumably all plant species [21,22]. These bioelectrical signals would be modified when plants perceive environmental signals or cues and prioritise one or some of these signals and cues over others [23–25]. The effect of these interactions on the electrophysiology of plants has been observed in different plant species under different conditions [26–32].

It might feel odd to discuss about attention in plants, as this phenomenon is traditionally studied in humans or animals phylogenetically very close to us. However, attention is an evolved characteristic, and hence, if it is present in humans and our evolutive close relatives, it should therefore be present in our common ancestors in some form. How far in the tree of life we can find this phenomenon remains to be elucidated. However, the fact that attention connects internal physiological states with environmental information to guide behaviour hints to a rather ancient origin, and attention could therefore be present in a range of organisms far wider than expected. Indeed, recent studies demonstrate attention not only in mammals but also in birds, fish, and invertebrates [33–36]. Other studies also pointed to correlations between the basal attentive system of insects and those of humans [8,37,38]. However, attention is not necessarily restricted to the animal kingdom.

Marder [20] attempted to define attention in non-zoocentric terms, so that hypotheses focusing on attention do not a priori reject non-traditional groups of organisms from their frameworks. In defining attention as a disproportionate investment of energy of a cell, tissue, or organism into a particular activity, or in the reception of a stimulus or set of stimuli, Marder was the first to propose that attention could be a phenomenon observed in plants [20]. A few years later, when studying the electrophysiology of the parasitic dodder plant (*Cuscuta racemosa*), Parise et al. [31] found what they identified as the first empirical evidence of a state of attention in a plant when *C. racemosa* twigs were near to their hosts. A hypothesis of plant attention was then further developed to combine the theory and practice in a framework of attention that would make sense to the biology of plants and other sessile organisms [9].

When reviewing the literature on plant electrophysiology, Parise et al. [9] observed that attention in plants could follow the same pattern in different species. In short, attention could be observed in plants through EPG analyses when the complexity of the electrome (meaning the collection of all the electrical activity of an organ or tissue) drops while the correlation between the signals increases. Following the definition of Marder [20], typically, an increase in the energy of the signals would also be measured. This happens because, when involved in an activity or receiving stimuli that likely requires attention by the whole plant, the modules would synchronise their bioelectrical behaviour to coordinate their actions in response to the stimuli. This behaviour was observed in soybean (*Glycine max*), common bean (*Phaseolus vulgaris*), and dodder plants (*Cuscuta racemosa*) [9]. Another recent study on *P. vulgaris* analysed the electrome of local leaves (i.e., leaves that received heat and/or wounding stimuli) and systemic leaves (i.e., leaves that did not). In both the local and systemic leaves, the electromes presented a transiently decreased complexity accompanied with an increased correlation [39], thereby supporting the hypothesis that attention in these cases requires the synchronisation of all the modules. Putative attention in plants most likely leads to the generation of specific EPG patterns and electromic signatures, which are a phenomenon greatly facilitated by the mesological plasticity of plants in their singular milieu [40].

The studies mentioned above corroborate the hypothesis of plant attention as formulated by Parise et al. [9] and suggest that electrophysiological analyses are valuable to empirically observe this phenomenon. However, the analyses proposed by Parise et al. [9] could be further improved to better reflect the presumed attentional dynamics. For example, such as in humans and non-human animals, plants could also have different ranges of wave frequencies related to different stimuli or activities in a form analogous to the brain waves (e.g., alpha, theta, gamma, and delta waves). Understanding how the energy of the signals is distributed across the wave frequencies could provide important insights into the electrical ecophysiology of plants, and yield clues of how the mechanisms for information processing operate in aneural organisms. Therefore, analyses that better capture the subtleties of the electrical signals in plants under states of active perception or (putative) attention should be developed to convey more precise information on the dynamics of this cognitive ecophysiological phenomenon.

In this study, we revisited the electrophysiological data obtained from the dodder plant, *C. racemosa*, a holoparasitic plant, when presented to two different hosts: the bean plant, a suitable host; and the wheat plant, a host that dodders cannot parasitise [31]. We assumed that the plants were under a state of attention in the presence of their hosts as their electromes followed the dynamics described above and described in detail in Parise et al. [9]. Upon host perception, the electrome changed its dynamics as the parasitic plants would be focusing their attention mainly on the suitable hosts and organising their physiology to grow towards the hosts and change their pigment content according to the species detected (see [31] for details). In other words, when a meaningful cue appeared in the environment, the plants invested more energy in the perception of this cue and in the actions required to respond adaptively to it.

In the present study, we hypothesised that there would be a difference in the energy of the electrophysiological waves of the dodder electromes. This difference would probably be more evident in the lowest frequencies, as according to how plant electrophysiological activity usually operates [22]. We employed a different and more reliable technique, time dispersion analysis of features (TDAF) to infer the increase of energy manifested through the difference of potential (DDP), and the average band power (ABP) to separate the electrophysiological activity of *C. racemosa* into their respective band frequencies.

2. Results

The results based on TDFA (time dispersion analysis of features) were obtained combining the data of all the dodder plants in each treatment, with the lines in Figure 1 representing the median for all the values. Therefore, the time series analysed in this study result from the analysis of the twenty-three individual time series of each plant in the same treatment. The shaded areas demarcate the range between the highest and lowest values obtained. The series shown in Figures 1 and 2 show a general behaviour where, despite the higher dispersion around the median in several moments, overall, the power distribution was mostly uniform.

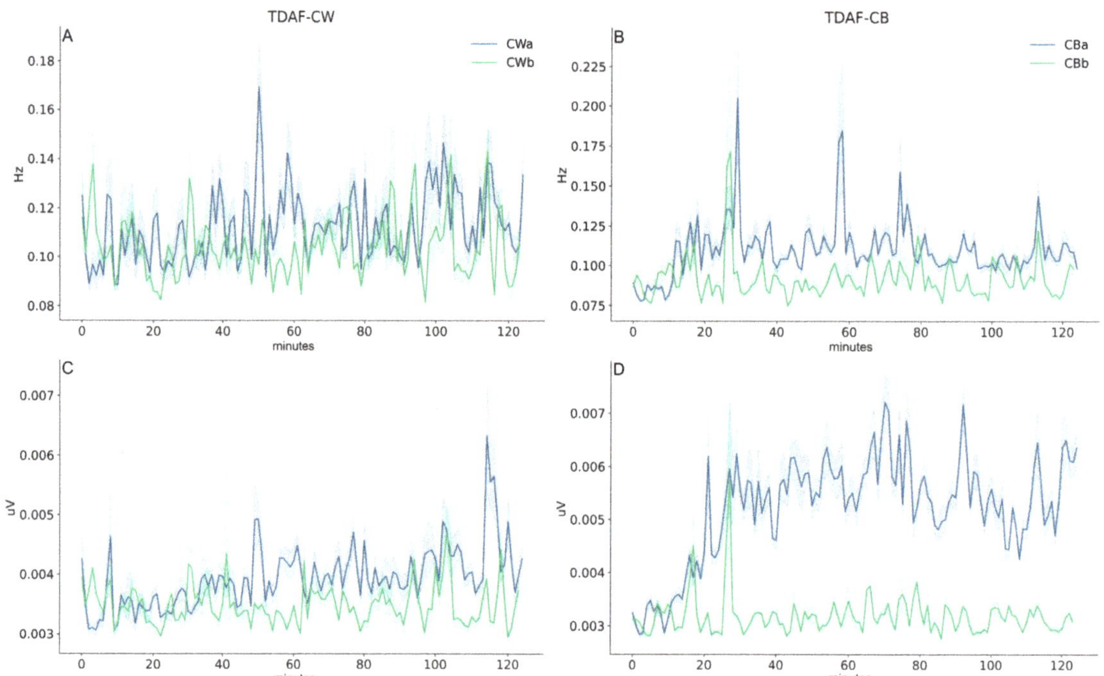

Figure 1. In (**A**,**B**), the mean of the frequencies of all the dodder plants analysed ($n = 23$) calculated for every minute are shown in the Y axis. (**C**,**D**) show the mean of the difference of potential of all the plants. The X axis shows the time of the recording in minutes. The data of dodders exposed to the wheat plants is displayed on the left side of the figure (**A**,**C**), and on the right side (**B**,**D**), the data of dodders exposed to the bean plants is shown. The darker lines indicate the median of the data, and the lower and upper limits of the shadow represent the minimal and maximum values obtained, respectively. The green line represents the values of the dodders before being exposed to the host, and the blue line, after exposure.

In Figure 1A,B, the mean of the frequencies decomposed by the fast Fourier transform (FFT) is shown at every minute during the entire duration of data recording (2 h). In Figure 1A, the mean of the frequencies before and after the dodder being exposed to wheat can be observed. In this case, there was no significant difference found in the general behaviour. Around 50 min post-presentation to the host, there was a spike observed, but the dispersion was also high in this moment, therefore suggesting that this effect could be due to one or a few plants that presented a different behaviour in this moment. However, when the dodders were exposed to the bean plant (Figure 1B), there was a clear increase in the frequencies observed around 20 min later, and this increase persisted until the end of the recordings. In this case, even though there was a high dispersion around the spikes, the median after exposure to the suitable host plant was kept at a higher level than before exposure throughout the entirety of the recording. Additionally, a clear micro-voltage increase and persistence in the difference of potential (DDP) after dodder exposure to the bean plants was observed (Figure 1D). However, this behaviour was not observed in dodders exposed to wheat plants (Figure 1C).

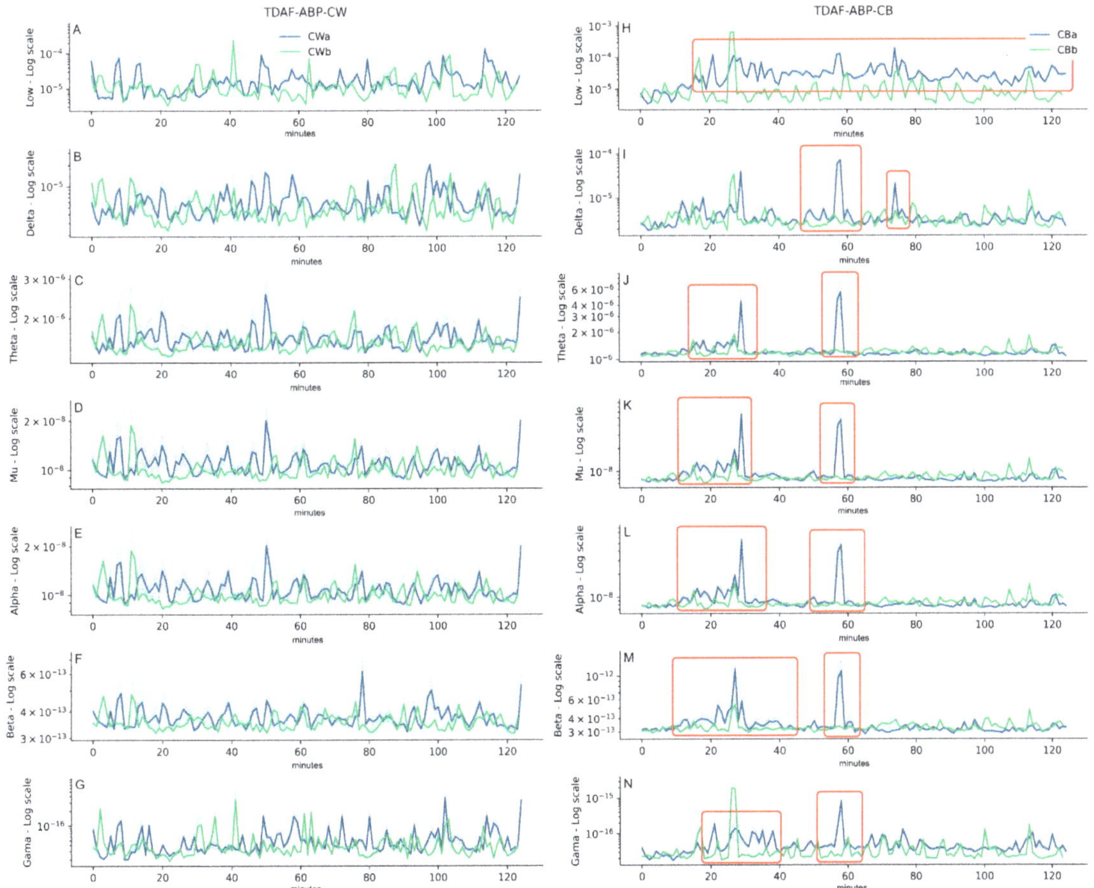

Figure 2. ABP analysis for the electrical signals of the dodders for all the plants studied (n = 23 per treatment). The ranges were named low (0.0–0.5 Hz; (**A**,**H**)), delta (0.5–4 Hz; (**B**,**I**)), theta (4–8 Hz; (**C**,**J**)), mu (9–11 Hz; (**D**,**K**)), alpha (8–13 Hz; (**E**,**L**)), beta (13–30 Hz; (**F**,**M**)), and gamma (20–100 Hz; (**G**,**N**)) waves. The Y axis shows the power values returned for each range of frequencies, and the X axis shows the time minute-by-minute during the recordings of the data (2 h before and 2 h after exposure to hosts). The darker lines indicate the median of the data, and the lower and upper limits of the shadow represent the minimal and maximum values obtained, respectively. The green line represents the values of the dodders before being exposed to the host, and the blue line, after the exposure. Relevant changes in the electrome dynamics were highlighted in red.

Figure 2 shows the ABP analyses for the signals sampled. Again, the left side of the figure shows the values of dodders presented to the wheat plants (Figure 2A–G), and on the right side, dodders presented to the bean plants (Figure 2H–N). The figure shows the band frequencies for low (0.0–0.5 Hz), delta (0.5–4 Hz), theta (4–8 Hz), mu (9–11 Hz), alpha (8–13 Hz), beta (13–30 Hz), and gamma (20–100 Hz) waves. The behaviour of the signals after the dodder being presented to the bean plant was quite distinct from before (Figure 2H–N, highlighted in red). After around 20 min, there was a sustained increase observed in the energy of the low waves throughout the recording (Figure 2H). The delta waves showed two peaks of energy investment, one at around 60 and the other around 80 min after exposure of the dodders to the bean plants (Figure 2I). In all the other band frequencies, there was a peak in energy investment observed at around 20 and 60 min after presentation to the host (Figure 2J–N)—except for gamma waves after 20 min, where the main peak occurred *before* presentation to the host. It was noticeable that the dispersion of the data was narrow, suggesting little variation in the behaviour of the plants that were analysed.

3. Discussion

Parise et al. [31] demonstrated that the bioelectrical activity of *C. racemosa* changes when the parasite is near a potential host, and that this change is not the same to both hosts. The results suggested that dodder plants recognise different host species from a distance, and this difference was observed already at the level of electrical signalling. The change in the dodders' electrome was starker in the plants presented to the suitable host, the bean plant. Additionally, the dynamics of the electrome suggested that the dodder was being attentive to the cues of the host plants. Here, we investigated whether there is a difference in the investment of the electrome energy, and how this energy is distributed in the different band frequencies. We observed that dodders that were presented to the bean (i.e., the suitable host) invested more energy in their electromes than the dodders presented to the wheat plant when it comes to the electrical signals produced. This was evident not only in the increase of the DDP (Figure 1) but also in the distribution of frequencies given by the ABP analysis (Figure 2).

In this work, the same protocol for power distribution in band frequencies used in neuroscience was employed. This was necessary, as we are unaware of protocols specific for plants that permit the categorisation of the bands for different plant behaviours, as indicated by changes in the electrome. Consequently, we must have several caveats when interpreting the results. In neuroscience, each band typically demonstrates a specific behaviour or activity of the brain. For example, delta waves indicate a low cognitive activity [41], and gamma waves are related to learning and attention [42]. Since plants are very distinct beings, their bioelectrical behaviour was expected to be quite different, and likely more related to lower frequencies [43]. Accordingly, we found higher activity in this range of frequencies, here called low (0.0–0.5 Hz). However, spikes of activity were found in all band frequencies, mainly when the dodders were presented to the bean plants. When combining the ABP technique with TDAF, it was evident there was a change in the energy of the frequencies when the dodder is perceiving a salient feature of the environment.

Our results corroborate the hypothesis that plants under putative states of attention or active perception increase the energy expended in this activity [9]. This was particularly clear in this study when a suitable host was presented to *C. racemosa*. After approximated 20 min, the dodder allocated more energy to the electrome, specifically to low wave frequencies, and after around 60 min, a peak in the energy of the frequencies in all the plants was observed in the dodders presented to the bean plant.

Although it would be important to further investigate the links between the cyclic and low-frequency activities in *C. racemosa* and many other plants, our results indicated that, whatever the cognitive value we attribute to the perceptual process involved (such as selective attention), the operating mode was, in the particular case of the dodder: (1) host-selective (operational choice), (2) energy-intensive (in proportion to the frequencies detected), (3) associated with a possible state of attention on transdisciplinary grounds [9], and (4) of a mesological order (*Umwelt* of plants, a direct link with their singular milieu [44,45]).

This is in line with other studies on electrome complexity [26,40,43], including some that demonstrated specific electromic signatures in pathogen-infected plants [28,46], bio-electric patterns of oscillations recorded in loco in *Miconia* sp. [30], and bioelectrical reaction of fruit petioles when fruits are chewed by caterpillars [32]. Our study now shows that low-frequency peaks likely predominate in these electromes, but higher frequency bands also exist, which is a novel finding. To our knowledge, gamma-like waves in plants were never recorded before this study.

Frequency bands of EPGs are different and much slower in plants. However, they fall within the range of classical EEG activities [47–49] while being clearly different from them, as was demonstrated early on [23]. Specifically, they behave similarly in three main aspects:

(1) The complex functioning of ion channels and dipoles in cell membranes (apart from differences in the ions involved in plant and animal signals), whose direct influence on the shape of the waves has been confirmed at least in human brains with computer models [50]. How plant ion channels influence bioelectrical oscillation in plants remains to be elucidated.

(2) The generation of spontaneous low-voltage bioelectrical signals (background EPG activity) through layers of tissue dipoles that correspond for brainwaves to extracellular EEG ionic currents [51,52], and intracellular electromagnetic field potentials that can be recorded by MEG [53]. Spontaneous EPGs, as part of the electrome, may synchronise when plant tissues are stimulated [23,24,54,55].

(3) The nature of the signals collected in the plant tissues, ranging within the microvolt amplitude (5–250 μV) and relatively low frequencies (0.5–15 Hz)—despite being, as shown here, lower, and less diversified than in humans and non-human animals.

Nevertheless, the electrogenesis observed in plants is different from local field potentials (LFPs) or cerebral potentials observed in animals. These bioelectrical events reflect the algebraic sum of time-synchronised synaptic potentials from large populations of neurons, including interneuronal architecture and specialised brain networks that function as intracerebral generators and pacemaker systems [56]. They contribute to the permanent generation of frequency bands that are classically linked to states of vigilance, attention, or sleep in humans or non-human animals [57–60].

On the other hand, the plant's perceptive system, including the spontaneous bioelectrical oscillations within plant bodies, involves ion channels that respond to chemical or physical stimuli [61]. As a result, there is a remarkable convergence in some plant and animal bioelectrical processes that suggest a deep evolutionary origin of the same mechanisms [24,45,62]. Plant intrinsic capacity to generate a permanent electromic activity is only now beginning to be explored [26,40]. There is an exciting potential for insights into the electrophysiological nature of plants and its relation to the establishment of scale-free states of self-organised criticality (SOC) to be made that can be quantified during 'sensory hook-ups' [26].

The electrome exhibits spontaneous and complex dynamics that is continuously generated and characterised by transition states. These transition states seem to underpin active perceptions and physiological modifications during complex tasks that require selective attention, memorisation, habituation, and associative learning [15,63,64], which require a cognitive apparatus [65].

The activity resulting from the electrome would be underpinned specifically by the dynamic non-linear organisation of cellular networks and bio-oscillators linked to certain tissue layers (e.g., meristems [66]). This is similar to the observed increase in the relative strength of certain brain rhythms at the neurobiological level, such as the theta frequency band of the parietal and temporal lobes during auditory stimulation [67], or during more complex phenomena such as binding or perceptual binding following visual stimulation. Different plant organs or modules also exhibit electromic activity in response to stimuli, similarly to how the brain exhibits increased activity in certain frequency bands following stimulation [24,54].

Temporal synchronisation of neural activity [68–70] involves the brain's ability to synchronise the oscillatory phases of neurons from different regions to reconstruct both the shape and colour of an object or image. Similarly, our working hypothesis was that synchronising the electrical activity of different plant parts is the basis of plant electrome plasticity [26,39,40], and is one of the major keys to understand complex cognitive behaviours at the plant–environment interface. Spontaneous variations in the EPGs can indicate the real-time reaction of plants to stress and environmental stimuli in the form of spatiotemporal patterns that are linked to a specific stimulus or action. The oscillatory behaviours of plants, often involving Ca^{2+} waves, have been studied and modelled [55,66,71,72]. *C. racemosa* and many other plants have predominantly low frequency ranges related to energy expenditure that might increase during operational choice and attention facing external cues [9]. This phenomenon is remarkable and may have similarities with information processing at the neural scale [24,25,45,73,74].

Low-frequency cortical oscillations (<1 Hz), such as the K-complex, have been observed in the brain during sleep, and similar low-frequency EPG emissions have been observed in plants [25,45,75–77]. Although these processes cannot be compared directly, the mechanisms underlying the electromes of living organisms present similarities that could reflect a strong mesological plasticity linked to the biology and evolution of living species and their environment [25,45].

The aim now is to link these behaviours to cognitive typologies specific to plants (e.g., dynamic coupling mechanisms, electromagnetic and ecoplastic interfaces, distributed, extended, or embodied cognition, and learning), and to show more specifically how these putative states of selective attention could enable dodders to identify which prey to parasitise, and how much energy to allocate to this operation. This same type of behaviour could also be found in vines when selecting and moving towards the supports they want to attach to [17–19].

Finally, our current band analysis was conducted based on methods from neuroscience, but future studies should develop analyses tailored to the biology of plants. This will enable the uncovering of plant-specific electrophysiological patterns and bring to light the minutiae of the bioelectrical behaviour not only of *Cuscuta racemosa*, but other plant species as well. Furthermore, studies on plant attention are at their infancy, and more research must be conducted to corroborate this hypothesis. Once this phenomenon becomes clearer, we will have a better understanding of the evolution of a cognitive phenomenon that may be widespread beyond the animal kingdom. Indeed, selective attention seems to be critical to the survival of every organism. We should pay more attention to it.

4. Materials and Methods

All the methodology for acquiring the data was described in detail in [31]. Here, a summary of what was done is presented. The difference is in the electrophysiological analyses realised and described in Section 4.3.

4.1. Plant Material and Experimental Setup

Twigs of *Cuscuta racemosa* were collected from a stock grown on basil plants (*Ocimus* sp.) and kept in greenhouse conditions. They were trimmed so that every twig had a 10 cm length with only one node at one extremity, and their mean weight was 0.172 mg $\pm$ 0.042. They were placed inside a Styrofoam box (20 $\times$ 25 $\times$ 17 cm) which contained a shelf (12 $\times$ 17 cm) placed 5 cm deep inside the box. The shelf covered part of the box interior, which left a pit in the other half where the host plants were placed. Four dodders per box were placed in parallel with the longer side of the box with their nodes pointed to the pit. A pair of stainless-steel needle electrodes (model EL-452, BIOPAC Systems $R^©$, Goleta, CA, USA) was inserted in the dodder's stem immediately below the node and 1 cm apart from each other. The boxes were placed inside a Faraday cage and illuminated by a white LED light (PPFD 230 μmol m^{-2} s^{-1}), with a photoperiod of 12.0 h and constant mean temperature of 25 °C $\pm$ 1. The boxes were sealed with transparent polyvinyl chloride (PVC) film to maintain humidity and then left in the laboratory overnight for acclimation.

4.2. Electrophysiological Recordings

In the following morning, the electrome of the dodders' was recorded for 2 h before and 2 h after a pot with a host was introduced in the box. The introduced host was either a bean plant (*Phaseolus vulgaris* L. cv. BRS-Expedito) in the V3 stage of development, with the first trifoliolate leaf developing (following the classification of de Oliveira et al. [78]); or a wheat plant (*Triticum aestivum* L. cv. BRS-Parrudo), with the third leaf emerging after germination. This species was chosen as dodders cannot parasitise wheat plants and avoid them when possible [79–81].

The data was recorded with the system Biopac Student Lab (BIOPAC Systems R$^{©}$, Goleta, CA, USA), model MP36 with four channels with a high input impedance (10 GΩ). Signals were collected with a sampling rate of fs = 62.5 Hz amplified with a gain of 1000-fold. The protocol used was ECG-AHA (0.05–100 Hz), with a notch frequency of 60.0 Hz to minimise the influence of the electrical network. The data obtained was a time series of micro-voltage variations as $\Delta\mathbb{V} = \{\Delta V_1, \Delta V_2, \ldots, \Delta V_n\}$, where ΔV_i is the difference of electrical potential between the electrodes, and n is the length of the time series. This length was derived from a sample of 2 h (7200 s) of data acquisition, and using fs = 62.5 Hz, N = 450,000 points. In the end, 23 time series for the test and 23 time series for the control were obtained and analysed. A detailed description of all the methods described until this point can be found in [31].

4.3. Electrophysiological Analyses

The fast Fourier transform (FFT) technique was used to understand the frequencies of the dodder's electrome. For this work, we used Bluestein's algorithm [82] and the rfft Hermitian-symmetric algorithm from scipy library [83] to silence imaginary values. The PSD (power spectral density) was calculated using the Welch method [84]. This method estimated the PSD by dividing the data in windows and calculating the periodogram for each window. Again, the scipy library was used [83] with the Welch method with 4 s windows. The PSD value returned was then used to calculate the ABP (average band power). This method consists of analysing the signals in specific frequencies. It is commonly used for the generation of EEG features in neuroscientific analyses [85–91], as well as in attention tests [92–94]. To calculate the ABP, the area determined by the interest must be integrated to the function provided by the PSD. This was done using the Simpson's compound rule [95] which, in general, decomposes the area in many parabolas and then sum them up returning the value of the area of interest. For this calculation, the method Simpson from scipy library was used [83]. The band frequencies calculated were 0.0–0.5 Hz (here named low waves), 0.5–4.0 Hz (delta), 4.0–8 Hz (theta), 9–11 Hz (mu), 8–13 Hz (alpha), 13–30 Hz (beta), and 20–100 Hz (gamma), respectively.

To visualise these analyses, the method time dispersion analysis of features (TDAF) was employed [39]. To generate the characteristics, the time series were divided into shorter series, and with TDAF, the temporal information of the position of each clipping of the time series was kept and taken to each feature. After completing the calculations, it was possible to aggregate each characteristic for each stretch of time and make a dispersion analysis (with the maximum value, minimum value, median, and quartiles) in a specific moment. When all the analyses were plotted in the correct temporal order, it was then possible to obtain the dynamics of the dispersion of the feature for each time series analysed. Therefore, a qualitative way of analysing chaotic series through time was obtained.

Author Contributions: Conceptualisation, A.G.P., T.F.d.C.O., G.M.S. and M.-W.D.; methodology, A.G.P., T.F.d.C.O. and G.M.S.; software, T.F.d.C.O.; validation, A.G.P., T.F.d.C.O., M.-W.D. and G.M.S.; data curation, T.F.d.C.O.; writing—original draft preparation, A.G.P., T.F.d.C.O., M.-W.D. and G.M.S.; writing—review and editing, A.G.P.; supervision, G.M.S.; funding acquisition, G.M.S. All authors have read and agreed to the published version of the manuscript.

Funding: This study was supported in part by the Coordenação de Aperfeiçoamento de Pessoal de Nível Superior, Brazil (CAPES) code 001. The authors are also grateful to the Conselho Nacional de Desenvolvimento Científico e Tecnológico (CNPq) for the financial support provided (G.M.S. is supported by the CNPq grant 302592/2021-0). A.G.P. is currently supported by the University of Reading International PhD Studentship.

Data Availability Statement: The raw data used in this work can be freely accessed in Parise et al. [31].

Acknowledgments: The authors are grateful to Michael Marder, Monica Gagliano, and Umberto Castiello for early discussions on the hypothesis of plant attention, to Hannah Rhiannon Hall for kindly reviewing the manuscript, and to all the LACEV members for their support during the collection of the data.

Conflicts of Interest: The authors declare no conflict of interest.

References

1. Neisser, U. *Cognition and Reality*; W. H. Freeman and Company: San Francisco, CA, USA, 1976.
2. Maturana, H.R.; Varela, F.J. *Autopoiesis and Cognition: The Realization of the Living*; D. Reidel Publishing Company: Dordrecht, The Netherlands, 1980.
3. Baluška, F.; Reber, A.S.; Miller, W.B., Jr. Cellular sentience as the primary source of biological order and evolution. *BioScience* **2022**, *218*, 104694. [CrossRef]
4. James, W. *The Principles of Psychology*; 1981 Reprint; Harvard University Press: Cambridge, MA, USA, 1890.
5. Driver, J. A selective review of selective attention research from the past century. *Br. J. Psychol.* **2001**, *92*, 53–78. [CrossRef] [PubMed]
6. Chun, M.M.; Golomb, J.D.; Turk-Browne, N.B. A taxonomy of external and internal attention. *Annu. Rev. Psychol.* **2011**, *62*, 73–101. [CrossRef] [PubMed]
7. Gibson, J.J. *The Ecological Approach to Visual Perception: Classic Edition*; Psychology Press: New York, NY, USA, 2014. [CrossRef]
8. Lev-Ari, T.; Beeri, H.; Gutfreund, Y. The ecological view of selective attention. *Front. Integr. Neurosci.* **2022**, *16*, 856207. [CrossRef] [PubMed]
9. Parise, A.G.; de Toledo, G.R.A.; Oliveira, T.F.d.C.; Souza, G.M.; Castiello, U.; Gagliano, M.; Marder, M. Do plants pay attention? A possible phenomenological-empirical approach. *Prog. Biophys. Mol. Biol.* **2022**, *173*, 11–23. [CrossRef]
10. Karban, R. *Plant Sensing and Communication*; The University of Chicago Press: Chicago, IL, USA, 2014.
11. Gagliano, M. Green symphonies: A call for studies on acoustic communication in plants. *Behav. Ecol.* **2013**, *24*, 789–796. [CrossRef]
12. Schwartz, A.; Koller, D. Diurnal phototropism in solar tracking leaves of *Lavatera cretica*. *Plant Physiol.* **1986**, *80*, 771–778. [CrossRef]
13. Trewavas, A. Aspects of plant intelligence. *Ann. Bot.* **2003**, *92*, 1–20. [CrossRef]
14. Lüttge, U. *Clusia*: Holy grail and enigma. *J. Exp. Bot.* **2008**, *59*, 1503–1514. [CrossRef]
15. Gagliano, M.; Vyazovskiy, V.V.; Borbély, A.A.; Grimonprez, M.; Depczynski, M. Learning by association in plants. *Sci. Rep.* **2016**, *6*, 38427. [CrossRef]
16. Munné-Bosh, S. Spatiotemporal limitations in plant biology research. *Trends Plant Sci.* **2022**, *27*, P346–P354. [CrossRef] [PubMed]
17. Guerra, S.; Peressotti, A.; Peressotti, F.; Bulgheroni, M.; Baccinelli, W.; D'Amico, E.; Gómez, A.; Massaccesi, S.; Ceccarini, F.; Castiello, U. Flexible control of movement in plants. *Sci. Rep.* **2019**, *9*, 16570. [CrossRef] [PubMed]
18. Raja, V.; Silva, P.L.; Holghoomi, R.; Calvo, P. The dynamics of plant nutation. *Sci. Rep.* **2020**, *10*, 19465. [CrossRef]
19. Wang, Q.; Guerra, S.; Bonato, B.; Simonetti, V.; Bulgheroni, M.; Castiello, U. Decision-making underlying support-searching in pea plants. *Plants* **2023**, *12*, 1597. [CrossRef]
20. Marder, M. Plant intelligence and attention. *Plant Signal. Behav.* **2013**, *8*, e23902. [CrossRef] [PubMed]
21. De Toledo, G.R.A.; Parise, A.G.; Simmi, F.Z.; Costa, A.V.L.; Senko, L.G.S.; Debono, M.-W.; Souza, G.M. Plant electrome: The electrical dimension of plant life. *Theor. Exp. Plant Physiol.* **2019**, *31*, 21–46. [CrossRef]
22. Volkov, A. *Plant Electrophysiology: Theory and Methods*, 1st ed.; Springer: Berlin, Germany, 2006.
23. Debono, M.-W.; Bouteau, F. Spontaneous and evoked surface potentials in *Kalanchoë* tissues. *Life Sci. Adv. Plant. Physiol.* **1992**, *11*, 107–117.
24. Debono, M.-W. Dynamic protoneural networks in plants: A new approach of extracellular spontaneous potential variations. *Plant Signal. Behav.* **2013**, *8*, e24207. [CrossRef]
25. Debono, M.-W. Perceptive levels in plants: A transdisciplinary challenge in living organism's plasticity. *Transdiscipl. J. Eng. Sci.* **2013**, *4*, 21–39. [CrossRef]
26. Souza, G.M.; Ferreira, A.S.; Saraiva, G.F.R.; Toledo, G.R.A. Plant "electrome" can be pushed toward a self-organized critical state by external cues: Evidences from a study with soybean seedlings subject to different environmental conditions. *Plant Signal. Behav.* **2017**, *12*, e1290040. [CrossRef]
27. Pereira, D.R.; Papa, J.P.; Saraiva, G.F.R.; Souza, G.M. Automatic classification of plant electrophysiological responses to environmental stimuli using machine learning and interval arithmetic. *Comput. Electron. Agric.* **2018**, *145*, 35–42. [CrossRef]
28. Simmi, F.; Dallagnol, L.; Ferreira, A.; Pereira, D.; Souza, G. Electrome alterations in a plant-pathogen system: Toward early diagnosis. *Bioelectrochemistry* **2020**, *133*, 107493. [CrossRef] [PubMed]
29. Najdenovska, E.; Dutoit, F.; Tran, D.; Plummer, C.; Wallbridge, N.; Camps, C.; Raileanu, L.E. Classification of plant electrophysiology signals for detection of spider mites infestation in tomatoes. *Appl. Sci.* **2021**, *11*, 1414. [CrossRef]
30. Gimenez, V.M.M.; Pauletti, P.M.; Silva, A.C.S.; Costa, E.J.X. Bioelectrical pattern discrimination of *Miconia* plants by spectral analysis and machine learning. *Theor. Exp. Plant Physiol.* **2021**, *33*, 329–342. [CrossRef]

31. Parise, A.G.; Reissig, G.N.; Basso, L.F.; Senko, L.G.S.; Oliveira, T.F.d.C.; de Toledo, G.R.A.; Ferreira, A.S.; Souza, G.M. Detection of different hosts from a distance alters the behaviour and bioelectrical activity of *Cuscuta racemosa*. *Front. Plant Sci.* **2021**, *12*, 409. [CrossRef]

32. Reissig, G.N.; Oliveira, T.F.C.; Oliveira, R.P.; Posso, D.A.; Parise, A.G.; Nava, D.E.; Souza, G.M. Fruit herbivory alters plant electrome: Evidence for fruit-shoot long-distance electrical signaling in tomato plants. *Front. Sustain. Food Syst.* **2021**, *5*, 657401. [CrossRef]

33. Mokeichev, A.; Segev, R.; Ben-Shahar, O. Orientation saliency without visual cortex and target selection in archer fish. *Proc. Natl. Acad. Sci. USA* **2010**, *107*, 16726–16731. [CrossRef] [PubMed]

34. Sridharan, D.; Schwarz, J.S.; Knudsen, E.I. Selective attention in birds. *Curr. Biol.* **2010**, *24*, R510–R513. [CrossRef]

35. Sareen, P.; Wolf, R.; Heisenberg, M. Attracting the attention of a fly. *Proc. Natl. Acad. Sci. USA* **2011**, *108*, 7230–7235. [CrossRef]

36. Paulk, A.C.; Stacey, J.A.; Pearson, T.W.; Taylor, G.J.; Moore, R.J.; Srinivasan, M.V.; van Swinderen, B. Selective attention in the honeybee optic lobes precedes behavioral choices. *Proc. Natl. Acad. Sci. USA* **2014**, *111*, 5006–5011. [CrossRef]

37. Van Swinderen, B. Attention-like processes in *Drosophila* require short-term memory genes. *Science* **2007**, *315*, 1590–1593. [CrossRef] [PubMed]

38. Zhang, K.; Guo, J.Z.; Peng, Y.; Xi, W.; Guo, A. Dopamine-mushroom body circuit regulates saliency-based decision-making in *Drosophila*. *Science* **2007**, *316*, 1901–1904. [CrossRef] [PubMed]

39. Costa, A.V.L.; Oliveira, T.F.C.; Posso, D.A.; Reissig, G.N.; Parise, A.G.; Barros, W.S.; Souza, G.M. Systemic signals induced by single and combined abiotic stimuli in common bean plants. *Plants* **2023**, *12*, 924. [CrossRef] [PubMed]

40. Debono, M.-W.; Souza, G.M. Plants as electromic plastic interfaces: A mesological approach. *Prog. Biophys. Mol. Biol.* **2019**, *146*, 123–133. [CrossRef]

41. Harmony, T. The functional significance of delta oscillations in cognitive processing. *Front. Integr. Neurosci.* **2013**, *7*, 0083. [CrossRef]

42. De Arcangelis, L.; Herrmann, H.J. Learning as a phenomenon occurring in a critical state. *Proc. Natl. Acad. Sci. USA* **2010**, *107*, 3977–3981. [CrossRef]

43. Saraiva, G.F.R.; Ferreira, A.S.; Souza, G.M. Osmotic stress decreases complexity underlying the electrophysiological dynamic in soybean. *Plant Biol.* **2017**, *19*, 702–708. [CrossRef]

44. Debono, M.-W. The plant-environment interface: A mesological approach to plant cognition. In Proceedings of the Third World Congress of Transdisciplinarity, CTU Weeks, Transdisciplinary Unesco Chair "Human Development and Culture of Peace", Florence, Italy, 13–15 October 2021.

45. Debono, M.-W. Mesological plasticity as a new model to study plant evolution, interactive ecosystems & self-organized evolutionary processes. In *Self-Organization as a New Paradigm in Evolutionary Biology: From Theory to Applied Cases in the Tree of Life*; Dambricourt-Malassé, A., Ed.; Springer-Nature: Cham, Switzerland, 2022; Volume 5, pp. 253–290. [CrossRef]

46. Simmi, F.Z.; Dallagnol, L.J.; Almeida, R.O.; Dorneles, K.R.; Souza, G.M. Barley systemic bioelectrical changes detect pathogenic infection days before the first disease symptoms. *Comput. Electron. Agric.* **2023**, *209*, 107832. [CrossRef]

47. Freeman, W.J. Origin, structure, and role of background EEG activity. Part 1. Analytic amplitude. *Clin. Neurophysiol.* **2004**, *115*, 2077–2088. [CrossRef]

48. Freeman, W.J. Origin, structure, and role of background EEG activity. Part 2. Analytic phase. *Clin. Neurophysiol.* **2004**, *115*, 2089–2107. [CrossRef]

49. Freeman, W.J. Origin, structure, and role of background EEG activity. Part 3. Neural frame classification. *Clin. Neurophysiol.* **2005**, *116*, 1118–1129. [CrossRef]

50. Reimann, M.W.; Anastassiou, C.A.; Perin, R.; Hill, S.L.; Markram, H.; Koch, C. A biophysically detailed model of neocortical local field potentials predicts the critical role of active membrane currents. *Neuron* **2013**, *79*, 375–390. [CrossRef] [PubMed]

51. Nunez, P.L.; Srinivasan, R. *Electric Fields of the Brain: The Neurophysics of EEG*; Oxford University Press: Oxford, UK, 2006.

52. Buzsáki, G.; Anastassiou, C.; Koch, C. The origin of extracellular fields and currents—EEG, ECoG, LFP and spikes. *Nat. Rev. Neurosci.* **2012**, *13*, 407–420. [CrossRef] [PubMed]

53. Amzica, F.; Lopes da Silva, F.H. Cellular substrates of brain rhythms. In *Niedermeyer's Electroencephalography: Basic Principles, Clinical Applications, and Related Fields*, 7th ed.; Schomer, D.L., Lopez da Silva, F.H., Eds.; Oxford University Press: Oxford, UK, 2017; Volume 1, pp. 20–62. [CrossRef]

54. Masi, E.; Ciszak, M.; Stefano, G.; Rena, L.; Azzarello, E.; Pandolfi, C.; Mugnai, S.; Baluška, F.; Arecchi, F.T.; Mancuso, S. Spatiotemporal dynamics of the electrical network activity in the root apex. *Proc. Natl. Acad. Sci. USA* **2009**, *106*, 4048–4053. [CrossRef]

55. Cabral, E.F.; Pecora, P.C.; Arce, A.I.C.; Tech, A.R.B.; Costa, E.J.X. The oscillatory bioelectrical signal from plants explained by a simulated electrical model and tested using Lempel-Ziv complexity. *Comput. Electron. Agric.* **2011**, *76*, 1–5. [CrossRef]

56. Kandel, E.R.; Koester, J.; Mack, S.; Siegelbaum, S. *Principles of Neural Science*, 6th ed.; McGraw Hill: New York, NY, USA, 2021.

57. Remmers, J.E.; Gautier, H. Neural and mechanical mechanisms of feline purring. *Respir. Physiol.* **1972**, *16*, 351–361. [CrossRef]

58. Da Silva, F.L. EEG: Origin and Measurement. In *EEG–fMRI*; Mulert, C., Lemieux, L., Eds.; Springer: Berlin/Heidelberg, Germany, 2009; pp. 19–38. [CrossRef]

59. Steriade, M. Grouping of brain rhythms in corticothalamic systems. *Neuroscience* **2006**, *137*, 1087–1106. [CrossRef] [PubMed]

60. Turner, D.C.; Bateson, P. (Eds.) *The Domestic Cat: The Biology of Its Behaviour*, 2nd ed.; The University of Cambridge Press: Cambridge, UK, 2000.

61. Canales, J.; Henriquez-Valencia, C.; Brauchi, S. The integration of electrical signals originating in the root of vascular plants. *Front. Plant Sci.* **2018**, *8*, 2173. [CrossRef]

62. Baluška, F.; Mancuso, S. Deep evolutionary origins of neurobiology: Turning the essence of 'neural' upside-down. *Commun. Integr. Biol.* **2009**, *2*, 60–65. [CrossRef]

63. Gagliano, M.; Renton, M.; Depczynski, M.; Mancuso, S. Experience teaches plants to learn faster and forget slower in environments where it matters. *Oecologia* **2014**, *175*, 63–72. [CrossRef]

64. Calvo, P.; Gagliano, M.; Souza, G.M.; Trewavas, A. Plants are intelligent: Here's how. *Ann. Bot.* **2020**, *125*, 11–28. [CrossRef]

65. Calvo Garzón, P.; Keijzer, F.A. Plants: Adaptive behavior, root-brains, and minimal cognition. *Adapt. Behav.* **2011**, *19*, 155–171. [CrossRef]

66. Damineli, D.S.C.; Portes, M.T.; Feijó, J.A. Electrifying rhythms in plant cells. *Curr. Opin. Cell Biol.* **2022**, *77*, 102113. [CrossRef] [PubMed]

67. Ala, T.S.; Ahmadi-Pajouh, M.A.; Nasrabadi, A.M. Cumulative effects of theta binaural beats on brain power and functional connectivity. *Biomed. Signal Process. Control.* **2018**, *42*, 242–252. [CrossRef]

68. Von der Malsburg, C. Synaptic plasticity as basis of brain organization. In *The Neural and Molecular Bases of Learning*; Changeux, J.P., Konishi, M., Eds.; John Wiley & Sons: Chichester, UK, 1987; pp. 411–431.

69. Engel, A.K.; Konig, P.; Kreiter, A.K.; Gray, C.M.; Singer, W. Temporal coding by coherent oscillations as a potential solution to the binding problem. In *NONLINEAR Dynamics and Neural Networks*; Schuster, H.G., Ed.; VCH Pub: New York, NY, USA, 1990.

70. Engel, A.K.; Singer, W. Temporal binding and the neural correlates of sensory awareness. *Trends. Cogn. Sci.* **2001**, *5*, 16–25. [CrossRef]

71. Hazledine, S.; Sun, J.; Wysham, D.; Downie, J.A.; Oldroyd, G.E.D.; Morris, R.J. Nonlinear time series analysis of nodulation factor induced calcium oscillations: Evidence for deterministic chaos? *PLoS ONE* **2009**, *4*, e6637. [CrossRef]

72. Sukhova, E.; Akinchits, E.; Sukhov, V. Mathematical models of electrical activity in plants. *J. Membr. Biol.* **2017**, *250*, 407–423. [CrossRef]

73. Debono, M.-W. Electrome & Cognition modes in plants: A transdisciplinary approach to the eco-sensitiveness of the world. *Transdiscipl. J. Eng. Sci.* **2020**, *11*, 213–239. Available online: https://www.atlas-tjes.org/index.php/tjes/article/view/169 (accessed on 10 May 2023).

74. Tian, W.; Wang, C.; Gao, Q.; Li, L.; Luan, S. Calcium spikes, waves and oscillations in plant development and biotic interactions. *Nat. Plants* **2020**, *6*, 750–759. [CrossRef]

75. Amzica, F.; Steriade, M. The K-complex: Its slow (<1-Hz) rhythmicity and relation to delta waves. *Neurology* **1997**, *49*, 952–959. [CrossRef]

76. Neske, G.T. The slow oscillation in cortical and thalamic networks: Mechanisms and functions. *Front. Neural Circuits* **2015**, *9*, 88. [CrossRef] [PubMed]

77. Achermann, P.; Borbély, A.A. Low-frequency (<1 Hz) oscillations in the human sleep electroencephalogram. *Neuroscience* **1997**, *81*, 213–222. [CrossRef] [PubMed]

78. De Oliveira, L.F.C.; Oliveira, M.G.C.; Wendland, A.; Heinemann, A.B.; Guimarães, C.M.; Ferreira, E.P.B.; Quintela, E.D.; Barbosa, F.R.; Carvalho, M.C.S.; Lobo Junior, M.; et al. *Conhecendo a Fenologia do Feijoeiro e Seus Aspectos Fitotécnicos*; Embrapa: Brasília, DF, Brazil, 2018.

79. Costea, M.; Tardif, F.J. The biology of Canadian weeds. 133. *Cuscuta campestris* Yuncker, C. *gronovii* Willd. ex Schult., *C. umbrosa* Beyr. ex Hook., C. *epithymum* (L.) L. and C. *epilinum* Weihe. *Can. J. Plant Sci.* **2006**, *86*, 293–316. [CrossRef]

80. Runyon, J.B.; Mescher, M.C.; de Moraes, C.M. Volatile chemical cues guide host location and host selection by parasitic plants. *Science* **2006**, *313*, 1964–1967. [CrossRef]

81. Albert, M.; Belastegui-Macadam, X.; Bleischwitz, M.; Kaldenhoff, R. Cuscuta spp: "parasitic plants in the spotlight of plant physiology, economy and ecology". In *Progress in Botany*; Lüttge, U., Beyschlag, W., Murata, J., Eds.; Springer: Berlin/Heidelberg, Germany, 2008; pp. 267–277. [CrossRef]

82. Bluestein, L.A. Linear filtering approach to the computation of discrete Fourier transform. *IEEE Trans. Audio Electroacoust.* **1970**, *18*, 451–455. [CrossRef]

83. Virtanen, P.; Gommers, R.; Oliphant, T.E.; Haberland, M.; Reddy, T.; Cournapeau, D.; Burovski, E.; Peterson, P.; Weckesser, W.; Bright, J.; et al. SciPy 1.0: Fundamental algorithms for scientific computing in Python. *Nat. Methods* **2020**, *17*, 261–272. [CrossRef]

84. Welch, P. The use of fast Fourier transform for the estimation of power spectra: A method based on time averaging over short, modified periodograms. *IEEE Trans. Audio Electroacoust.* **1967**, *15*, 70–73. [CrossRef]

85. Evans, J.R. Neurofeedback. In *Encyclopedia of the Human Brain*; Elsevier: Amsterdam, The Netherlands, 2002; pp. 465–477.

86. Heraz, A.; Razaki, R.; Frasson, C. Using machine learning to predict learner emotional state from brainwaves. In Proceedings of the Seventh IEEE International Conference on Advanced Learning Technologies (ICALT 2007), Niigata, Japan, 18–20 July 2007. [CrossRef]

87. Abhang, P.A.; Gawali, B.W.; Mehrotra, S.C. Technical aspects of brain rhythms and speech parameters. In *Introduction to EEG- and Speech-Based Emotion Recognition*; Abhang, P.A., Gawali, B.W., Mehrotra, S.C., Eds.; Elsevier: London, UK, 2016; pp. 51–79.

88. Abhang, P.A.; Gawali, B.W.; Mehrotra, S.C. Technological basics of EEG recording and operation of apparatus. In *Introduction to EEG- and Speech-Based Emotion Recognition*; Abhang, P.A., Gawali, B.W., Mehrotra, S.C., Eds.; Elsevier: London, UK, 2016; pp. 19–50.

89. Delimayanti, M.K.; Purnama, B.; Nguyen, N.G.; Faisal, M.R.; Mahmudah, K.R.; Indriani, F.; Kubo, M.; Satou, K. Classification of brainwaves for sleep stages by high-dimensional FFT characteristics from EEG signals. *Appl. Sci.* **2020**, *10*, 1797. [CrossRef]

90. Savadkoohi, M.; Oladunni, T.; Thompson, L. A machine learning approach to epileptic seizure prediction using electroencephalogram (EEG) signal. *Biocybern. Biomed. Eng.* **2020**, *40*, 1328–1341. [CrossRef]

91. Kora, P.; Meenakshi, K.; Swaraja, K.; Rajani, A.; Raju, M.S. EEG based interpretation of human brain activity during yoga and meditation using machine learning: A systematic review. *Complement. Ther. Clin. Pract.* **2021**, *43*, 101329. [CrossRef]

92. Zhong, G. Analysis of healthy people's attention based on EEG spectrum. In Proceedings of the 2015 International Conference on Mechatronics, Electronic, Industrial and Control Engineering, Shenyang, China, 1–3 April 2015. [CrossRef]

93. Kirmizi-Alsan, E.; Bayraktaroglu, Z.; Guruvit, H.; Keskin, Y.H.; Emre, M.; Demiralp, T. Comparative analysis of event-related potentials during Go/NoGo and CPT: Decomposition of electrophysiological markers of response inhibition and sustained attention. *Brain Res.* **2006**, *1104*, 114–128. [CrossRef] [PubMed]

94. Karamacoska, D.; Barry, R.J.; De Blasio, F.M.; Steiner, G.Z. EEG-ERP dynamics in a visual continuous performance test. *Int. J. Psychophysiol.* **2019**, *146*, 249–260. [CrossRef] [PubMed]

95. Press, W.H.; Flannery, B.P.; Teukolsky, S.A.; Vetterling, W.T. *Numerical Recipes in Pascal: The Art of Scientific Computing*; Cambridge University Press: Cambridge, UK, 1989.

Article

Water Cannot Activate Traps of the Carnivorous Sundew Plant *Drosera capensis*: On the Trail of Darwin's 150-Years-Old Mystery

Andrej Pavlovič [1,*], Ondřej Vrobel [2,3] and Petr Tarkowski [2,3]

[1] Department of Biophysics, Faculty of Science, Palacký University, Šlechtitelů 27, CZ-783 71 Olomouc, Czech Republic
[2] Czech Advanced Technology and Research Institute, Palacký University, Šlechtitelů 27, CZ-783 71 Olomouc, Czech Republic; ondrej.vrobel@upol.cz (O.V.); petr.tarkowski@upol.cz (P.T.)
[3] Centre of the Region Haná for Biotechnological and Agricultural Research, Department of Genetic Resources for Vegetables, Medicinal and Special Plants, Crop Research Institute, Šlechtitelů 29, CZ-783 71 Olomouc, Czech Republic
* Correspondence: andrej.pavlovic@upol.cz; Tel.: +420-585-634-831

Abstract: In his famous book *Insectivorous plants*, Charles Darwin observed that the bending response of tentacles in the carnivorous sundew plant *Drosera rotundifolia* was not triggered by a drop of water, but rather the application of many dissolved chemicals or mechanical stimulation. In this study, we tried to reveal this 150-years-old mystery using methods not available in his time. We measured electrical signals, phytohormone tissue level, enzyme activities and an abundance of digestive enzyme aspartic protease droserasin in response to different stimuli (water drop, ammonia, mechanostimulation, chitin, insect prey) in Cape sundew (*Drosera capensis*). Drops of water induced the lowest number of action potentials (APs) in the tentacle head, and accumulation of jasmonates in the trap was not significantly different from control plants. On the other hand, all other stimuli significantly increased jasmonate accumulation; the highest was found after the application of insect prey. Drops of water also did not induce proteolytic activity and an abundance of aspartic protease droserasin in contrast to other stimuli. We found that the tentacles of sundew plants are not responsive to water drops due to an inactive jasmonic acid signalling pathway, important for the induction of significant digestive enzyme activities.

Keywords: abscisic acid; aspartic protease; carnivorous plant; digestive enzyme; jasmonic acid; sundew

Citation: Pavlovič, A.; Vrobel, O.; Tarkowski, P. Water Cannot Activate Traps of the Carnivorous Sundew Plant *Drosera capensis*: On the Trail of Darwin's 150-Years-Old Mystery. *Plants* **2023**, *12*, 1820. https://doi.org/10.3390/plants12091820

Academic Editor: Anis Limami

Received: 29 March 2023
Revised: 21 April 2023
Accepted: 25 April 2023
Published: 28 April 2023

1. Introduction

Carnivorous plants of the genus sundew (*Drosera* sp.) capture insect prey by using sticky tentacles. The tentacles exhibit remarkably complex behaviour when they capture and digest insect prey. The heads of the marginal tentacles rapidly bend inward after mechanical stimulation. Rapid movement of the tentacles serves to hold captured insects on the leaf and is mediated by action potentials (APs) which are initiated by a receptor potential [1–5]. The APs are initiated just below the swollen head of the tentacle and propagated to its base, but not to the trap tissue [1–3,5]. A series of several APs are necessary to initiate an inflection of the tentacle [2]. Inflection brings the prey to the centre of the leaf and in contact with the short inner tentacles. The contact with the inner tentacles triggers the oscillation of membrane potentials and localized accumulation of jasmonates in the trap tissue [5,6]. Recent study revealed that this oscillation probably represents Ca^{2+} waves from individual tentacles [7], by means of which the individual tentacles communicate with each other. This results in the bending of all marginal tentacles that have no previous contact with the prey, thus safely fixing the prey [5,6]. The plant phytohormones from the group of jasmonates play an indispensable role in the bending reaction and initiation of digestive enzyme secretion [5,6,8]. The secretome of the carnivorous sundew plant consists of cysteine and aspartic proteases, chitinase, ribonuclease, phosphatase, alpha-galactosidase and β-1,3-glucanase [5,9–13].

The peculiarity of the insect-trapping behaviour of sundew plants attracted the attention of many early biologists. The first detailed experimental investigations on the common sundew (*Drosera rotundifolia*) were conducted by Charles Darwin and were described in his book *Insectivorous plants* [14]. Darwin showed that the tentacles are sensitive to both chemical and mechanical stimulation. He applied more than one hundred different organic and inorganic chemicals on the leaf and carefully observed the bending reactions of the sticky tentacles. He tested various nitrogenous and non-nitrogenous compounds in solution, and he found that the plant detects with almost unerring certainty the presence of nitrogen. In relation to these experiments, it was necessary first to ascertain the effect of distilled water, and he found that the more sensitive leaves are affected by it, but only slightly in comparison to nitrogenous compounds. Thus, the glands are insensible to the weight and repeated blows of drops of heavy rain, and the plants are thus likewise saved from much useless movement. On the other hand, the pressure exerted by the lighter human hair is sufficient for tentacle bending reactions. Darwin wrote about these observations: "*It would obviously have been a great evil to the plant if the tentacles were excited to bend by every shower of rain; but this evil has been avoided by the glands either having become through habit insensible to the blows and prolonged pressure of drops of water, or to their having been originally rendered sensitive solely to the contact of solid bodies*" [14].

Now, 150 years later, we can investigate the tentacle behaviour via more sophisticated methods in comparison to just observing its inflection. Recent studies have shown that the amount and type of enzymes secreted in sundew plants is regulated by different stimuli [5,9,15]. Already Darwin noticed that secretion induced by nitrogenous chemicals, but not mechanical stimuli, had the power to digest proteins, and that the glands of *Drosera* secrete: "*some ferment analogous to pepsin*" [14]. Recently, Darwin's assumption was confirmed, and the enzyme in digestive fluid related to pepsin was discovered—aspartic protease droserasin [5,12,16]. We also know that even mechanical stimulation of tentacles and subsequent generation of APs is essential for the initial release of low amounts of digestive enzymes; however, their production is boosted after chemical sensing. Chitin and proteins are sensed the best and trigger the expression of digestive enzymes [5,9,15]. Here, we investigated the effect of water drops on the secretion of digestive enzymes in Cape sundew (*Drosera capensis*) to complement Darwin's assertion of the inefficiency of water to induce the tentacle bending reaction. In addition, we also applied mechanical stimuli, ammonium chloride, chitin and live prey for comparison.

2. Results

2.1. Electrical Signals

Application of a drop of distilled water on the marginal tentacle head resulted in membrane hyperpolarization (positive voltage shift recorded extracellularly, representing intracellular hyperpolarization). In many cases, a receptor potential with few APs (7.4 ± 11.7, mean $\pm$ S.D., min = 0, max = 49, n = 22) was also triggered (Figure 1A). On the other hand, application of 50 mM of NH_4Cl resulted in the depolarization of membrane potentials (negative voltage shift recorded extracellularly, representing intracellular depolarization) which trigger the series of APs (10.7 ± 10.0, mean $\pm$ S.D., min = 0, max = 42, n = 21, Figure 1B). Touch or contact with any solid bodies induced the depolarization of a membrane potential (negative voltage shift recorded extracellularly, representing intracellular depolarization), probably a receptor potential (RP). The RP was accompanied by a series of APs (21.9 ± 20.2, mean $\pm$ S.D., min = 1, max = 87, n = 17). After that, the membrane potential repolarized (Figure 1C). The number of APs triggered by all these stimuli was highly variable (Figure 1D) and can be explained by the different sensitivities of the individual tentacles. Measurements of electrical signals in response to insect application was not measured due to the interference with the locomotion of the live insects.

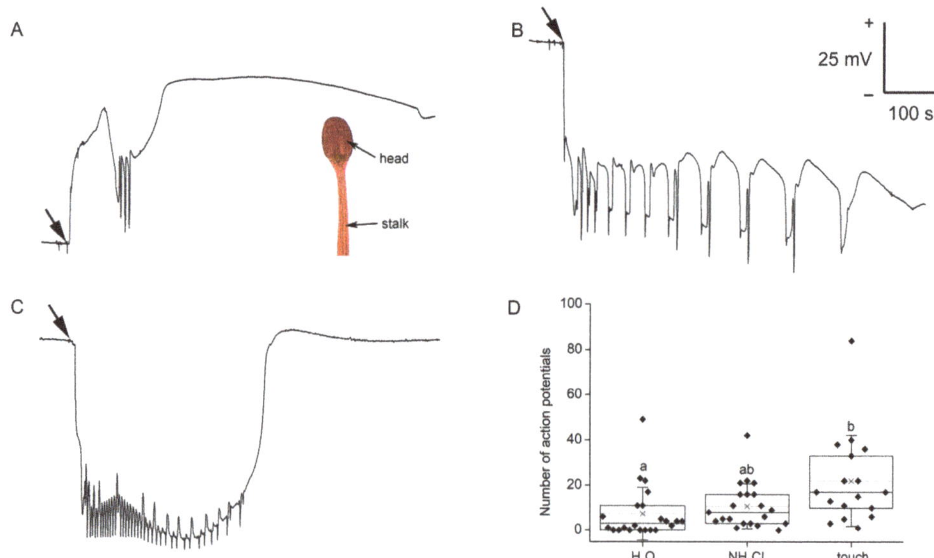

Figure 1. Extracellular recording of membrane potential from the tentacle head of the carnivorous sundew plant (*Drosera capensis*). (**A**) response to water drop application, with inset showing tentacle head and stalk; (**B**) response to drop of 50 mM of NH_4Cl; (**C**) response to touch; and (**D**) box plots of the number of action potentials triggered by different stimuli recorded during a 700-s timespan. Boxplots show the individual measurements (diamonds), 25th percentile, median and 75th percentile of the data points. Thin vertical lines with crosses represent the means. Whiskers indicate ± 1 S.D. Different letters indicate significant differences at $p < 0.05$ (ANOVA, Tukey's test), $n = 17$–22.

2.2. Accumulation of Phytohormones

Drops of distilled water increased the average accumulation of jasmonic acid (JA) two-fold in comparison to control plants; however, the increase was not significant. Jasmonoyl-L-isoleucine (JA-Ile) remained below the limit of detection (LOD), similar to the control plants (<0.022 pmol g^{-1} FW). In contrast, the application of 50 mM of NH_4Cl in the same drop of solution induced a significant, eight times higher accumulation of JA and at least a ten-fold increase in JA-Ile (calculated from LOD). Mechanical stimulation induced a 13-fold significant increase in JA and at least a 15-fold increase in JA-Ile. The flakes of chitin induced a similar response, a 16-fold higher accumulation of JA and at least a 10-fold increase in JA-Ile; however, due to the high variability of the samples, the increase was not significantly different in comparison to the control. A huge boost of JA and JA-Ile were detected after the application of live insect prey: a 220-fold and at least a 1100-fold increase, respectively (Figure 2A,B). The accumulation of abscisic acid (ABA) was increased fourfold only in response to chitin application (Figure 2C). The concentration of salicylic acid (SA) was not significantly affected at all (Figure 2D).

We were interested in whether the drop of water inducing membrane hyperpolarization could bring the tentacles into the state of insensibility as suggested by Darwin [14]. When the drop of water had been applied 5 min before the live insect prey, we did not observe any differences in the tentacle and trap bending reaction in comparison to traps with applied live prey without water. There were not any significant changes in jasmonate accumulation after 2 h; only the content of ABA significantly decreased in water drop pre-treated plants, probably due to better water status (Figure 3).

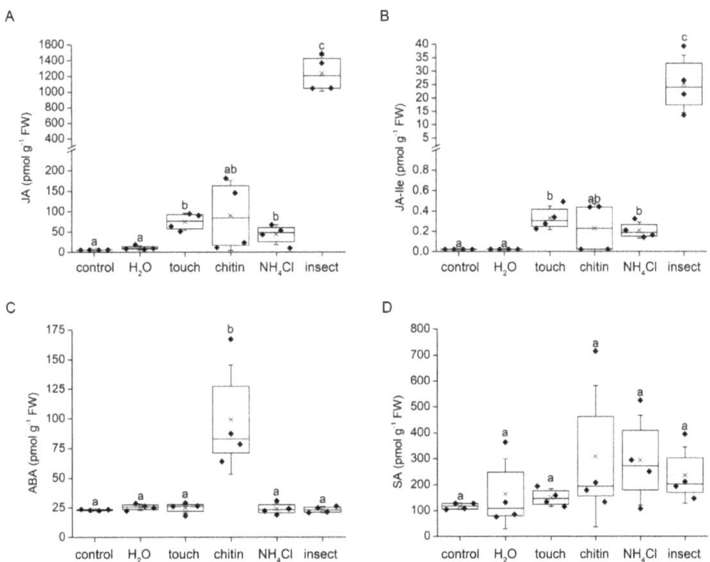

Figure 2. Trap tissue phytohormone accumulation in response to different stimuli in the carnivorous sundew plant (*Drosera capensis*) after 2 h. (**A**) jasmonic acid, (**B**) jasmonoyl-L-isoleucine, (**C**) abscisic acid, (**D**) salicylic acid. Boxplots show the individual measurements (diamonds), 25th percentile, median, and 75th percentile of the data points. Thin vertical lines with crosses represent the means. Whiskers indicate ± 1 S.D. Different letters indicate significant differences at $p < 0.05$ (ANOVA, Tukey's test or if non-homogeneity was present using multiple comparison via Welch's test), $n = 4$.

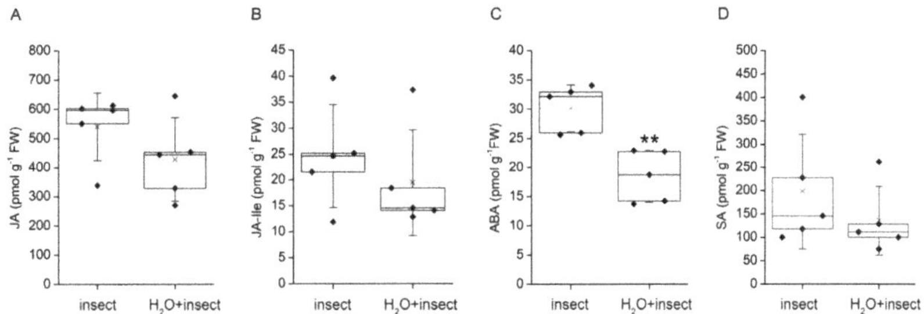

Figure 3. Trap tissue phytohormone accumulation in response to pre-treatment with water drops in the carnivorous sundew plant (*Drosera capensis*) after 2 h. The sundew traps had been pre-treated with three water drops 5 min before insect prey was applied (H_2O + insect) and compared with traps fed on insect prey (insect). (**A**) jasmonic acid, (**B**) jasmonoyl-L-isoleucine, (**C**) abscisic acid, (**D**) salicylic acid. Boxplots show the individual measurements (diamonds), 25th percentile, median, and 75th percentile of the data points. Thin vertical lines with crosses represent the means. Whiskers indicate ± 1 S.D. Asterisks denote significant differences at $p < 0.01$ (**), Student's *t*-test, $n = 5$.

2.3. Enzyme Activity

Phosphatase activity was not significantly increased in response to water drop application on the trap surface. However, the same drop of water containing 50 mM of NH_4Cl significantly increased phosphatase activities. Mechanostimulation also increased phosphatase activities, but the flakes of chitin did not cause any additional effect. The best activator of phosphatase activities was the living insect prey (Figure 4A).

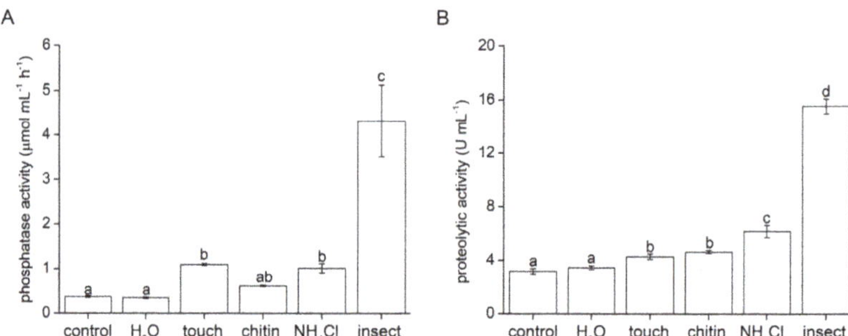

Figure 4. Enzyme activities in digestive fluids in response to different stimuli applied on the trap of the carnivorous sundew plant (*Drosera capensis*) after 24 h. (**A**) phosphatase activity, (**B**) proteolytic activity. Different letters denote significant differences at $p < 0.05$ (ANOVA, Tukey's test), $n = 4$.

Proteolytic activity was also not significantly increased in response to water drop application on the trap surface (Figure 4B). All other stimuli significantly increased proteolytic activity and insect prey was the most effective one. The stimulatory effect of chitin was comparable with mechanical stimulation, indicating no additional chitin specific response on proteolytic activities.

2.4. Abundance of Digestive Enzyme

Aspartic and cysteine proteases are responsible for proteolytic activities in digestive fluid in sundew plants [5,12]. We successfully immunodetected aspartic protease droserasin. The overall trend in its accumulation in digestive fluid is in accordance with proteolytic activity measurements. The signal intensities were comparable between the control and water drop treated plants. Application of NH_4Cl increased its abundance and mechanostimulation and chitin application had comparable signal intensities. Live insect prey induced the highest abundance of droserasin in digestive fluid (Figure 5).

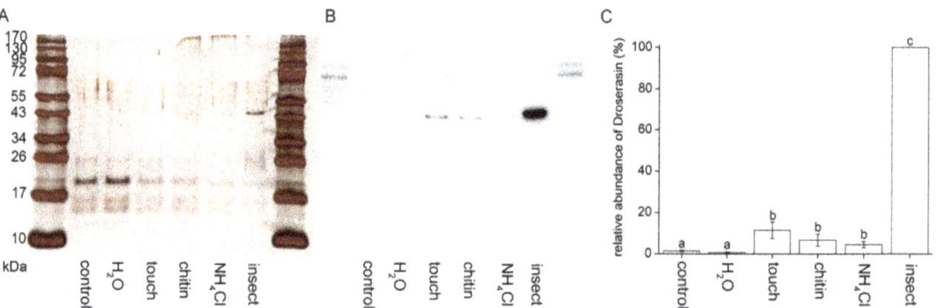

Figure 5. Immunodetection of aspartic protease (droserasin) in digestive fluid in response to different stimuli applied on the trap of the carnivorous sundew plant (*Drosera capensis*) after 24 h. The proteins were separated in 10% (v/v) sodium dodecyl sulphate–polyacrylamide gel electrophoresis (SDS-PAGE) and subjected to Western blot analysis. The same volume of digestive fluid was loaded (20 μL). (**A**) Silver-stained SDS-PAGE of digestive fluid in response to different stimuli. (**B**) Western blot analysis using antibodies against droserasin in response to different stimuli. (**C**) Quantification of chemiluminescence signal intensity. The results shown are representative of three independent experiments. Different letters denote significant differences at $p < 0.05$ (ANOVA, Tukey's test), $n = 3$.

3. Discussion

Charles Darwin in his famous book *Insectivorous plants* was the first to realize that the tentacles of sundew plants responded to water only very slightly, if at all [14]. We found that a drop of water induced the lowest number of APs in the tentacle. Because the tentacle bending reaction is proportional to the number of APs triggered [2], the weak reaction observed by Darwin could be caused by the low number of APs triggered. In contrast to the Venus flytrap, where a single touch induced a single AP [17], the sundew plant generates a series of APs in response to a single touch or any other stimuli [1,2,5]. None or a low number of APs triggered can be caused by a preceding rapid and long-lasting membrane hyperpolarization. This finding is consistent with the observations of Williams and Pickard [2], who observed the same response and found out that the probability of AP generation is decreased with membrane hyperpolarization and vice versa. Moreover, the number of APs generated by water drop application in our study can be overestimated due to the mechanical contact with the measuring electrode. In contrast to water, Charles Darwin found that the salts of ammonia are a powerful inducer of the tentacle bending reaction [14]. Application of NH_4Cl induced rapid long-lasting membrane depolarization and triggered more APs in accordance with Williams and Pickard's observations [2], which may also explain Darwin's observations. Mechanical stimulation of tentacles with solid bodies induced a strong and rapid bending response, short-lasting membrane depolarization called RP and the highest number of APs (Figure 1).

Membrane depolarization and generation of electrical signals in plants is related to the accumulation of a plant defence hormone from the group of jasmonates [18]. In the carnivorous plants, the jasmonate signalling pathway was co-opted for the regulation of carnivorous response, e.g., expression and secretion of digestive enzymes [19–22]. Water application did not induce a significant accumulation of jasmonates (JA, JA-Ile, Figure 2), probably due to water-induced membrane hyperpolarization. However, this hyperpolarization had no significant effect on jasmonate accumulation after subsequent prey capture (Figure 3), in contrast to Darwin's observation that a drop of water may bring the tentacles to a state of decreased responsiveness to mechanical stimuli [14]. Despite the fact that the weight of the applied water drop was ten-fold higher than polystyrene balls (treatment touch), the polystyrene balls induced a seven-fold and sixteen-fold higher accumulation of JA and JA-Ile, respectively (Figure 2). This is in line with Darwin's observations that tiny and much lighter human hair can induce the tentacle bending reaction in contrast to a water drop [14]. All other stimuli also induced a several-fold increase in jasmonates. On the other hand, only chitin induced a significant accumulation of ABA in accordance with our recent study on the Venus flytrap [23].

In our previous study we found that the digestive enzyme aspartic protease droserasin is under the control of jasmonates [5], and we used it as a proxy for the quantification of the physiological response. The accumulation of droserasin in tentacle droplets is not enhanced by water application, but all other stimuli induced its accumulation. The living prey triggered the highest accumulation of droserasin and corresponding proteolytic activity. Previous studies on different carnivorous plant genera showed that the living insect prey is indeed the best inducer of enzymatic activities [5,24,25]. This is probably caused by the combination of mechanical and chemical stimulations delivered to digestive organs of carnivorous plants. Chitin, a typical component of insect exoskeletons, did not trigger any additional accumulation of droserasin and proteolytic activity, in comparison to mechanical stimulation (Figure 5). This weak response of chitin on the secretion of proteolytic enzymes was also documented previously in *Drosera rotundifolia*, *Dionaea muscipula*, *Nepenthes alata* and *Nepenthes x Mixta* [9,23–25]. On the other hand, the presence of protein in insect bodies is a much better inducer of proteolytic activities [23,25].

Although the application of water did not induce the digestive ability of the carnivorous sundew, it was recently documented that water spray applied to *Arabidopsis thaliana* plants is sufficient to activate gene expression changes through JA signalling [26]. In a subsequent study it was demonstrated that the trichomes are true mechanosensors and that a rain drop initiated Ca^{2+} waves concentrically propagated away from trichomes, triggering a downstream sequence of responses. However, the accumulation of JA and JA-Ile was rather low, and only 11.8% of raindrop-induced genes were JA-responsive [27]. This is in accordance with our study, where a water drop did not induce significant accumulation of jasmonates (Figure 2) and thus enzymatic activities, which are under their control and therefore were not upregulated (Figure 4). This indicates that weak activation of the JA signalling pathway by a water drop was already pre-established in non-carnivorous plants and co-opted during the evolution of jasmonate signalling by carnivorous plants [28]. Matsumura et al. [27] suggested that rain drop mechanostimulation activates JA signalling only partially, and that mechanosensitive genes are regulated rather by calmodulin-binding transcription activator (CAMTA) directly through an increased intracellular Ca^{2+} concentration $[Ca^{2+}]_{cyt}$. Procko and his group [7] recently succeeded in the transformation of *Drosera spathulata* with Ca^{2+} sensor GCaMP3, observing that the Ca^{2+} wave propagated away from the tentacles in response to insect prey, mechanical stimulation and also in response to a water drop; however, a direct comparison of signal intensity in response to different treatments is not available (Procko, pers. comm.). Thus, only the increase in $[Ca^{2+}]_{cyt}$ is probably not sufficient for the induction of high jasmonate accumulation and enzyme synthesis in carnivorous plants, and the combination with another cellular factor is also important.

4. Materials and Methods

4.1. Plant Material, Culture Conditions and Experimental Setup

The carnivorous plant *Drosera capensis* L. (Cape sundew) is a small, erect perennial sundew native to the Cape region of South Africa. Experimental plants were grown from seeds and cultivated in a growth chamber at the Department of Biophysics in Olomouc (Czech Republic). Well-drained peat moss in plastic pots, placed in a tray filled with distilled water to a depth of 1–2 cm, was used. Daily temperatures fluctuated in the range of 23–26 °C, the relative air humidity was between 60–80% and there was a light/dark regime (12 h of light at 100-μmol photons m^{-2} s^{-1} per 12 h of dark). The traps were treated with 3 drops of distilled water or a 50-mM solution of NH_4Cl. The mechanical stimulation was performed by placing 3 small polystyrene balls on the trap surface. Three flakes of chitin from shrimp shells (95% acetylated, Sigma Aldrich, St. Louis, MO, USA) were placed in the same manner on the trap surface. The last treatment was feeding the plants with three living fruit flies (*Drosophila melanogaster*). They were cultured from eggs in a carbohydrate-rich medium (1 L composed of 87 g of corn meal, 15 g of agar, 25 g of dry yeast, 50 g of crystal sugar, 40 mL desinfection solution) at 23 °C, and before application, they were paralyzed with a cold treatment (4 °C for 10 min) for better manipulation. In an additional experiment, three drops of water had been applied 5 min before application of the fruit flies.

4.2. Extracellular Recording of Electrical Signals from Tentacle Head

The electrical signals were measured on the tentacle head using nonpolarizable Ag–AgCl surface electrodes (Scanlab Systems, Prague, Czech Republic) using a non-invasive device inside a Faraday cage as described previously in [5,29]. First, we established good electrical contact between the electrode and the head of a marginal tentacle. Sometimes, the mechanical contact with the electrode triggered a depolarization and series of APs, and in such a case, we had to wait for the restoration of the membrane potential. The tentacle head region was stimulated by a drop of distilled water, 50 mM of NH_4Cl or a gentle mechanical touch simulating a contact with solid bodies. The reference electrodes were submerged in a dish filled with 1–2 cm of water beneath the pot. The electrodes were connected to channels of an amplifier that had been made in-house (gain: 1–1000, noise: 2–3 mV, bandwidth (−3 dB): 10^5 Hz, response time: 10 μs, input impedance: 10^{12} Ω). The signals from the amplifier were transferred to an analogue-digital PC data converter (12-bit converter, ±10 V, PCA-7228AL supplied by TEDIA, Plzeň, Czech Republic), and the data were collected every 30 ms. The sensitivity of the device was 13 μV.

4.3. Quantitative Analysis of Phytohormones

Endogenous levels of phytohormones were quantified using the isotope dilution LC-MS/MS method [30] 2 h after treatments. The collected plant tissues were frozen in liquid nitrogen and ground using a mortar and pestle. The homogenized material was extracted and purified as described in [31]. Briefly, 25 mg of sample were extracted with 1 mL of ice cold 50% acetonitrile (ACN) containing a mixture of stable isotope-labelled standards. Unlabelled (SA, JA-Ile, ABA) and labelled standards (D4-SA, D2-JA-Ile, D6-ABA) were purchased from OlChemIm (Olomouc, Czech Republic); JA and D5-JA were purchased from Merck (Darmstadt, Germany). The extraction was performed with assistance of an ice-cold ultrasonic bath for 30 min. After centrifugation ($20,000\times$ g, 15 min, 4 °C) samples were purified using SPE. Waters (USA) Oasis™ HLB columns (30 mg, 1 mL cartridge) were activated by 1 mL of MeOH and equilibrated by 1 mL of H_2O and 1 mL of 50% ACN. During sample loading, the flow-through fraction was collected and pooled with the fraction from a single-washing step of 1 mL 30% ACN. Collected fractions were evaporated under a vacuum. If necessary, the dried samples were stored at -20 °C prior to analysis. For analysis, samples were resuspended in 40 µL of mobile phase, filtered through 0.2-µm microspins (Ciro, Deerfield Beach, FL, USA) and analyzed via LC-(+)ESI-MS/MS in the multiple reaction monitoring (MRM) mode. LC-MS/MS analysis was performed using a Nexera X2 modular liquid chromatograph system coupled to an MS 8050 triple quadrupole mass spectrometer (Shimadzu, Kyoto, Japan) via an electrospray interface. Chromatographic separation was performed using a reversed-phase analytical column, Waters CSH™ C18, 2.1 mm × 150 mm, 1.7 µm. The aqueous solvent A consisted of 15 mM of formic acid adjusted to a pH of 3.0 with ammonium hydroxide. Solvent B was pure ACN. Separation was achieved with gradient elution at a flow rate of 0.4 mL min^{-1} at 40 °C. Then, 0–1 min 20% B and 1–11 min 80% B linear gradients followed by washing and equilibration to initial conditions for a further 7 min were applied. If possible, up to 3 MRM transitions (1 quantitative, the others qualitative) were monitored for each analyte to ensure as much confidence as possible in the correct identification of analytes in the different plant matrices. Raw data were processed via Shimadzu software LabSolutions ver. 5.97 SP1. All data were log transformed to calculate the results. To reduce experimental biases, procedures included a randomized sample list, and the blinding was imposed on the analyst (OV).

4.4. Measurements of Enzyme Activities

After 24 h, the digestive fluid was collected from 8 leaves by submerging the part of leaves covered by tentacles into 1.5 mL of a 50-mM Na-acetate buffer (pH 5.0). To measure the activity of acid phosphatases, we used the chromogenic substrate 4-nitrophenyl phosphate (Sigma-Aldrich, St. Louis, MO, USA). The substrate was prepared in a 50-mM acetate buffer (pH 5.0), and the concentration was 5 mM. Then, 50 µL of the collected fluid was added to 500 µL of a 50-mM Na-acetate buffer (pH 5.0) and mixed with 400 µL of the substrate. As a control, 400 µL of the substrate solution was mixed with 550 µL of the Na-acetate buffer. Mixed samples were incubated at 25 °C for 1 h and then 160 µL of 1.0 N NaOH was added to terminate the reaction. Absorbance was measured at 405 nm with a Specord 250 Plus double-beam spectrophotometer (Analytik Jena, Germany). A calibration curve was determined using 4-nitrophenol.

The proteolytic activity of the leaf exudates was determined by incubating 150 µL of a sample with 150 µL of 2% (w/v) bovine serum albumin (BSA) in 200 mM of glycine-HCl (pH 3.0) at 37 °C for 1 h. The reaction was stopped by the addition of 450 µL of 5% (w/v) trichloroacetic acid (TCA). Samples were incubated on ice for 10 min and centrifuged at $20,000\times$ g for 10 min at 4 °C. The amount of released non-TCA-precipitable peptides was used as a measure of proteolytic activity, which was determined by comparing the absorbance of the supernatant at 280 nm to a blank sample with a Specord 250 Plus double-beam spectrophotometer (Analytik Jena, Jena, Germany). One unit of proteolytic activity was defined as an increase of 0.001 per min in the absorbance at 280 nm [9].

4.5. SDS-PAGE and Western Blots

To detect and quantify aspartic protease (droserasin), a polyclonal antibody against this protein was raised in rabbits by Genscript (Piscataway, NJ, USA). The following amino acid sequence (epitope) based on data from mass spectrometry [5] was synthesized: (NH$_2$-) SAIMDTGSDLIWTQC (-CONH$_2$) by Genscript, Piscataway, NJ, USA. The sequence was coupled to a carrier protein (keyhole limpet haemocyanin, KLH) and injected into two rabbits each. The terminal cysteine of the peptide was used for conjugation. The rabbit serum was analyzed for the presence of antigen-specific antibodies using an ELISA test.

The digestive fluid collected for the enzyme assays was subjected to Western blotting. The samples were heated and denatured for 30 min at 70 °C and mixed with a modified Laemmli sample buffer to a final concentration of 50 mM of Tris-HCl (pH 6.8), 2% SDS, 10% glycerol, 1% β-mercaptoethanol, 12.5 mM of EDTA, and 0.02% bromophenol blue. The same volume of digestive fluid was electrophoresed in 10% (v/v) SDS polyacrylamide gel [32]. The proteins in the gels were either visualized via silver staining (ProteoSilver; Sigma Aldrich) or transferred from the gel to a nitrocellulose membrane (Bio-Rad, Hercules, CA, USA) using a Trans-Blot SD Semi-Dry Electrophoretic Transfer Cell (Bio-Rad). After blocking in TBS-T containing 5% BSA overnight, the membranes were incubated with the primary antibody for 1 h at room temperature, and after washing, the membrane was incubated with the secondary antibody: the goat antirabbit IgG (H + L)-horseradish peroxidase conjugate (Bio-Rad). Blots were visualized and chemiluminescence was quantified using an Amersham Imager 600 gel scanner (GE HealthCare Life Sciences, Tokyo, Japan).

4.6. Statistical Analyses

Before statistical analyses, the data were tested for homogeneity of variance (Brown-Forsythe test). If the homogeneity was fulfilled, Student's t-test and one-way analysis of variance (ANOVA) with Tukey's post hoc test was used (Origin 8.5.1, Northampton, MA, USA). If homogeneity was not present multiple comparisons via Welch's test were used (Microsoft Excel).

5. Conclusions

In conclusion, water drops cannot activate the digestive process in the carnivorous sundew plant *D. capensis*. The reason is a subthreshold accumulation of jasmonates which control the expression of digestive enzymes. This is in line with evidence from experiments on non-carnivorous plants [27].

Author Contributions: Conceptualization, A.P.; methodology, A.P., O.V. and P.T.; software, A.P., O.V. and P.T; validation, A.P., O.V. and P.T.; formal analysis, A.P., O.V. and P.T.; investigation, A.P., O.V. and P.T.; resources, A.P., O.V. and P.T.; data curation, A.P., O.V. and P.T.; writing—original draft preparation, A.P.; writing—review and editing, A.P., O.V. and P.T.; visualization, A.P.; supervision, A.P. and P.T.; project administration, P.T.; funding acquisition, P.T. All authors have read and agreed to the published version of the manuscript.

Funding: This work was funded by project No. RO0423 funded by the Ministry of Agriculture, Czech Republic.

Data Availability Statement: Upon request, the data will be provided by the corresponding author.

Conflicts of Interest: The authors declare no conflict of interest.

References

1. Williams, S.E.; Pickard, B.G. Properties of action potentials in *Drosera* tentacles. *Planta* **1972**, *103*, 222–240. [CrossRef] [PubMed]
2. Williams, S.E.; Pickard, B.G. Receptor potentials and action potentials in *Drosera* tentacles. *Planta* **1972**, *103*, 193–221. [CrossRef] [PubMed]
3. Williams, S.E.; Pickard, B.G. The role of action potentials in the control of capture movements of *Drosera* and *Dionaea*. In *Plant Growth Substances*; Skoog, F., Ed.; Springer: Berlin/Heidelberg, Germany, 1980; pp. 470–480.
4. Williams, S.E.; Spanswick, R.M. Propagation of the neuroid action potential of the carnivorous plant *Drosera*. *J. Comp. Physiol.* **1976**, *108*, 211–223. [CrossRef]

5. Krausko, M.; Perutka, Z.; Šebela, M.; Šamajová, O.; Šamaj, J.; Novák, O.; Pavlovič, A. The role of electrical and jasmonate signalling in the recognition of captured prey in the carnivorous sundew plant *Drosera capensis*. *New Phytol.* **2017**, *213*, 1818–1835. [CrossRef] [PubMed]

6. Nakamura, Y.; Reichelt, M.; Mayer, V.E.; Mithöfer, A. Jasmonates trigger prey-induced formation of 'outer stomach' in carnivorous sundew plants. *Proc. R. Soc. B Biol. Sci.* **2013**, *280*, 20130228. [CrossRef] [PubMed]

7. Procko, C.; Radin, I.; Hou, C.; Richardson, R.A.; Haswell, E.S.; Chory, J. Dynamic calcium signals mediate the feeding response of the carnivorous sundew plant. *Proc. Natl. Acad. Sci. USA* **2022**, *119*, e2206433119. [CrossRef] [PubMed]

8. Mithöfer, A.; Reichelt, M.; Nakamura, Y. Wound and insect-induced jasmonate accumulation in carnivorous *Drosera capensis*: Two sides of the same coin. *Plant Biol.* **2014**, *5*, 982–987. [CrossRef]

9. Matušíková, I.; Salaj, J.; Moravčíková, J.; Mlynárová, L.; Nap, J.P.; Libantová, J. Tentacles of in vitro-grown round-leaf sundew (*Drosera rotundifolia* L.) show induction of chitinase activity upon mimicking the presence of prey. *Planta* **2005**, *222*, 1020–1027. [CrossRef]

10. Okabe, T.; Yoshimoto, I.; Hitoshi, M.; Ogawa, T.; Ohyama, T. An S-like ribonuclease gene is used to generate a trap-leaf enzyme in the carnivorous plant *Drosera Adelae*. *FEBS Lett.* **2005**, *579*, 5729–5733. [CrossRef]

11. Renner, T.; Specht, C.D. Molecular and functional evolution of class I chitinases for plant carnivory in the Caryophyllales. *Mol. Biol. Evol.* **2012**, *29*, 2971–2985. [CrossRef]

12. Takahashi, K.; Nishii, W.; Shibata, C. The digestive fluid of *Drosera indica* contains a cysteine endopeptidase ("Droserain") similar to Dionain from *Dionaea muscipula*. *Carniv. Plant Newsl.* **2012**, *41*, 132–134. [CrossRef]

13. Nishimura, E.; Jumyo, S.; Arai, N.; Kanna, K.; Kume, M.; Nishikawa, J.; Tanase, J.; Ohyama, T. Structural and functional characteristics of S-like ribonucleases from carnivorous plants. *Planta* **2014**, *240*, 147–159. [CrossRef] [PubMed]

14. Darwin, C. *Insectivorous Plants*; John Murray: London, UK, 1875.

15. Jopcik, M.; Moravcikova, J.; Matusikova, I.; Bauer, M.; Rajninec, M.; Libantova, J. Structural and functional characterisation of a class I endochitinase of the carnivorous sundew (*Drosera rotundifolia* L.). *Planta* **2017**, *245*, 313–327. [CrossRef]

16. Sprague-Piercy, M.A.; Bierma, J.C.; Crosby, M.G.; Carpenter, B.P.; Takahashi, G.R.; Paulino, J.; Hung, I.; Zhang, R.; Kelly, J.E.; Kozlyuk, N.; et al. The Droserasin 1 PSI: A Membrane-Interacting Antimicrobial Peptide from the Carnivorous Plant *Drosera capensis*. *Biomolecules* **2020**, *10*, 1069. [CrossRef] [PubMed]

17. Pavlovič, A.; Jakšová, J.; Novák, O. Triggering a false alarm: Wounding mimics prey capture in the carnivorous Venus flytrap (*Dionaea muscipula*). *N. Phytol.* **2017**, *216*, 927–938. [CrossRef]

18. Farmer, E.E.; Gao, Y.-Q.; Lenzoni, G.; Wolfender, J.-L.; Wu, Q. Wound- and mechanostimulated electrical signals control hormone responses. *New Phytol.* **2020**, *227*, 1037–1050. [CrossRef]

19. Pavlovič, A.; Saganová, M. A novel insight into the cost–benefit model for the evolution of botanical carnivory. *Ann. Bot.* **2015**, *115*, 1075–1092. [CrossRef]

20. Bemm, F.; Becker, D.; Larisch, C.; Kreuzer, I.; Escalante-Perez, M.; Schulze, W.X.; Ankenbrand, M.; Van de Weyer, A.-L.; Krol, E.; Al-Rasheid, K.A.; et al. Venus flytrap carnivorous lifestyle builds on herbivore defense strategies. *Genome Res.* **2016**, *26*, 812–825. [CrossRef] [PubMed]

21. Pavlovič, A.; Mithöfer, A. Jasmonate signalling in carnivorous plants: Copycat of plant defence mechanisms. *J. Exp. Bot.* **2019**, *70*, 3379–3389. [CrossRef]

22. Adamec, L.; Matušíková, I.; Pavlovič, A. Recent ecophysiological, biochemical and evolutionary insights into plant carnivory. *Ann. Bot.* **2021**, *128*, 241–259. [CrossRef]

23. Jakšová, J.; Libiaková, M.; Bokor, B.; Petřík, I.; Novák, O.; Pavlovič, A. Taste for protein: Chemical signal from prey stimulates enzyme secretion through jasmonate signalling in the carnivorous plant Venus flytrap. *Plant Physiol. Biochem.* **2020**, *146*, 90–97. [CrossRef] [PubMed]

24. Yilamujiang, A.; Reichelt, M.; Mithöfer, A. Slow food: Insect prey and chitin induce phytohormone accumulation and gene expression in carnivorous *Nepenthes* plants. *Ann. Bot.* **2016**, *118*, 369–735. [CrossRef]

25. Saganová, M.; Bokor, B.; Stolárik, T.; Pavlovič, A. Regulation of enzyme activities in carnivorous pitcher plants of the genus *Nepenthes*. *Planta* **2018**, *248*, 451–464. [CrossRef] [PubMed]

26. Van Moerkercke, A.; Duncan, O.; Zander, M.; Šimura, J.; Broda, M.; Bossche, R.V.; Lewsey, M.G.; Lama, S.; Singh, K.B.; Ljung, K.; et al. A MYC2/MYC3/MYC4-dependent transcription factor network regulates water spray-responsive gene expression and jasmonate levels. *Proc. Natl. Acad. Sci. USA* **2019**, *116*, 23345–23356. [CrossRef]

27. Matsumura, M.; Nomoto, M.; Itaya, T.; Aratani, Y.; Iwamoto, M.; Matsuura, T.; Hayashi, Y.; Mori, T.; Skelly, M.J.; Yamamoto, Y.Y.; et al. Mechanosensory trichome cells evoke a mechanical stimuli–induced immune response in *Arabidopsis thaliana*. *Nat. Commun.* **2022**, *13*, 1216. [CrossRef]

28. Pavlovič, A. How the sensory system of carnivorous plants has evolved. *Plant Commun.* **2022**, *3*, 100462. [CrossRef]

29. Ilík, P.; Hlaváčková, V.; Krchňák, P.; Nauš, J. A low-noise multi-channel device for the monitoring of systemic electrical signal propagation in plants. *Biol. Plant.* **2010**, *54*, 185–190. [CrossRef]

30. Ljung, K.; Sandberg, G.; Moritz, T. Methods of plant hormone analysis. In *Plant Hormones*; Davies, P.J., Ed.; Springer: Berlin/Heidelberg, Germany, 2010; pp. 717–740.

31. Šimura, J.; Antoniadi, I.; Široká, J.; Tarkowská, D.; Strnad, M.; Ljung, K.; Novák, O. Plant hormonomics: Multiple phytohormone profiling by targeted metabolomics. *Plant Physiol.* **2018**, *177*, 476–489. [CrossRef]
32. Schägger, H. Tricine-SDS-PAGE. *Nat. Protoc.* **2006**, *1*, 16–22. [CrossRef]

Article

Decision-Making Underlying Support-Searching in Pea Plants

Qiuran Wang [1,*], Silvia Guerra [1], Bianca Bonato [1], Valentina Simonetti [1,2], Maria Bulgheroni [2] and Umberto Castiello [1]

[1] Department of General Psychology, University of Padova, 35131 Padova, Italy; silvia.guerra@unipd.it (S.G.); bianca.bonato.1@phd.unipd.it (B.B.); valentinasimonetti@ab-acus.eu (V.S.); umberto.castiello@unipd.it (U.C.)
[2] Ab.Acus srl, 20155 Milan, Italy; mariabulgheroni@ab-acus.com
* Correspondence: qiuran.wang@gmail.com

Abstract: Finding a suitable support is a key process in the life history of climbing plants. Those that find a suitable support have greater performance and fitness than those that remain prostrate. Numerous studies on climbing plant behavior have elucidated the mechanistic details of support-searching and attachment. Far fewer studies have addressed the ecological significance of support-searching behavior and the factors that affect it. Among these, the diameter of supports influences their suitability. When the support diameter increases beyond some point, climbing plants are unable to maintain tensional forces and therefore lose attachment to the trellis. Here, we further investigate this issue by placing pea plants (*Pisum sativum* L.) in the situation of choosing between supports of different diameters while their movement was recorded by means of a three-dimensional motion analysis system. The results indicate that the way pea plants move can vary depending on whether they are presented with one or two potential supports. Furthermore, when presented with a choice between thin and thick supports, the plants showed a distinct preference for the former than the latter. The present findings shed further light on how climbing plants make decisions regarding support-searching and provide evidence that plants adopt one of several alternative plastic responses in a way that optimally corresponds to environmental scenarios.

Keywords: decision-making; plant movement; kinematics; plant behavior

Citation: Wang, Q.; Guerra, S.; Bonato, B.; Simonetti, V.; Bulgheroni, M.; Castiello, U. Decision-Making Underlying Support-Searching in Pea Plants. *Plants* **2023**, *12*, 1597. https://doi.org/10.3390/plants12081597

Academic Editors: Frantisek Baluska and Gustavo Maia Souza

Received: 28 February 2023
Revised: 6 April 2023
Accepted: 7 April 2023
Published: 10 April 2023

1. Introduction

Scientists have long been intrigued by the specialized adaptations of climbing plants that enable them to compete for necessary resources, such as sunlight [1]. However, despite this prolonged fascination, we know surprisingly little about how climbing plants make 'decisions' with regard to stimulus searching and attachment behaviors. Indeed, climbing plants can be an ideal model system for studying the decision-making in plants because they show rapid changes in response to environmental cues [2]. For them, finding a suitable support upon which they can climb is among the most important factors affecting their growth and development [3].

The study of climbing plant behavior is chiefly based on Darwin's observations on the oscillatory movements of exploring stems and tendrils (i.e., *circumnutation*) [4]. He noted that vines are not only able to locate potential supports and grow towards them, but they can even show an aversive response [4]. He first described this effect with regard to the *Bignonia capreolata* L. tendrils that initially seized and then let go of sticks that were inappropriate in terms of size. If, because of its thickness, a stimulus was perceived as 'inadequate', after initially seizing it, the tendrils let go of it [4]. This case provides a degree of support to speculative claims that some climbing plants can judge the thickness of potential supports and modify their circumnutation patterns to a greater or lesser extent, depending on the features of potential supports with respect to what would be expected by chance movement. To date, most studies on climbing plants focus on the attachment stage, which is the final coiling step [5], whereas the approaching stage, which occurs before any physical contact with the support, is rarely examined. Experimental evidence demonstrates that this stage can be anticipatory and adapted on the basis of the physical properties of the support, and hence it has the potential to be highly informative [6–10]. For instance, Guerra and colleagues demonstrated that pea plants (*Pisum sativum* L.) are able to perceive a support and modulate the kinematics of the tendrils' aperture depending on its thickness [8]. The aperture of the tendrils refers to the maximum distance between the tips of the tendrils reached during movements leaning towards a support. The average and maximum velocities of the tendrils were found to be higher for thinner supports compared to thicker ones. In temporal terms, it took more time for the tendrils to reach peak velocity and maximum aperture when the supports were thinner [6,8]. Further, they modulate the production of a number of secondary velocity peaks (i.e., submovements) as a function of the support's thickness, suggestive of "on-line adjustments" [7]. The frequency of submovements tends to increase when the support is thick. This signifies that they need to make more adjustments in order to establish contact points along the support [7].

The above results are in line with Darwin's previous observations highlighting that thinner and thicker supports are different for climbing plants [4,5,11–13], with the touching and grasping of thick supports being more 'difficult' since it is more energy-demanding with respect to the thinner ones. In fact, it implies that the plant not only needs to increase the length of its tendrils in order to wrap itself around the stimulus efficiently [14], but it also has to strengthen its tensional forces to counteract gravity [2,15] and modulate kinematics [8]. Still, could climbing plants "choose" between thinner and thicker supports? Should they manifest a preference for the thinner? Would they perform according to their choice?

In light of these considerations, the aim of the current study is twofold. First, to ascertain what pea plants do when confronted with differently sized supports. To test this, after germination, pea plants were exposed to both a thin and thick support. We hypothesized that if pea plants inevitably prefer thinner supports, then we should observe a significantly higher frequency of movements directed toward them. Second, to ascertain whether such a decisional process impacts the kinematics of the tendrils' circumnutations, we compared a 'choice' condition termed the "double-support" (DS) condition, in which a thin and thick support were present in the environment with a "single-support" (SS) condition, where only a thin support was present in the environment. We foresee differences across the conditions evident at the level of movement kinematics. Although the plants would prefer the thinner support, they might still keep into account the thicker one as a potential option for an ever-changing environment. If so, a hybrid kinematical patterning accounting for differently sized supports would be evident.

2. Results

2.1. Qualitative Results

For all plants and in both experimental conditions (i.e., DS and SS), the tendrils displayed a circumnutating growing pattern. As soon as a plant sensed the support, it strategically altered the tendril's movement trajectory so as to bend towards the support (Figure 1a,b). For the DS condition, plants exhibited a very strong preference for the thin support and grew less than the plants for the SS condition by the time they touched the support (Figure 1c,d). Eight of the nine plants for the DS condition began to grow and move toward the thin support relatively early, even though they were too tiny to reach out for any support. These plants were able to aim fairly precisely toward the thin support and touch it by modulating/twisting the angles of the new petiole, and this is visible to the naked eye (see Video S1). Only one plant made an attempt to cling onto the thick support but ultimately failed and fell. The data for this plant have not been analyzed further. In this connection, given that we could not observe any full movement towards the thicker support for the DS condition, here, we consider only the thin SS condition as a comparison with the DS condition.

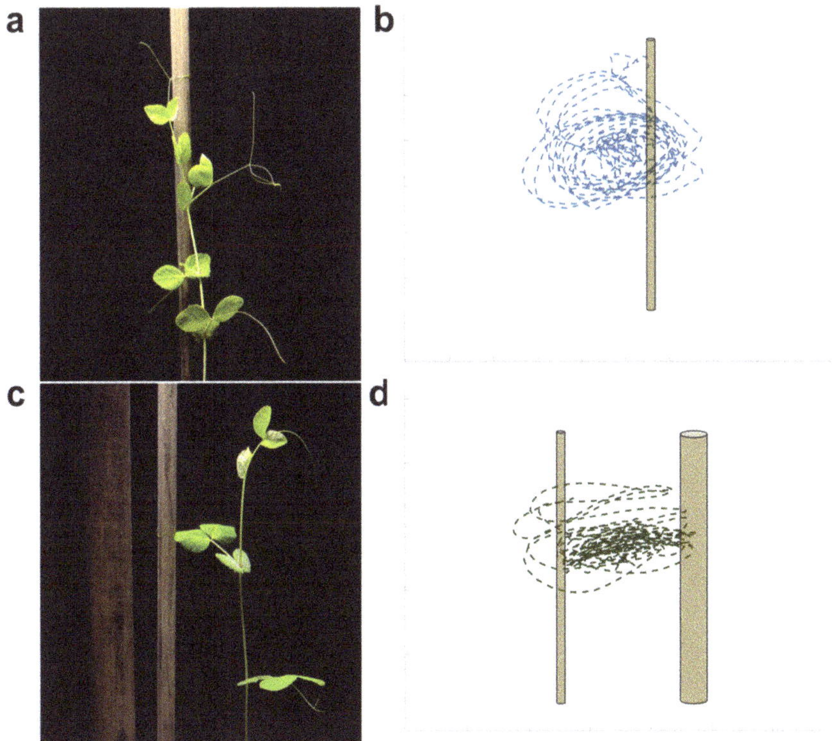

Figure 1. A frame representing an exemplar plant approaching the support for (**a**) the single-support (SS) condition with (**b**) a graphical representation of its trajectory. A plant approaching the thinner support for (**c**) the double-support (DS) condition with (**d**) a graphical representation of its trajectory.

Among the eight plants in the SS condition: two circumnutated clockwise and two circumnutated counterclockwise. The remaining four exhibited both a clockwise and counterclockwise circumnutation pattern during the entire movement. As for the DS condition, four plants circumnutated clockwise, one plant circumnutated counterclockwise and three circumnutated in a mixed manner.

2.2. Kinematic Results

The descriptive statistics and the kinematic results, when comparing the DS with the SS conditions, are provided below (Tables 1 and 2). The comparison is between the thin support for the SS condition and the thin support for the DS condition. This is because, for the DS condition, plants always choose the thinner support.

2.2.1. Number of Circumnutations

For the DS condition, subjects performed, on average, 24.924 (SD = 4.247, SE = 0.303, 95% CI: [24.327, 25.521]) circumnutations, whereas for the SS condition, they performed, on average, 26.553 (SD = 6.156, SE = 0.439, 95% CI: [25.688, 27.418]) circumnutations. The Bayesian Mann–Whitney U analysis revealed a Bayes factor (BF_{10}) of 314.656, suggesting that there is a decisive difference between the SS and the DS conditions with respect to the number of circumnutations ($BF_{10} = 314.656$, $BF_{01} = 0.003$, W = 14220, R-hat = 1.008, 95% CI: [−0.657, −0.229]).

Table 1. Descriptive statistics for the considered dependent measures.

	Group	Mean	SD	SE	Coefficient of Variation	95% CI Lower	95% CI Upper
Number of circumnutations	DS	24.924	4.247	0.303	0.170	24.327	25.521
	SS	26.553	6.156	0.439	0.232	25.688	27.418
Circumnutation duration (min)	DS	66.746	13.190	0.940	0.198	64.893	68.600
	SS	69.000	14.451	1.030	0.209	66.969	71.031
Distance from the circumnutation gravity center to the origin (cm)	DS	15.899	10.429	0.743	0.656	14.434	17.364
	SS	27.895	24.340	1.734	0.873	24.475	31.315
Length of the circumnutation major axis (mm)	DS	91.214	38.929	2.774	0.427	85.744	96.684
	SS	72.908	43.538	3.102	0.597	66.791	79.026
Circumnutation length (mm)	DS	243.403	124.957	8.903	0.513	225.846	260.961
	SS	188.148	115.972	8.263	0.616	171.853	204.443
Circumnutation area (mm^2)	DS	4992.504	4634.422	330.189	0.928	4341.325	5643.684
	SS	3217.099	3505.097	249.728	1.090	2724.601	3709.598
Amplitude of maximum peak velocity (mm/min)	DS	6.541	5.650	0.403	0.864	5.748	7.335
	SS	4.660	2.840	0.202	0.610	4.260	5.059

Note. DS = double-support condition; SS = single-support condition; SD = standard deviation; SE = standard error; CI = credible interval.

Table 2. Two-sided Bayesian Mann–Whitney U test for the DS and the SS conditions.

	BF$_{10}$	W	R-Hat
Number of circumnutation	314.656	14,220.000	1.008
Circumnutation duration	0.387	17,083.000	1.000
Distance from the circumnutation gravity center to the origin	136.096	15,132.000	1.031
Length of the circumnutation major axis	734.705	24,455.000	1.016
Circumnutation length	980.421	24,433.000	1.015
Circumnutation area	1267.886	24,611.500	1.008
Amplitude of maximum peak velocity	4137.588	25,438.000	1.014

Note. Results based on data augmentation algorithm with five chains of 1000 iterations.

2.2.2. Circumnutation Duration

The duration of the circumnutation was, on average, 66.746 min for a single circumnutation (SD = 13.190, SE = 0.940, 95% CI: [64.893, 68.600]) for the DS condition, whereas for the SS condition, it was, on average, 69 min (SD = 14.451, SE = 1.030, 95% CI: [66.969, 71.031]). The Bayesian Mann–Whitney U analysis revealed a Bayes factor (BF$_{10}$) of 0.387, suggesting that there is no difference between the SS and the DS conditions with respect to the circumnutation duration (BF$_{10}$ = 0.387, BF$_{01}$ = 2.584, W = 17083, R-hat = 1.000, 95% CI: [−0.354, 0.029]).

2.2.3. Distance from the Circumnutation Gravity Center to the Origin

The distance from the circumnutation gravity center to the origin was 15.899 cm (SD = 10.429, SE = 0.743, 95% CI: [14.434, 17.364]) for the DS condition, whereas it was 27.895 cm (SD = 24.340, SE = 1.734, 95% CI: [24.475, 31.315]) for the SS condition. The Bayesian Mann–Whitney U analysis revealed a Bayes factor (BF$_{10}$) of 136.096, suggesting that there is a decisive difference between the SS and the DS conditions with respect to the distance from the circumnutation gravity center to the origin (BF$_{10}$ = 136.096, BF$_{01}$ = 0.007, W = 15132, R-hat = 1.031, 95% CI: [−0.575, −0.169].

2.2.4. Length of the Circumnutation Major Axis

The length of the circumnutation major axis was 91.214 mm (SD = 38.929, SE = 2.774, 95% CI: [85.744, 96.684]) for the DS condition, whereas it was 72.908 mm (SD = 43.538, SE = 3.102, 95% CI: [66.791, 79.026]) for the SS condition. The Bayesian Mann–Whitney U analysis revealed a Bayes factor (BF$_{10}$) of 734.705, suggesting that there is a decisive difference between the SS and the DS conditions with respect to the length of the circumnutation major axis (BF$_{10}$ = 734.705, BF$_{01}$ = 0.001, W = 24455, R-hat = 1.016, 95% CI: [0.275, 0.676]).

2.2.5. Circumnutation Length

The circumnutation length for the DS condition was 243.403 mm (SD = 124.957, SE = 8.903, 95% CI: [225.846, 260.961]), whereas, for the SS condition, it was 188.148 mm (SD = 115.972, SE = 8.263, 95% CI: [171.853, 204.443]). The Bayesian Mann–Whitney U analysis revealed a Bayes factor (BF_{10}) of 980.421, suggesting that there is a decisive difference between the SS and DS conditions with respect to the circumnutation length (BF_{10} = 980.421, BF_{01} = 0.001, W = 24433, R-hat = 1.015, 95% CI: [0.290, 0.693]).

2.2.6. Circumnutation Area

The area of circumnutation for the DS condition is, on average, 4992.504 mm^2 (SD = 4634.422, SE = 330.189, 95% CI: [4341.325, 5643.684]), whereas for the SS condition is 3217.099 mm^2 (SD = 3505.097, SE = 249.728, 95% CI: [2724.601, 3709.598]). The Bayesian Mann–Whitney U analysis revealed a Bayes factor (BF10) of 1267.886, suggesting that there is a decisive difference between the SS and DS conditions with respect to the area of circumnutation (BF_{10} = 1267.886, BF_{01} = 0.0008, W = 24611.5, R-hat = 1.008, 95% CI: [0.299, 0.697]).

2.2.7. Amplitude of Maximum Peak Velocity

The amplitude of maximum peak velocity was, on average, 6.541 mm/min (SD = 5.650, SE = 0.403, 95% CI: [5.748, 7.335]) for the DS condition, whereas it was 4.660 mm/min (SD = 2.840, SE = 0.202, 95% CI: [4.260, 5.059]) for the SS condition. The Bayesian Mann–Whitney U analysis revealed a Bayes factor (BF_{10}) of 4137.588, suggesting that there is a decisive difference between the SS and DS conditions with respect to the amplitude of maximum peak velocity (BF_{10} = 4137.588, BF_{01} = 0.0002, W = 25438, R-hat = 1.014, 95% CI: [0.380, 0.780]).

2.2.8. Correlational Analyses

We noticed a non-significant difference for the circumnutation duration across conditions, while the amplitude of peak velocity increased for the DS with respect to the SS condition. We felt that this might indicate the plants putting in place a sort of isochrony principle [16] (see the Discussion section). To test this, we performed Pearson's correlation analysis [17] between the circumnutation length and the amplitude of peak velocity [18]. The results indicate a significant correlation between these measures (Pearson's r = 0.715, p-value = 0.000, 95% CI: [0.663, 0.760]; Figure 2).

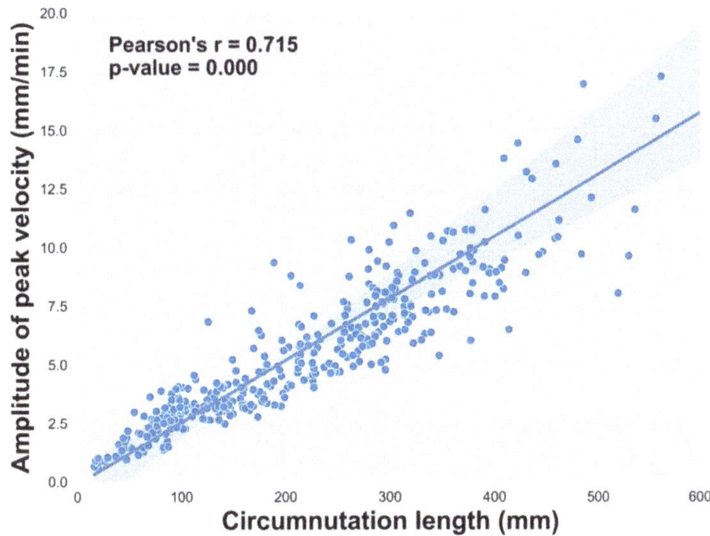

Figure 2. Pearson's correlation coefficient between the "circumnutation length" and the "amplitude of peak velocity".

3. Discussion

In this study, we have examined the kinematics of pea plants' tendrils' circumnutation from the beginning of circumnutation till they touched and grasped the support. Our findings show that most of the considered dependent measures differed markedly between the DS and SS conditions, indicating that pea plants exhibit distinct movement patterns depending on the conditions. For instance, plants perform fewer and larger circumnutations, as evidenced by a lower "number of circumnutations", a longer "length of circumnutation major axis", and a longer "circumnutation length" for the DS than the SS condition. Further, the "circumnutation area" is greater for the DS than the SS condition. To achieve all this, plants increased the "amplitude of maximum peak velocity" for the DS condition. Altogether, this pattern of results might imply a more active and exploratory patterning for the plants facing a "choice" scenario. The "circumnutation duration", on the other hand, remains the same for both conditions. In this respect, the correlational analysis indicates that the "circumnutation length" and the "amplitude of the peak velocity" are strongly correlated. This suggests that the pea plants' movement is based on the isochrony principle [16]. The isochrony principle refers to a spontaneous tendency to increase the velocity of a movement depending on the linear extent of its trajectory to maintain the execution time as approximately constant [19]. In our circumstances, plants maintain constant movement duration and scale velocity in order to cover longer distances, as witnessed by the longer circumnutation lengths. This appears to be an easy and appropriate organizational option adopted by the plant to program the patterning of circumnutation when a decision based on alternatives has to be taken.

At this stage, the question is more about how climbing plants avoid an unsuitable host and choose a suitable one. A common belief is that the physiological mechanisms underlying behavioral responses in plants tend to be caused by simple, local reactions [20]. As proposed by Saito, these 'reactions' might also be the basis of the decision-making processes related to the support diameter characterizing tendrils' coiling [5]. In this view, changes in the coiling responses may be caused by local reactions in the tendrils. For instance, in many climbing plants, the coiling of tendrils is thought to be caused by the contraction of the gelatinous fibers (G fibers) after stimuli have been contacted [5,21]. That is to say, when a suitable support is detected and recognized, the tendril shows a reflex behavior and rapidly bends in the stimulated direction [22]. Put simply, at the basis of plants' support selection, there might be a mechanism that makes it possible to select a support with an appropriate diameter.

The emerging picture from the "choice" that the plants made might suggest a trade-off in terms of metabolic use. Touching and grasping a thicker support would imply the growth of longer tendrils, which, in turn, would be more demanding in terms of energy exploitation. This metabolically based decision would also reflect on movement kinematics. The movement towards thicker supports is much slower than for thinner supports [8] and shows a great deal of online adjustments, visible as submovements along the velocity profiles [7]. Therefore, plants might have the ability to monitor, detect, and process information that determines the preference for a thin support. These aspects are particularly evident when comparing circumnutation between the thin support for the SS and the DS conditions. Plants move faster and execute less but larger circumnutations for the latter than for the former. This signifies that, despite that the plants are aiming at supports of the same size, being exposed to an alternative (the thicker support for the DS condition) determines a decisional complexity that is played out in the kinematics of circumnutation. Therefore, it appears that circumnutation is not only affected by a complex occurrence of factors, such as light, gravity, touch, and hormonal signals [23], but also by the presence of alternative supports in the environment.

Decision-making implies making choices from several alternatives to achieve a desired result [24]. In recent years, decision-making has been studied on a variety of organisms [25], including plants [26,27]. Dener and colleagues investigated decision-making in the root development of the pea plant (*Pisum sativum*) using the risk sensitivity theory (RST) [26].

According to RST, the rational decision is the one that maximizes fitness [28]. In the study, root growth displayed both risk-prone and risk-averse behaviors, which better support the RST hypothesis than previous animal testing. It appears that pea plants make "rational" economic decisions in terms of risk sensitivity [26,29]. Plant decision-making is also explored in the context of the social environment. Gruntman and colleagues compared the responses of *Potentilla reptans*, centered on their ability to out-compete their neighbors for accessing light [27]. Observed shifts in the responses between vertical growth, shade tolerance, and lateral growth suggest that plants can choose adaptively from several alternatives under light-competition scenarios [27].

Altogether, these findings suggest that plants possess the ability to make decisions and adjust their behavior in response to their surroundings. Our findings contribute to the literature, demonstrating that a plant's behavior is flexible, as opposed to rigid and mechanical [30], reinforcing the idea that plants are systems with a remarkable ability to deal with the complexities of an ever-changing environment [31].

At this stage, the question is how and at which level pea plants implement such decisions that then translate into specific behavioral patterns. One possible mechanism could be light acquisition at the level of the stomata [32,33], which might allow them to distinguish the light reflections determined by differently sized supports. Alternatively, Souza and colleagues introduced the concept of "plant electrome", describing the totality of the ionic dynamics at different scales of plant organization, engendering a constant electrical activity [34,35]. Souza and colleagues demonstrated that, rather than pure random noise, the amount of complexity characterizing environmental stimuli might alter several characteristics of the temporal dynamics of the plant electrome [34,36,37]. It was reported that some frequencies (the higher ones) exhibited by non-stimulated plants faded after stimulation. Only the lowest frequencies remain, allowing for low-energy-cost long-distance signaling [35]. In this view, the electrome could be considered a unifying factor of whole plant reactivity in a constantly changing environment and, therefore, might be a good candidate for understanding the flexible behavior of plants [35].

A caveat of the present results at the observational level is that the direction of the circular movements could be either clockwise or counterclockwise, and it could change within the same plant. Whether climbing plants are right- or left-handers is an aspect tackled in the previous literature [38], and that may be pursued in connection with decision-making. Further research is required to establish such a link.

In conclusion, the results of this study offer a contextual framework for the different well-known responses of climbing plants when searching for a support. More importantly, we have demonstrated a decision-making ability in plants, which allows them to adaptively 'choose' between responses according to the diameter of the available supports. Overall, the results of our study suggest that plants are capable of acquiring and integrating complex information about their environment in order to modify their extent of plastic responses adaptively. Such complex decision-making in plants could have important implications for our understanding of the processes that govern plant behavior.

4. Materials and Methods

4.1. Subjects

A total of 17 snow peas (*Pisum sativum* var. *saccharum* cv Carouby de Maussane) were chosen as study plants. Cylindrical pots (40 cm in diameter, 20 cm in depth) were filled with river sand (type 16SS, dimension 0.8/1.2 mm, weight 1.4). Seeds were potted at 8 cm from the pot's border and sowed at a depth of 2.5 cm.

4.2. Type of Support

Two types of wooden support were considered: a 'thin' support of 13 mm in diameter (Koto -13 mm) and a 'thick' support of 40 mm in diameter (Koto -40 mm; Figure 3a). Both supports were 54 cm in height. The supports were inserted 7 cm below the soil surface (Figure 3b). The supports were made available to the plants immediately after germination.

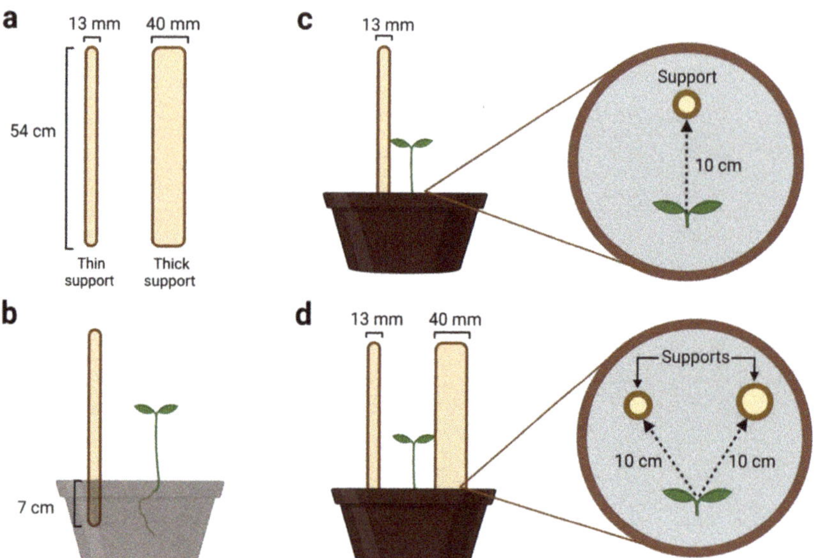

Figure 3. Graphical depiction of the (**a**) "thin" and "thick" supports; (**b**) the location of the support in the pot and how it was inserted in the soil. The single-support and double-support conditions are represented in panels (**c**) and (**d**), respectively.

4.3. Experimental Conditions

The subjects were randomly assigned to two experimental conditions termed single- (SS) and double-support (DS) conditions. For the SS condition, 8 plants were raised individually in the presence of the 'thin' support (Figure 3c). For the DS condition (Figure 3d), 8 plants were raised individually in the presence of both the 'thin' and the 'thick' support. The location of the differently sized supports was counterbalanced across subjects to avoid a potential bias due to the direction of circumnutation (clockwise or counterclockwise). The supports were positioned so that the first leaf developed by a sprout faced the midpoint between the two supports. This was done to prevent a growing bias in favor of either one or the other support. It should be noted that here, we did not include a 'thick' single-support condition. This decision was based on the observation that, during data acquisition for the DS condition, none of the plants successfully touched or grasped the thick support—they all went for the thin support. Consequently, it would be impossible to compare trials for a potentially thick SS condition with trials for the DS condition. Moreover, the differences between the thin and thick supports have been previously reported [6–8], and it has been established that the thicker support is not the best option for climbing plants [4,11–15]. Therefore, we confined our comparison to plants that achieved the same outcome of touching and grasping the thin support under the SS and DS conditions.

In addition, our setting considered an equal distance between the plant and the surface of the supports and not necessarily the center of the support (Figure 3c,d). This appears to be a suitable positioning solution, given that we are focusing on the approaching phase preceding the grasping of the support and not on the coiling phase of the support. Note, however, that in the studies concerned with the measurements related to the coiling pattern, the equal distance between the plant and the exact center of the support has been considered [5].

4.4. Experimental Setup

The plants grew individually in a thermo-light-controlled growth chamber (Cultibox SG combi 80 × 80 × 160 cm; Figure 4). The temperature was set at 26 °C by means of an extractor fan equipped with a thermo-regulator (TT125 vents; 125 mm-diameter;

max 280 mc/h) and an input-ventilation fan (Blauberg Tubo 100–102 m^3/h). The two-fan combination allowed for a steady air flow rate into the growth chamber with a mean air residence time of 60 s. The fan was carefully placed so that the circulation of air did not affect the plants' movements. Each plant was exposed for 12 h (6 a.m. to 6 p.m.) to a cool white LED lamp (V-TAC innovative LED lighting, VT-911-100W, Des Moines, IA, USA) that was positioned 50 cm above each seedling. The photosynthetic photon flux density at 50 cm under the lamp in correspondence with the seedling was 350 μmol$_{ph}$/(m^{2}s) (quantum sensor LI-190R, Lincoln, NE, USA). At the beginning of each experiment, the pots were fertilized using a half-strength solution culture (Murashige and Skoog Basal Salt Micronutrient Solution; see https://www.sigmaaldrich.com/RO/en/technical-documents/technical-article/cell-culture-and-cell-culture-analysis/plant-tissue-culture/murashige-skoog, accessed on 20 February 2023). The pots were watered with 1 L a week using distilled water (Sai Acqua Demineralizzata, Parma, Italy).

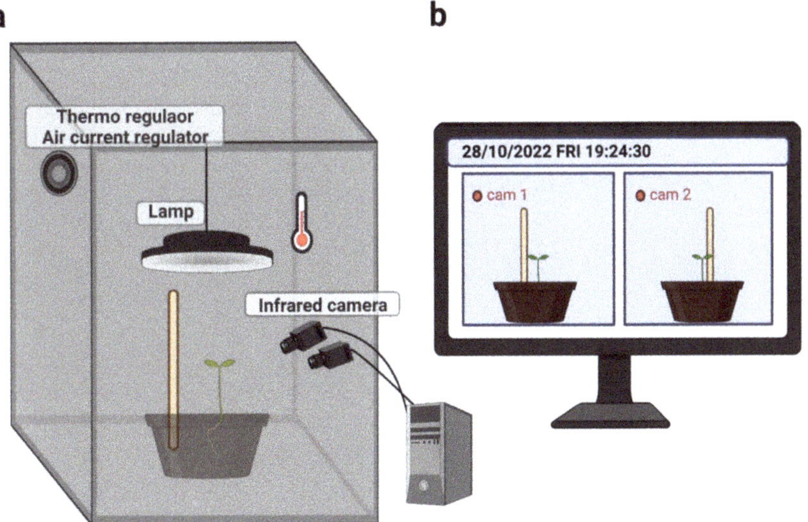

Figure 4. Graphical illustration of (**a**) experimental setup and (**b**) demonstration of how plants were captured by the infrared cameras.

4.5. Kinematic Acquisition and Data Processing

For each growth chamber, a pair of RGB-infrared cameras (IP 2.1 Mpx outdoor varifocal IR 1080P) were placed 110 cm above the ground, spaced at 45 cm to record the stereo images of the plant (Figure 4). The cameras were connected via ethernet cables to a 10-port wireless router (D-Link Dsr-250n) connected via Wi-Fi to a PC. The frame acquisition and saving processes were controlled by CamRecorder software (Ab.Acus s.r.l., Milan, Italy; Figure 4). Each camera's intrinsic, extrinsic, and lens distortion parameters were estimated using a Matlab Camera Calibrator application. Depth extraction from the single images was carried out by taking 20 pictures of a chessboard (squares' size of 18 × 18 mm, 10 columns × 7 rows) from multiple angles and distances in natural non-direct light conditions. For the stereo calibration, the same chessboard used for the single-camera calibration process was placed in the middle of the growth chamber. The two cameras synchronously acquired the frame every 180 s (frequency 0.0056 Hz). RGB images were acquired during the daylight cycle, and infrared images during the night cycle. The anatomical landmarks of interest were the tendrils developing from the considered leaf. We considered the initial frame as the one corresponding to the appearance of the tendrils for the considered leaf. The end frame was defined as the frame in which the tendrils start to coil the support. Images from both the left and right cameras were used in order to reconstruct 3D trajectories. An ad hoc software (Ab.Acus s.r.l., Milan, Italy), developed in Matlab, was used to identify the anatomical points to be investigated by means of markers and to track their position frame-by-frame on the images acquired by the two cameras to reconstruct the 3D trajectory of each marker. The markers on the anatomical landmarks of interest (i.e., the tendrils) were inserted post hoc. The tracking procedures were performed automatically throughout the time of the movement sequence using the Kanade–Lucas–Tomasi (KLT) algorithm on the frames acquired by each camera after distortion removal. The tracking was manually verified by the experimenter, who checked the position of the markers frame-by-frame. The 3D trajectory of each tracked marker was computed by triangulating the 2D trajectories obtained from the two cameras. Finally, the trajectory was reconstructed with a series of coordinates in 3D (x, y, z), where the x-z plane is the horizontal plane, and the x-y plane and z-y plane are the vertical planes perpendicular to each other.

4.6. Dependent Measure

The considered dependent measures were the following [39]:

(i)   The number of circumnutations: the number of circumnutations performed by a plant from the time it was potted to the time it touched the support.
(ii)   The circumnutation duration: the time taken by a plant to complete a single circumnutation.
(iii)   Distance from the circumnutation gravity center to the origin (Figure 5. Segment a): the distance between the circumnutation gravity center and the plant origin.
(iv)   The length of the circumnutation major axis (Figure 5. Segment b): the maximum distance between two points of the circumnutation trajectory.
(v)   The circumnutation length (Figure 5. Segment c): the length of the overall path computed as the sum of all the Euclidean distances between the subsequent points during a single circumnutation.
(vi)   The circumnutation area (Figure 5. Segment d): the sum of pixels with a value equal to 1, obtained from the binarization of the circumnutation trajectory.
(vii)   The amplitude of peak velocity: the values for the average of the maximum velocity.

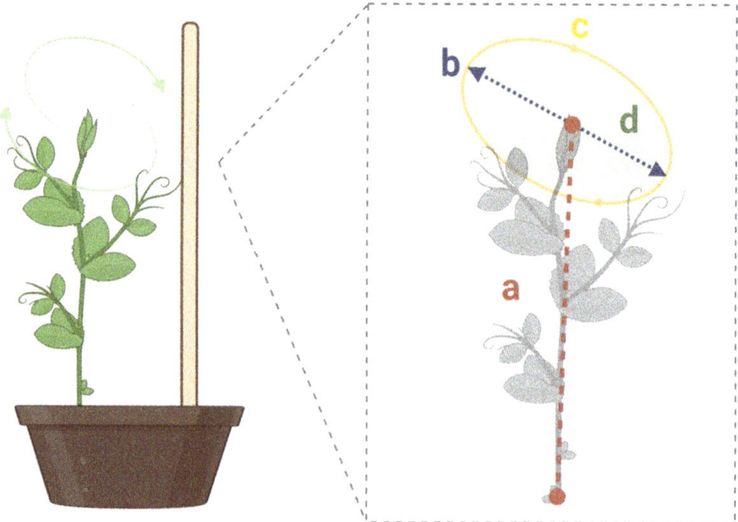

Figure 5. Graphical representation for some of the considered dependent measures: (**a**) the distance from the circumnutation gravity center to the origin is represented as a red/dash line; (**b**) the length of the circumnutation major axis is represented as a blue/dash line; (**c**) the circumnutation length is represented as a yellow/solid line; (**d**) the circumnutation area is represented in green.

4.7. Statistical Analysis

The descriptive statistics, including the mean, standard deviation (SD), standard error (SE), and coefficient of variation, were calculated. Statistical analyses were conducted using the Bayesian approach. The objective of Bayesian estimation is to allocate credibility to a distribution of alternative parameter values (posterior distribution) that is consistent with the observed data by generating a large number of samples using the Markov chain Monte Carlo approach (MCMC). In this study, we adopted the two-sided Bayesian Mann–Whitney U test, given that the dependent variables are not normally distributed. The Mann–Whitney U test is a non-parametric test that does not require the assumption of normality. The analysis was performed using JASP [40], which was nested within the environment R (see the https://jasp-stats.org/r-package-list/, accessed on 20 February 2023) [41]. We choose the default that was prior defined by a Cauchy distribution, which was centered on a zero-effect size (δ) and a scale of 0.707 because prior knowledge regarding the exposition of plants to a double-support condition is absent [42,43]. Data augmentation was generated with five chains of 1000 iterations, allowing for a simpler and more feasible simulation from a posterior distribution. In the analysis, W was calculated in the Mann–Whitney U test as the smaller of the rank total between the two conditions. The Bayes factor (BF) was obtained to quantify the relative predictive performance of two hypotheses [42]. The BF quantifies evidence for the presence or absence of the difference between the DS and SS conditions. Here, the null hypothesis (H0) is that there is no difference in kinematics between the DS and SS conditions. The alternative hypothesis (H_1) is that there is a difference. The BF_{10} value is the likelihood given H_1 divided by H_0. The BF_{01} value is calculated as H_0 divided by H_1. The results are reported based on Jeffery's scheme, which proposes a series of labels for which specific Bayes factor values can be considered as either "no evidence (0–1)", "anecdotal (1–3)", "moderate (3–10)", "strong (10–30)", "very strong (30–100)", or "decisive (>100)" relative evidence for alternative hypotheses [44]. R-hat is also reported to check the degree of convergence of the MCMC algorithms based on outcome stability. The closer the value of R-hat is to 1, the better convergence to the underlying distribution. Credible intervals (CI) are set as 95%, which is simply the central portion of the posterior distribution that contains 95% of the values.

Supplementary Materials: The following supporting information can be downloaded at: https://www.mdpi.com/article/10.3390/plants12081597/s1, Video S1: Plant in DS condition.

Author Contributions: Conceptualization, Q.W. and U.C.; methodology, Q.W.; software, Q.W. and V.S.; validation, Q.W.; formal analysis, Q.W.; investigation, Q.W., S.G. and B.B.; resources, M.B.; data curation, Q.W.; writing—original draft preparation, Q.W. and U.C.; writing—review and editing, U.C. and Q.W.; visualization, Q.W.; supervision, U.C.; project administration, Q.W. All authors have read and agreed to the published version of the manuscript.

Funding: This research received no external funding.

Institutional Review Board Statement: Not applicable.

Informed Consent Statement: Not applicable.

Data Availability Statement: The data is available online: http://doi.org/10.6084/m9.figshare.22574443.

Acknowledgments: Figures 3–5 were created with BioRender.com. Qiuran Wang is funded by the China Scholarship Council (CSC). We thank Matthew Sims for his invaluable comments on this research project.

Conflicts of Interest: The authors declare no conflict of interest.

References

1.   Niklas, K.J. Climbing plants: Attachment and the ascent for light. *Curr. Biol.* **2011**, *21*, R199–R201. [CrossRef] [PubMed]
2.   Gianoli, E. The behavioural ecology of climbing plants. *AoB Plants* **2015**, *7*, plv013. [CrossRef] [PubMed]
3.   Gianoli, E.; Gonzalez-Teuber, M. Effect of support availability, mother plant genotype and maternal support environment on the twining vine Ipomoea purpurea. *Plant Ecol.* **2005**, *179*, 231–235. [CrossRef]

4.   Darwin, C. *The Movements and Habits of Climbing Plants*; John Murray: London, UK, 1875.
5.   Saito, K. A study on diameter-dependent support selection of the tendrils of Cayratia japonica. *Sci. Rep.* **2022**, *12*, 1–8. [CrossRef] [PubMed]
6.   Ceccarini, F.; Guerra, S.; Peressotti, A.; Peressotti, F.; Bulgheroni, M.; Baccinelli, W.; Bonato, B.; Castiello, U. Speed–accuracy trade-off in plants. *Psychon. Bull. Rev.* **2020**, *27*, 966–973. [CrossRef] [PubMed]
7.   Ceccarini, F.; Guerra, S.; Peressotti, A.; Peressotti, F.; Bulgheroni, M.; Baccinelli, W.; Bonato, B.; Castiello, U. On-line control of movement in plants. *Biochem. Biophys. Res. Commun.* **2020**, *564*, 86–91. [CrossRef] [PubMed]
8.   Guerra, S.; Peressotti, A.; Peressotti, F.; Bulgheroni, M.; Baccinelli, W.; D'Amico, E.; Gomez, A.; Massaccesi, S.; Ceccarini, F.; Castiello, U. Flexible control of movement in plants. *Sci. Rep.* **2019**, *9*, 16570. [CrossRef]
9.   Guerra, S.; Bonato, B.; Wang, Q.; Peressotti, A.; Peressotti, F.; Baccinelli, W.; Bulgheroni, M.; Castiello, U. Kinematic Evidence of Root-to-Shoot Signaling for the Coding of Support Thickness in Pea Plants. *Biology* **2022**, *11*, 405. [CrossRef]
10. Castiello, U. (Re) claiming plants in comparative psychology. *J. Comp. Psychol.* **2020**, *135*, 127–141. [CrossRef]
11. Carrasco-Urra, F.; Gianoli, E. Abundance of climbing plants in a southern temperate rain forest: Host tree characteristics or light availability? *J. Veg. Sci.* **2009**, *20*, 1155–1162. [CrossRef]
12. Goriely, A.; Neukirch, S. Mechanics of climbing and attachment in twining plants. *Phys. Rev. Lett.* **2006**, *97*, 184302. [CrossRef] [PubMed]
13. Putz, F.E.; Holbrook, N.M. Biomechanical studies of vines. In *The Biology of Vines*; Cambridge University Press: Cambridge, UK, 1992; pp. 73–98.
14. Rowe, N.P.; Isnard, S.; Gallenmüller, F.; Speck, T. 2 Diversity of Mechanical Architectures in Climbing Plants: An Ecological Perspective. In *Ecology and Biomechanics: A Mechanical Approach to the Ecology of Animals and Plants*; Taylor & Francis: New York, NY, USA, 2006; p. 35.
15. Sousa-Baena, M.S.; Hernandes-Lopes, J.; Van Sluys, M.-A. Reaching the top through a tortuous path: Helical growth in climbing plants. *Curr. Opin. Plant Biol.* **2021**, *59*, 101982. [CrossRef] [PubMed]
16. Viviani, P.; McCollum, G. The relation between linear extent and velocity in drawing movements. *Neuroscience* **1983**, *10*, 211–218. [CrossRef]
17. Cohen, I.; Huang, Y.; Chen, J.; Benesty, J.; Benesty, J.; Chen, J.; Huang, Y.; Cohen, I. Pearson correlation coefficient. In *Noise Reduction in Speech Processing*; Springer: Cham, Switzerland, 2009; pp. 1–4.
18. Van Rossum, G.; Drake, F.L., Jr. *Python Tutorial*; Centrum voor Wiskunde en Informatica: Amsterdam, The Netherlands, 1995; Volume 620.
19. Sartori, L.; Camperio-Ciani, A.; Bulgheroni, M.; Castiello, U. Reach-to-grasp movements in Macaca fascicularis monkeys: The Isochrony Principle at work. *Front. Psychol.* **2013**, *4*, 114. [CrossRef] [PubMed]
20. Karban, R. Plant behaviour and communication. *Ecol. Lett.* **2008**, *11*, 727–739. [CrossRef]
21. Bowling, A.J.; Vaughn, K.C. Gelatinous fibers are widespread in coiling tendrils and twining vines. *Am. J. Bot.* **2009**, *96*, 719–727. [CrossRef] [PubMed]
22. Vidoni, R.; Mimmo, T.; Pandolfiy, C.; Valentinuzzi, F.; Cesco, S. SMA bio-robotic mimesis of tendril-based climbing plants: First results. In Proceedings of the 2013 16th International Conference on Advanced Robotics (ICAR), Montevideo, Uruguay, 25–29 November 2013; pp. 1–6.
23. Stolarz, M. Circumnutation as a visible plant action and reaction: Physiological, cellular and molecular basis for circumnutations. *Plant Signal. Behav.* **2009**, *4*, 380–387. [CrossRef]
24. Eisenfuhr, F. Decision making. *Acad. Manag. Rev.* **2011**, *19*, 312–330.
25. Kacelnik, A.; El Mouden, C. Triumphs and trials of the risk paradigm. *Anim. Behav.* **2013**, *86*, 1117–1129. [CrossRef]
26. Dener, E.; Kacelnik, A.; Shemesh, H. Pea plants show risk sensitivity. *Curr. Biol.* **2016**, *26*, 1763–1767. [CrossRef]
27. Gruntman, M.; Groß, D.; Májeková, M.; Tielbörger, K. Decision-making in plants under competition. *Nat. Commun.* **2017**, *8*, 2235. [CrossRef]
28. McNamara, J.M.; Houston, A.I. Risk-sensitive foraging: A review of the theory. *Bull. Math. Biol.* **1992**, *54*, 355–378. [CrossRef]
29. Schmid, B. Decision-making: Are plants more rational than animals? *Curr. Biol.* **2016**, *26*, R675–R678. [CrossRef]
30. Calvo, P. The philosophy of plant neurobiology: A manifesto. *Synthese* **2016**, *193*, 1323–1343. [CrossRef]
31. Souza, G.M.; Toledo, G.R.; Saraiva, G.F. Towards systemic view for plant learning: Ecophysiological perspective. In *Memory and Learning in Plants*; Springer: Cham, Switzerland, 2018; pp. 163–189.
32. Chen, C.; Xiao, Y.-G.; Li, X.; Ni, M. Light-regulated stomatal aperture in Arabidopsis. *Mol. Plant* **2012**, *5*, 566–572. [CrossRef] [PubMed]
33. Sharkey, T.D.; Raschke, K. Separation and measurement of direct and indirect effects of light on stomata. *Plant Physiol.* **1981**, *68*, 33–40. [CrossRef] [PubMed]
34. Souza, G.M.; Ferreira, A.S.; Saraiva, G.F.; Toledo, G.R. Plant "electrome" can be pushed toward a self-organized critical state by external cues: Evidences from a study with soybean seedlings subject to different environmental conditions. *Plant Signal. Behav.* **2017**, *12*, e1290040. [CrossRef]
35. Debono, M.W.; Souza, G.M. Plants as electromic plastic interfaces: A mesological approach. *Prog. Biophys. Mol. Biol.* **2019**, *146*, 123–133. [CrossRef] [PubMed]

36. Saraiva, G.; Ferreira, A.; Souza, G. Osmotic stress decreases complexity underlying the electrophysiological dynamic in soybean. *Plant Biol.* **2017**, *19*, 702–708. [CrossRef]
37. Bak, P.; Tang, C.; Wiesenfeld, K. Self-organized criticality: An explanation of the 1/f noise. *Phys. Rev. Lett.* **1987**, *59*, 381. [CrossRef]
38. Schuster, J.; Engelmann, W. Circumnutations of Arabidopsis thaliana seedlings. *Biol. Rhythm. Res.* **1997**, *28*, 422–440. [CrossRef]
39. Simonetti, V.; Bulgheroni, M.; Guerra, S.; Peressotti, A.; Peressotti, F.; Baccinelli, W.; Ceccarini, F.; Bonato, B.; Wang, Q.; Castiello, U. Can Plants Move Like Animals? A Three-Dimensional Stereovision Analysis of Movement in Plants. *Animals* **2021**, *11*, 1854. [CrossRef] [PubMed]
40. JASP Team. JASP (Version 0.17) [Computer Software]. 2023. Available online: https://jasp-stats.org/ (accessed on 20 February 2023).
41. R Development Core Team, R. *A Language and Environment for Statistical Computing*; R Foundation for Statistical Computing: Vienna, Austria, 2010.
42. van Doorn, J.; van den Bergh, D.; Böhm, U.; Dablander, F.; Derks, K.; Draws, T.; Etz, A.; Evans, N.J.; Gronau, Q.F.; Haaf, J.M. The JASP guidelines for conducting and reporting a Bayesian analysis. *Psychon. Bull. Rev.* **2021**, *28*, 813–826. [CrossRef] [PubMed]
43. Ly, A.; Verhagen, J.; Wagenmakers, E.-J. Harold Jeffreys's default Bayes factor hypothesis tests: Explanation, extension, and application in psychology. *J. Math. Psychol.* **2016**, *72*, 19–32. [CrossRef]
44. Jeffreys, H. *The Theory of Probability*; OuP Oxford: New York, NY, USA, 1998.

Article

Kin Recognition in an Herbicide-Resistant Barnyardgrass (*Echinochloa crus-galli* L.) Biotype

Le Ding [1], Huan-Huan Zhao [2], Hong-Yu Li [1], Xue-Fang Yang [3] and Chui-Hua Kong [1,*]

[1] College of Resources and Environmental Sciences, China Agricultural University, Beijing 100193, China; s20223030381@cau.edu.cn (L.D.); lihongyulq2011@163.com (H.-Y.L.)
[2] College of Geography and Environmental Science, Henan University, Kaifeng 475004, China; zhaohh@henu.edu.cn
[3] College of Life Science, Hebei University, Baoding 071000, China; yang_xue_fang1@126.com
* Correspondence: kongch@cau.edu.cn

Abstract: Despite increasing evidence of kin recognition in natural and crop plants, there is a lack of knowledge of kin recognition in herbicide-resistant weeds that are escalating in cropping systems. Here, we identified a penoxsulam-resistant barnyardgrass biotype with the ability for kin recognition from two biotypes of penoxsulam-susceptible barnyardgrass and normal barnyardgrass at different levels of relatedness. When grown with closely related penoxsulam-susceptible barnyardgrass, penoxsulam-resistant barnyardgrass reduced root growth and distribution, lowering belowground competition, and advanced flowering and increased seed production, enhancing reproductive effectiveness. However, such kin recognition responses were not occurred in the presence of distantly related normal barnyardgrass. Root segregation, soil activated carbon amendment, and root exudates incubation indicated chemically-mediated kin recognition among barnyardgrass biotypes. Interestingly, penoxsulam-resistant barnyardgrass significantly reduced a putative signaling (–)-loliolide production in the presence of closely related biotype but increased production when growing with distantly related biotype and more distantly related interspecific allelopathic rice cultivar. Importantly, genetically identical penoxsulam-resistant and -susceptible barnyardgrass biotypes synergistically interact to influence the action of allelopathic rice cultivar. Therefore, kin recognition in plants could also occur at the herbicide-resistant barnyardgrass biotype level, and intraspecific kin recognition may facilitate cooperation between genetically related biotypes to compete with interspecific rice, offering many potential implications and applications in paddy systems.

Keywords: allelopathy; biomass allocation; flowering and reproduction; herbicide resistance; kin recognition; (–)-loliolide; rice–barnyardgrass interactions; root behavior

Citation: Ding, L.; Zhao, H.-H.; Li, H.-Y.; Yang, X.-F.; Kong, C.-H. Kin Recognition in an Herbicide-Resistant Barnyardgrass (*Echinochloa crus-galli* L.) Biotype. *Plants* **2023**, *12*, 1498. https://doi.org/10.3390/plants12071498

Academic Editors: Frantisek Baluska and Gustavo Maia Souza

Received: 8 March 2023
Revised: 28 March 2023
Accepted: 28 March 2023
Published: 29 March 2023

1. Introduction

Two or more plants occur together, resulting in a series of intraspecific and interspecific interactions. Plant–plant interactions, either positive or negative, are important driving forces for plant coexistence and community assembly [1–3]. Plant–plant negative interactions mainly include competition and allelopathy that are well-known in natural and managed ecosystems [4–7]. Positive or beneficial interactions involve in plant facilitation and kin recognition [8–10]. In contrast to competition and allelopathy, plant facilitation and kin recognition within and among species have received less attention, particularly for kin recognition [11,12]. Kin recognition allows plants to assess kinship to discriminate neighboring kin (collaborators) from non-kin (competitors), and then plants display morphological and biochemical plasticity toward reduced competition and defense when growing with kin, leading to increased fitness in kin groups [13,14]. Therefore, plant–plant beneficial interactions can result from kin recognition.

Kin recognition is a cooperative behavior among plants acting exclusively at the intraspecific level [15,16]. The ability to recognize and respond appropriately to close relatives may allow plants to tailor response strategies and optimize their competitive traits. Such kin recognition and cooperation have been observed at the individual, population, accession and cultivar levels in natural and cropping systems [17–22]. In natural systems, kin recognition in plants has dealt with conspecific individuals. Different from naturally occurring species where kin represent plants sharing the same mother, as either full or half siblings [17,23,24], in cropping systems, kin recognition occurs at the cultivar level because artificial selection generates crop cultivars with genetical and morphological uniform. Therefore, kin in crop plants are across closely related cultivars rather than siblings within a natural species [21,22].

Rice (*Oryza sativa* L.) is one of the principal grain crops. Barnyardgrass (*Echinochloa crusgalli* L.) is an intractable weed that coexists with rice in paddy systems for millennia. Rice–barnyardgrass represents a well-characterized model system to understand the combined roles of plant–plant intraspecific and interspecific interactions. However, barnyardgrass has evolved several herbicide-resistant biotypes duo to continuous use of herbicides in the past decades [25–27]. The incidence of herbicide-resistant barnyardgrass is escalating in rice fields. Accordingly, rice-barnyardgrass interactions should be altered by herbicide-resistant barnyardgrass biotype. Rice has to co-occur and interact with herbicide-resistant and -susceptible barnyardgrass, as well as normal barnyardgrass biotypes in paddies. However, the local coexistence and interactions among these biotypes are largely unknown. In fact, herbicide-resistant and -susceptible barnyardgrass usually come from a population with the same genetic background [28]. It is thought that kin recognition may occur between herbicide-resistant and -susceptible barnyardgrass biotypes.

Herbicide-resistant barnyardgrass has been a serious problem in paddy systems. Numerous studies have documented that allelopathic rice cultivars can release allelochemicals to suppress barnyardgrass and provide a competitive advantage for their own growth [29–32]. In particular, allelopathic rice cultivars interfered with both herbicide-susceptible and -resistant barnyardgrass while herbicide-resistant and -susceptible biotypes responded differently to allelopathic rice cultivars [28]. However, relatedness-mediated interactions between herbicide-resistant and -susceptible barnyardgrass, as well as their consequence for the interference of allelopathic rice cultivars with barnyardgrass biotypes retain obscure.

In the present study, we test whether kin recognition occurs at the herbicide-resistant barnyardgrass biotype level. To achieve this, we identified a penoxsulam-resistant barnyardgrass biotype with the ability for kin recognition from two biotypes of penoxsulam-susceptible barnyardgrass and normal barnyardgrass at different levels of relatedness. Furthermore, we determined the kin recognition responses of penoxsulam-resistant barnyardgrass with or without root segregation, soil activated carbon amendment and root exudates incubation. In addition, we examined the role of a putative root exudate signal in kin recognition among barnyardgrass biotypes. Finally, we assessed relatedness-mediated impacts on the interference of allelopathic rice cultivars with barnyardgrass biotypes. Together, these efforts provide a new insight into the kin recognition in plants, with a further understanding of rice-barnyardgrass interactions underlying the escalation of herbicide-resistant biotypes in paddy systems.

2. Results

2.1. Phenotypic Profile of Penoxsulam-Resistant Barnyardgrass in the Presence of Penoxsulam-Susceptible Barnyardgrass or Normal Barnyardgrass

When penoxsulam-resistant barnyardgrass (focal biotype) was paired with itself, penoxsulam-susceptible barnyardgrass (closely related biotype, kin) or normal barnyardgrass (distantly related biotype, non-kin), phenotypic profile of penoxsulam-resistant barnyardgrass varied with neighbor relatedness (Figures 1 and 2). There were not significant differences between penoxsulam-resistant and -susceptible barnyardgrass. However, the presence of distantly related barnyardgrass biotype significantly increased root biomass, reduced shoot biomass and seed production, and delayed flowering of penoxsulam-resistant barnyardgrass (Figure 1a–d). Similar changes with neighbor relatedness were observed in root measurements. Mixed-culture with the closely related penoxsulam-susceptible barnyardgrass did not alter the root measurements of penoxsulam-resistant barnyardgrass. However, distantly related barnyardgrass altered root measurements of penoxsulam-resistant barnyardgrass, and significant changes occurred in increased root biomass and total root length (Figure 2a,b). These results indicated that penoxsulam-resistant barnyardgrass adjusted phenotypic profile based on neighbor identity from different levels of relatedness, particularly for making plants with closely related biotype shifted biomass allocation from competitive roots to flowering and seed reproduction. Such relatedness-based phenotypic responses and biomass allocation indicated kin recognition in penoxsulam-resistant barnyardgrass biotype.

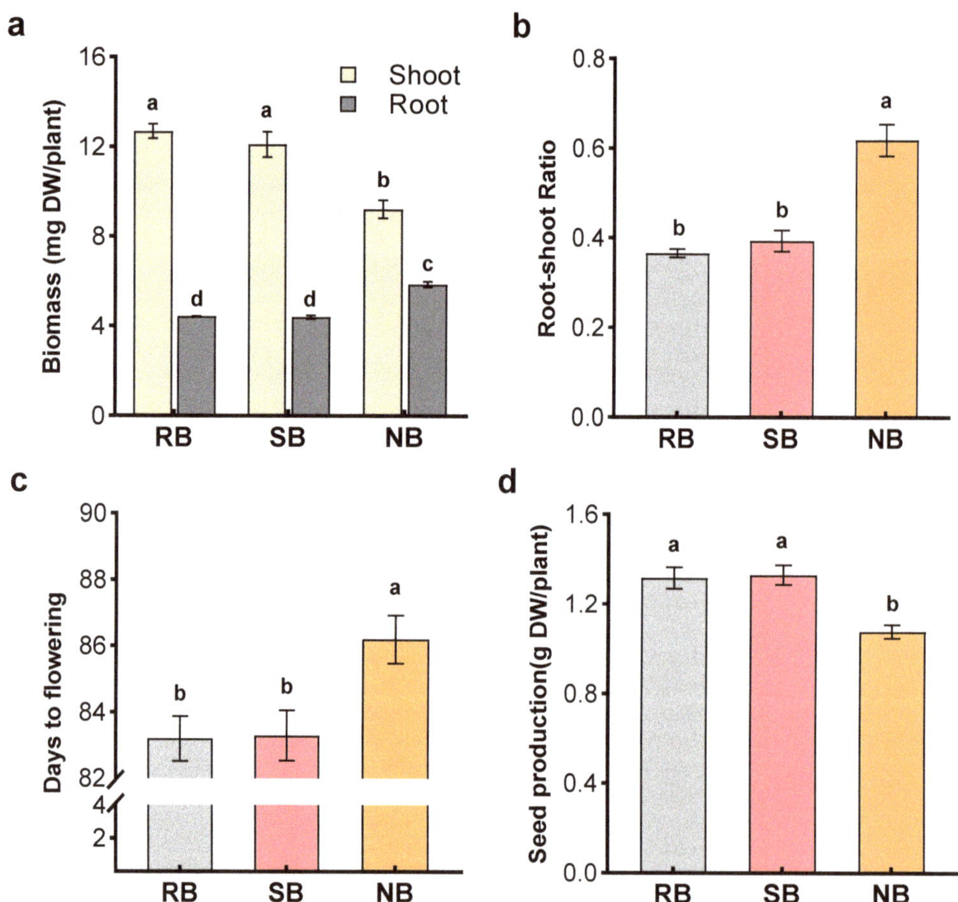

Figure 1. Biomass (**a**), Root-shoot radio (**b**), flowering time (**c**), and seed production (**d**) of penoxsulam-resistant barnyardgrass (**RB**) in the presence of penoxsulam-susceptible barnyardgrass (**SB**) and normal barnyardgrass (**NB**). Values plotted are means plus/minus SE. Columns with the same letter are not significantly different at $p < 0.05$ according to ANOVA, followed by Tukey HSD tests.

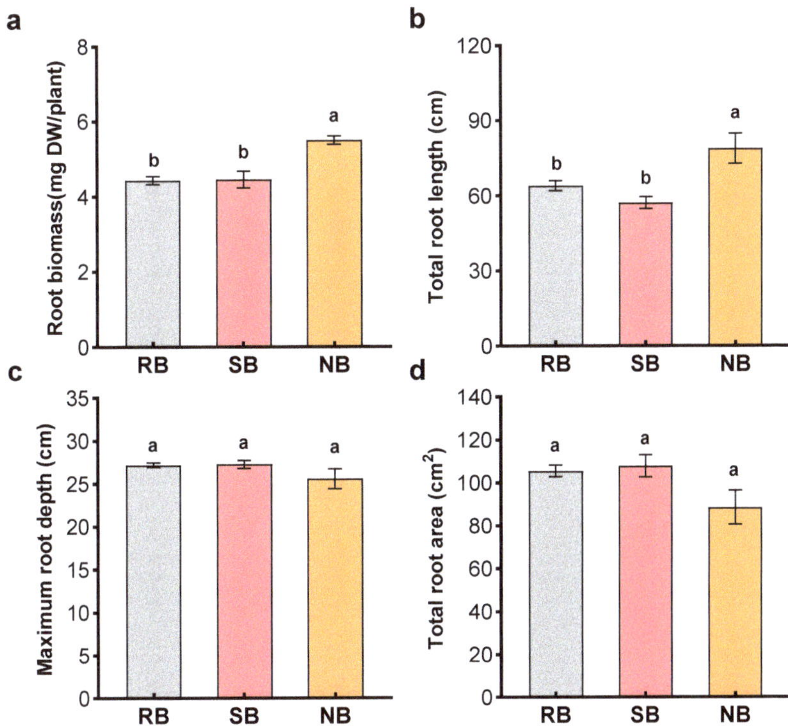

Figure 2. Root biomass (**a**), total root length (**b**), maximum root depth (**c**) and total root area (**d**) of penoxsulam-resistant barnyardgrass (**RB**) in the presence of penoxsulam-susceptible barnyardgrass (**SB**) and normal barnyardgrass (**NB**). Values plotted are means plus/minus SE. Columns with the same letter are not significantly different at $p < 0.05$ according to ANOVA, followed by Tukey HSD tests.

2.2. Chemically Mediated Kin Recognition in Barnyardgrass Biotypes

When penoxsulam-resistant barnyardgrass and penoxsulam-susceptible barnyardgrass or normal barnyardgrass occurred together and interacted, root biomass of penoxsulam-resistant barnyardgrass were changed with and without root segregation. Root biomass consistently increased when grown with distantly related normal barnyardgrass under root contact and root segregation with 30 μm nylon mesh. However, the increased root biomass was not observed under root segregation with plastic film (Figure 3a). The plastic film blocked belowground physical, chemical, and biological interactions, limiting all interactions to aboveground. There were not significant differences on root biomass under root segregation with plastic film regardless of neighbor biotypes, indicating that the kin recognition responses relied on belowground interactions rather than aboveground interactions. Furthermore, the root measurements of penoxsulam-resistant barnyardgrass were significantly different in soil amended with and without activated carbon. Soil amended with activated carbon greatly changed root biomass, total root length, and total root area of penoxsulam-resistant barnyardgrass regardless of neighbor biotypes. In particular, significant changes in the root measurements of penoxsulam-resistant barnyardgrass induced by distantly related normal barnyardgrass were attenuated or terminated after soil activated carbon amendment (Figure 3b–d). When penoxsulam-resistant barnyardgrass was exposed to the root exudates of distantly related barnyardgrass biotype, the seedling produced larger root systems. However, root biomass and morphology were not significantly changed in the root exudates of closely related barnyardgrass biotype (Figure 4a,b). The results indicated the importance of belowground chemical interactions in mediating kin recognition among barnyardgrass biotypes.

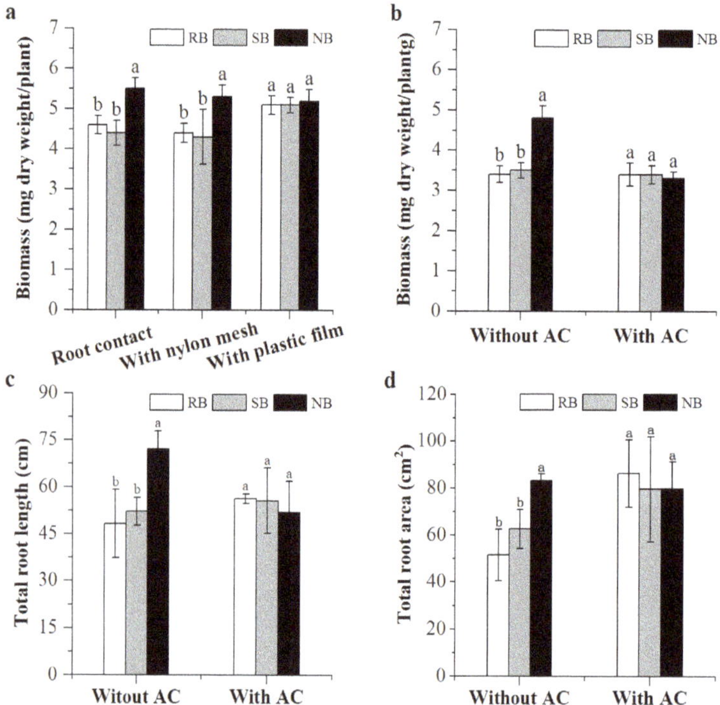

Figure 3. The effects of root segregation and soil activated carbon (**AC**) amendment on root kin recognition responses of penoxsulam-resistant barnyardgrass (**RB**) in the presence of penoxsulam-susceptible barnyardgrass (**SB**) and normal barnyardgrass (**NB**). (**a**), Root biomass under root segregation; (**b**), Root biomass with and without soil activated carbon amendment; (**c**), Total root length with and without soil activated carbon amendment; (**d**), Total root area with and without soil activated carbon amendment. Values plotted are means plus/minus SE. Columns with the same letter within identical treatments are not significantly different at $p < 0.05$ according to ANOVA, followed by Tukey HSD tests.

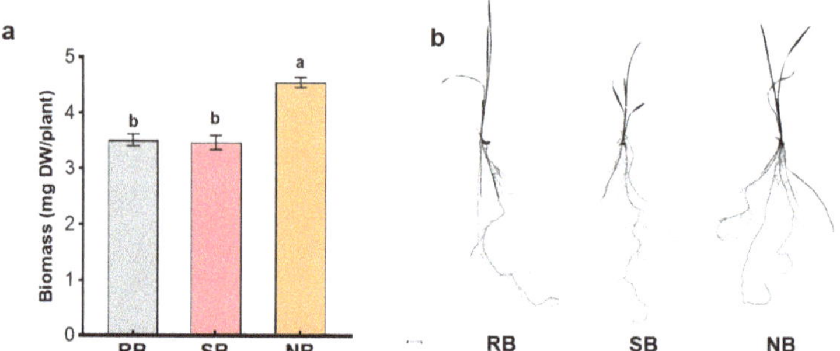

Figure 4. Root biomass (**a**) and morphology (**b**) of penoxsulam-resistant barnyardgrass (**RB**) in the root exudates of penoxsulam-susceptible barnyardgrass (**SB**) and normal barnyardgrass (**NB**). Values plotted are means plus/minus SE. Columns with the same letter are not significantly different at $p < 0.05$ according to ANOVA, followed by Tukey HSD tests.

2.3. The Role of (–)-Loliolide in Kin Recognition among Barnyardgrass Biotypes

Kin recognition indicates relatedness-mediated neighbor discrimination, mainly reducing intraspecific competition. While (–)-loliolide is a common signal for plant neighbor detection and competition [33,34]. This plant–plant signaling is reminiscent of the role of (–)-loliolide in kin recognition among barnyardgrass biotypes. Thus, we examined the level of (–)-loliolide in penoxsulam-resistant barnyardgrass in response to neighbors of differing relatedness. As expected, the concentration of (–)-loliolide in penoxsulam-resistant barnyardgrass significantly varied with neighbor relatedness (Figure 5). The more distant allelopathic rice cultivar led to the highest (–)-loliolide concentration in both root and shoot, followed by the distantly related barnyardgrass biotype. Penoxsulam-resistant barnyardgrass monocultures and those with closely related barnyardgrass biotype contained the lowest (–)-loliolide concentrations (Figure 5). This suggests that penoxsulam-resistant barnyardgrass may produce more (–)-loliolide in response to the presence of unrelated neighbors regardless of whether they are intraspecific or interspecific competitors, indicating (–)-loliolide's role as a key component of penoxsulam-resistant barnyardgrass response to kinship interactions.

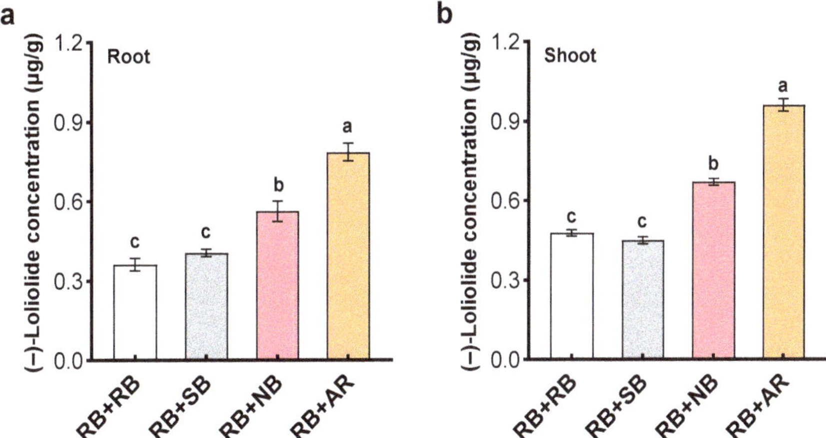

Figure 5. (–)-Loliolide levels in root (**a**) and shoot (**b**) of penoxsulam-resistant barnyardgrass (**RB**) in response to the presence of penoxsulam-susceptible barnyardgrass (**SB**) and normal barnyardgrass (**NB**), as well as allelopathic rice cultivar (**AR**). Values plotted are means plus/minus SE. Columns with the same letter are not significantly different at $p < 0.05$ according to ANOVA, followed by Tukey HSD tests.

2.4. Rice-Barnyardgrass Allelopathic Interactions in Response to Kin Recognition among Barnyardgrass Biotypes

The presence of allelopathic rice cultivar significantly inhibited the growth of all barnyardgrass biotype mixtures, but the inhibition was dependent on the relatedness of mixed with barnyardgrass biotypes. There was more significant inhibition in penoxsulam-resistant barnyardgrass mixed with distantly related barnyardgrass biotype than in one mixed with closely related barnyardgrass biotype (Figure 6a). Similarly, relatedness-mediated inhibition occurred in the interference of barnyardgrass with allelopathic rice cultivar. Compared with rice monoculture, barnyardgrass biotype mixtures reduced rice biomass. However, significant reduction was observed in penoxsulam-resistant barnyardgrass mixed with the same and closely related biotypes rather than in one mixed with closely related barnyardgrass biotype (Figure 6b). These results showed that kin recognition among barnyardgrass biotypes affected the interactions between allelopathic rice and barnyardgrass. In particular, relatedness-mediated barnyardgrass biotypes cooperated resistance to inhibition of allelopathic rice cultivar, and enhanced the interference of barnyardgrass with allelopathic rice cultivar.

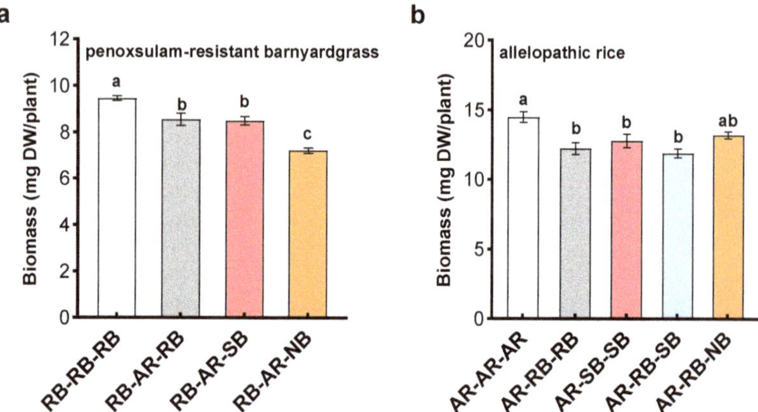

Figure 6. Effects of relatedness-mediated rice–barnyardgrass allelopathic interactions on biomass of penoxsulam-resistant barnyardgrass biotype (**a**) and allelopathic rice cultivar (**b**). **RB**, penoxsulam-resistant barnyardgrass biotype; **SB**, penoxsulam-susceptible barnyardgrass biotype; **NB**, normal barnyardgrass biotype; **AR**, allelopathic rice cultivar. Values plotted are means plus/minus SE. Columns with the same letter are not significantly different at $p < 0.05$ according to ANOVA, followed by Tukey HSD tests.

3. Discussion

Kin recognition in plants, as well as the evolutionary and ecological mechanisms underlying such recognition, has received a great deal of attention in the past decade [12,13,16,35]. However, there is still controversial issues in this fascinating area, particularly for the concept and definition of 'kin' in natural and managed ecosystems [36]. In natural ecosystems, 'kin' of wild species is limited to siblings or half-siblings and sometimes can be extended to plants sharing the same population [17,19,24]. In managed ecosystems, crops and weeds have undergone intense artificial selection and herbicide pressure, resulting in crop cultivars with genetic and morphological uniformity and the biotype of herbicide-resistant weeds [37]. Kin recognition at the cultivar or accession level has been observed in several crop species [18,21,22]. This study presents the first case of kin recognition at the herbicide-resistant weed biotype level.

Herbicide-resistant and -susceptible weeds simultaneously occur in cropping systems and evolve under the pressure of herbicides. The evolution of herbicide-resistant weeds is driven by the use of herbicides in large quantities. Most studies have focused on the incidence and gene mutation of herbicide-resistant weeds [26,27,37]. There is a lack of information on the interactions between herbicide-resistant and -susceptible biotypes or normal weeds. Such information is critical for understanding the coexistence, evolution, and mechanisms of weed biotypes and their consequences in cropping systems. From a system of penoxsulam-resistant and -susceptible barnyardgrass as well as normal barnyardgrass biotypes, we found the interactions among these biotypes. In particular, differential interactions were observed at varying relatedness level, indicating kin recognition of penoxsulam-resistant barnyardgrass at the biotype level. The direct evidence was that, when grown with closely related penoxsulam-susceptible barnyardgrass, there were (1) reduced root growth and distribution, lowering belowground competition and (2) advanced flowering and increased seed production, enhancing reproductive effectiveness. However, such kin recognition responses were not occurred when penoxsulam-resistant barnyardgrass was grown with a distantly related barnyardgrass biotype.

The root behavior is a key kin recognition response. Kin recognition can reduce competitive root systems, allowing greater allocation to reproduction [16,38,39]. Through window rhizobox and root segregation, this study clearly demonstrated the ability of penoxsulam-resistant barnyardgrass to respond to closely related penoxsulam-susceptible barnyardgrass, reducing competitive root systems. Penoxsulam-resistant barnyardgrass significantly increased their root growth and distribution in the presence of distantly related biotype but avoid the root growth and distribution in the presence of the same and closely related biotypes. Such relatedness-mediated root behavior to minimize root competition is key in creating more allocation to flowering and reproduction.

Flowering time is key to a plant reproductive strategy, and reproductive success depends on the flowering behavior of immediate conspecific neighbors [24,40,41]. Kin selection plays a role in the evolution of placentation and reproductive traits in flowering plants [42,43]. In particular, early flowering plants are favored by phenotypic selection on flowering phenology [44]. In the current study, penoxsulam-resistant barnyardgrass growing with closely related biotype accelerated flowering relative to those growing with distantly related biotype, providing a linkage to belowground root behavior and aboveground flowering and reproduction. In this manner, the penoxsulam-resistant barnyardgrass biotype may maximize their own growth to fit individual and population.

Kin recognition in plants involves both physical and chemical signals [45,46]. Most evidence suggests root exudates as the signal of relatedness [18,21,47]. In this study, we found consistent root measures of penoxsulam-resistant barnyardgrass regardless of neighbor biotypes in the presence of activated carbon. The tremendous adsorptive capacity of activated carbon for functional metabolites in soil alleviated root-secreted the signal of relatedness, resulting in kin recognition responses hard to occur. Furthermore, the exposure of penoxsulam-resistant barnyardgrass seedlings to the root exudates of distantly related biotype induced larger root systems than exposure of seedlings to the root exudates of their own or closely related penoxsulam-susceptible barnyardgrass biotype. These results agree with previous studies showing that root exudates mediate kin recognition in plants [18,21,47].

Cooperation arising from kin recognition allows cooperative behaviors to be directed preferentially toward kin that promotes the increase of offspring from related families, which is a mechanism to ensure the continuation and evolution of the population [48,49]. Kin recognition in plants can reduce intraspecific competition, allowing greater allocation to reproduction in kin groups or relatives mixtures. However, how to define the distinction between 'kin' and 'non-kin', actually quantifying the level of genetic relatedness is challenging particularly at the cultivar and biotype levels [36]. Several studies have used genetic distances by molecular approach to identify closely versus distantly related populations or cultivars [14,21,50]. In the current study, penoxsulam-resistant barnyardgrass significantly reduced (–)-loliolide production in the presence of closely related penoxsulam-susceptible barnyardgrass but increased production when growing with a distantly related barnyardgrass biotype and more distantly related interspecific allelopathic rice cultivar. In fact, kin recognition may allow plants to optimize competitive strategies, resulting in less intraspecific competition and more cooperation among plants [8]. Therefore, competition or cooperation are the most important, intraspecific traits between kin and non-kin. Relatedness allows plants to discriminate their neighboring collaborators (kin) or competitors (non-kin), either interspecifically or intraspecifically, and adjust their growth and competitiveness accordingly [14]. It appears from the results that the level of (–)-loliolide may indicate neighbor kinship, discriminating kin from non-kin and even differentiating interspecific competitors from conspecific competitors. (–)-Loliolide is a general signaling chemical for plant neighbor detection and plant competition, as well as other biotic and abiotic stressors [33,34]. (–)-Loliolide response to neighbor kinship may provide a chemical approach to quantify the level of genetic relatedness and define the distinction between 'kin' and 'non-kin' among crop cultivars and weed biotypes. Of course, for such (–)-loliolide-based interactions, the distinction between 'kin' and 'non-kin' needs to be verified in other plant systems.

Most kin recognition studies mainly focus on the occurrence and the extent to which plants express their cooperative behaviors towards conspecific neighbors based on the extent of genetic similarity between them [17,19,20,22,23]. In fact, plants often grow in mixtures of kin, non-kin conspecifics, and other species, and thus, intraspecific kin recognition and interspecific interactions may often occur simultaneously. Accordingly, an increasing number of studies have investigated how kin recognition mediates the context of cooperation between genetically related plants to fight against herbivores, to attract pollinators, and to compete with interspecific plants [14,24,50–52]. In particular, intraspecific kin recognition contributes to interspecific allelopathy in allelopathic rice interference with paddy weeds [14]. This study showed that penoxsulam-resistant barnyardgrass plants could modulate their root systems, flowering time and seed production in response to closely or distantly related biotypes. In this manner, penoxsulam-resistant barnyardgrass with the ability for kin recognition maximize inclusive fitness and stand performance. Importantly, genetically identical penoxsulam-resistant and -susceptible barnyardgrass biotypes synergistically interact to influence the action of allelopathic rice cultivar, providing evidence that kin recognition may also facilitate cooperation between genetically related plants to compete with interspecific intruders. The discovery of kin recognition in herbicide-resistant barnyardgrass at the biotype level, as well as a further understanding of its potential mechanism, may lead to evolutionary and ecological insights into herbicide-resistant weeds. A thorough understanding of intraspecific and interspecific interactions between crops and weeds in cropping systems will offer potential implications for applications and the development of sustainable agriculture strategies.

4. Materials and Methods

4.1. Plant Materials, Soil, and Chemicals

An allelopathic rice (*Oryza sativa* L.) cultivar (Huagan-3) and two penoxsulam-resistant and -susceptible barnyardgrass (*Echinochloa crus-galli* L.) biotypes were used in this study. Huagan-3 is the first commercially approved allelopathic rice cultivar against paddy weeds in China [53]. The penoxsulam-resistant barnyardgrass seeds were originally collected from a rice field subjected to penoxsulam treatment for several consecutive years at Lujiang Experimental Station of Rice Research, Anhui province of East China (30.47° N, 117.38° E), which located on the side of the Yangtze River. Seed samples were taken for the experiments to identify and segregate penoxsulam-resistant and -susceptible individuals as described in a previous study [28]. Finally, the homozygous barnyardgrass seeds of penoxsulam-resistant and -susceptible biotypes were obtained by a 3-year experiment of continuous selection of penoxsulam application. The penoxsulam-resistant and -susceptible biotypes within a population and location possess the same genetic background, resulting in their relatedness. In addition, normal barnyardgrass seeds from a distant population without herbicide application were collected at Shenyang Experimental Station of Chinese Academy of Sciences, Liaoning Province of Northeast China (41°31′ N, 123°24′ E), which was distantly related biotype for both penoxsulam-resistant and -susceptible biotypes. Accordingly, penoxsulam-resistant barnyardgrass was selected as the focal biotype while penoxsulam-susceptible barnyardgrass was kin biotype and normal barnyardgrass was used as a more distant non-kin biotype.

Soil was collected randomly from the surface (0–10 cm) of a paddy field of Lujiang Experimental Station as described above. The soil is a typical fluvaquent, Etisol (US taxonomy) with pH 5.8, organic matter of 25.1 g/kg, total nitrogen of 1.6 g/kg, available phosphourus of 30.9 mg/kg, and available potassium of 60.4 mg/kg. Soil samples were air-dried, mixed, and then sieved (2 mm mesh) to remove plant tissues for a series of experiments as described below.

(–)-Loliolide, a putative root-secreted chemical signal involved in rice–barnyardgrass allelopathic interactions [30], was isolated and identified from root exudates using previously developed methods [33] and verified with its authentic standard obtained from Yuanye Biology Corporation (Shanghai, China). Other organic solvents and chemicals were purchased from China National Chemical Corporation (Beijing, China).

4.2. Pot-Culture Experiments of Mixed-Biotype Barnyardgrass

Two pot-culture experiments were conducted in a greenhouse at 20–30 °C night and daytime temperatures and 65–90% relative humidity maintained. Each experiment was conducted in a completely randomized design with three replicates for each treatment or control. The sterilized barnyardgrass seeds for each biotype were separately sown to Petri dishes (9 cm diameter) with moistened filter paper for pre-germination in a chamber set at a temperature of 28 °C.

The first experiment investigated the performance of penoxsulam-resistant barnyardgrass (focal biotype) in the presence of penoxsulam-susceptible barnyardgrass (kin biotype) or normal barnyardgrass (non-kin biotype) in a series of 15 (diameter) × 12 cm (height) plastic pots containing 2 kg of soil as described above. Four pregerminated seeds of penoxsulam-resistant barnyardgrass were spaced uniformly in the center of each pot while four seeds of penoxsulam-susceptible barnyardgrass or normal barnyardgrass were sown in the surrounding area. Pots with penoxsulam-resistant barnyardgrass monocultures in the same planting pattern served as controls. All pots were placed in the greenhouse, watered daily, and their positions randomized weekly. The penoxsulam-resistant barnyardgrass in 1/3 of pots were sampled at the seedling stage, and their shoots and roots were determined biomass and quantified for (–)-loliolide as described below (Section 4.6). In the remaining 2/3 of pots, flowering time and seed biomass of penoxsulam-resistant barnyardgrass were recorded at the flowering and mature stages, respectively.

A second experiment was run to evaluate the impact of root segregation on the performance of penoxsulam-resistant barnyardgrass in the presence of penoxsulam-susceptible barnyardgrass or normal barnyardgrass. A series of 11 cm (diameter) × 12 cm (height) plastic pots that contained a central cylinder (7.5 cm diameter, 12 cm height) where a barrier could be inserted were divided into three groups. The cylinders in the first group were not modified while the cylinders in the second and third group were covered with 30 μm nylon mesh or plastic film. The open cylinders were full contact. The 30 μm nylon mesh prevented penetration of root systems but allowed chemical and microbial interactions in the pots. The plastic film completely blocked root–soil interactions [33]. Four pre-germinated seeds of penoxsulam-resistant barnyardgrass were uniformly sown in the cylinder of each pot containing 800 g of soil; four pre-germinated seeds of penoxsulam-susceptible barnyardgrass or normal barnyardgrass were sown outside the cylinder of each pot. Monocultures of penoxsulam-resistant barnyardgrass (4:4) in a pot for each group with or without root segregation served as the controls. All the pots were placed in the greenhouse, watered daily, and randomized once a week. The seedlings of penoxsulam-resistant barnyardgrass were harvested after four weeks, and their biomass was measured.

4.3. Rhizobox Experiments of Mixed-Biotype Barnyardgrass

Root behavior of penoxsulam-resistant barnyardgrass in the presence of penoxsulam-susceptible barnyardgrass or normal barnyardgrass were determined using a window rhizobox method in a completely randomized design with three replications. The window rhizobox was made of a 20 cm (length) × 2 cm (width) × 30 cm (height) polyvinyl chloride box with a clear plexiglass sheet [54]. The penoxsulam-resistant barnyardgrass was grown in monoculture, or paired with penoxsulam-susceptible barnyardgrass or normal barnyardgrass in the window rhizoboxes containing 500 g of soil. The soils in half of the rhizoboxes were amended with 2% activated carbon. Each window rhizobox was vertically divided into two equal parts. A single penoxsulam-resistant barnyardgrass seed or an interacting barnyardgrass seed for each biotype was sown into each half. Monocultures of barnyardgrass for each biotype served as the controls. Window rhizoboxes were placed in racks and set to a 40° angle with the clear plexiglass sheet facing down and away from the light source in the greenhouse with 20–30 °C night and day temperatures and 65–90% relative humidity. The window rhizoboxes were opened after 4 weeks when root systems reached the horizontal or vertical margin of the rhizoboxes. The roots were scanned with an Epson Perfection V700 scanner (Seiko Epson, Nagano-ken, Japan) to yield a grey scale TIFF image [32]. The image was processed with the WinRHIZO system (Regent Instruments Inc., Quebec, Canada) to obtain three root measurements, including a size-related metric (total root length) and two measures of habitat occupancy (total root area and maximum root depth). Finally, the roots were freeze-dried for biomass determination.

4.4. Root Exudates Incubation

Root behavior of penoxsulam-resistant barnyardgrass in the root exudates of each biotype was evaluated with an incubation experiment. The root exudates of barnyardgrass for each biotype were collected hydroponically [28]. Sterilized and germinated seeds were sown in nursery seedling plates. Thirty barnyardgrass seedlings at the 2-leaf stage of each biotype were inserted into holes in a Styrofoam float and transplanted into a container (6 cm × 10 cm × 12 cm) containing 250 mL distilled water in a sterile growth chamber at 25 ± 1 °C with a 16 h light and 8 h dark photoperiod. After 7 days, the solution was filtered with sterile filter paper (GE Healthcare Whatman) and yielded the root exudates of each barnyardgrass biotype that were used for subsequent experiment.

A uniform penoxsulam-resistant barnyardgrass seedling at the 2-leaf stage was each transplanted into a sponge plug to stabilize the seedling that was then inserted into a series of transparent bottles with 100 mL root exudates of barnyardgrass for each biotype. All bottles in a completely randomized design with three replications were placed in a sterile environmental chamber at 28 °C with a 12 h photoperiod. After penoxsulam-resistant barnyardgrass seedlings in their own exudates reached the 3-leaf stage, all bottles were removed from the chamber. The seedling was each scanned to obtain a grey scale TIFF image (Figure 4b), and then, their root biomass was recorded.

4.5. Rice–Barnyardgrass Mixes-Species Experiment

The interactions between allelopathic rice cultivar and barnyardgrass biotypes were evaluated by mixes-species experiment. A series of plastic pots (12 cm diameter × 10 cm height) containing 1000 g of soil were used for the experiment. A total of eight pre-germinated seeds were sown into each pot. Four rice seeds were spaced uniformly in the central area (6 cm diameter) while four barnyardgrass seeds were sown in the outer circle in two biotypes cross manner. Monocultures of eight rice plants and four barnyardgrass plants within each biotype surrounding four rice plants served as the controls. The experiment was conducted in a completely randomized design with three replicates for treatment and control. All pots were placed in the greenhouse with 20–30 °C night and day temperatures and 65–90% relative humidity, watered daily, and randomized once a week. Both rice and barnyardgrass were harvested after four weeks, and their plants were separately freeze-dried for determining dry weight.

4.6. Quantification of (−)-Loliolide

Quantitative analysis of (−)-loliolide was performed by liquid extraction/solid-phase extraction followed by high-performance liquid chromatography (HPLC). Root and shoot of penoxsulam-resistant barnyardgrass were each freeze-dried and ground with liquid nitrogen. The resulting powder (250 mg) was extracted with 10 mL of a MeCN (acetonitrile)-H2O-HOAC mixture (90:9:1, $v/v/v$), vortexed for 5 min at 25 °C, and centrifuged at 2800× g for 10 min. The supernatant was each filtered with a 0.22 μm nylon syringe filter (Sterlitech, Kent, WA, USA). The filtrates were evaporated to dryness individually under vacuum. Dry residues were dissolved in 50% aqueous methanol and loaded onto reversed phase C_{18} Sep-Pak cartridges (Waters, Co., Milford, MA, USA), equilibrated with water, and eluted with MeOH. The MeOH fraction was concentrated with nitrogen gas to a final volume of 100 μL. The concentrated samples were subsequently subjected to a Waters 152 HPLC equipped with a C_{18} reverse-phase column (Hypersil 100 mm $\times$ 4.0 mm, 5 µm) and a diode array UV detector at 220 nm. Elution was performed with a mixture of 1 % acetic acid and MeOH (70:30, v/v) at a constant flow rate of 1.0 mL min^{-1} at 35 °C. The peak of (–)-loliolide was identified by its retention time (ca. 9.8 min) and coelution with an authentic standard (Yuanye Biology Co., Shanghai, China). (–)-Loliolide was quantified by regression analysis of the peak areas against standard concentrations.

4.7. Statistical Analysis

The normality and homogeneity of variances were verified in all statistical analyses. The data were analyzed using a one-way ANOVA followed by Tukey's post hoc tests (HSD) to compare significant difference between treatments. All statistical procedures were carried out using SPSS 22.0 (IBM Corp., Armonk, NY, USA).

Author Contributions: Conceptualization, C.-H.K.; methodology, C.-H.K. and H.-H.Z.; investigation, L.D., H.-H.Z. and H.-Y.L.; validation, L.D. and H.-Y.L.; resources, X.-F.Y.; formal analysis, L.D. and X.-F.Y.; data curation, L.D. and H.-Y.L.; writing—original draft preparation, L.D. and C.-H.K.; writing—review and editing, C.-H.K.; supervision, C.-H.K.; funding acquisition, X.-F.Y. and C.-H.K. All authors have read and agreed to the published version of the manuscript.

Funding: This research was funded by the National Natural Science Foundation of China (grant number 32271593).

Data Availability Statement: All data supporting the findings of this research are available within the paper. Raw data used here can be obtained directly from the authors.

Acknowledgments: The authors sincerely thank anonymous referees for their constructive comments and suggestions.

Conflicts of Interest: The authors declare no conflict of interest.

References

1. Bennett, J.A.; Riibak, K.; Tamme, R.; Lewis, R.I.; Partel, M. The reciprocal relationship between competition and intraspecific trait variation. *J. Ecol.* **2016**, *104*, 1410–1420. [CrossRef]
2. Bennett, T. Plant–plant interactions. *Plant Cell Environ.* **2021**, *44*, 995–996. [CrossRef]
3. Yu, R.P.; Lambers, H.; Callaway, R.M.; Wright, A.J.; Li, L. Belowground facilitation and trait matching: Two or three to tango? *Trends Plant Sci.* **2021**, *26*, 1227–1235. [CrossRef] [PubMed]
4. Pierik, R.; Mommer, L.; Voesenek, L.A.C.J.; Robinson, D. Molecular mechanisms of plant competition: Neighbor detection and response strategies. *Funct. Ecol.* **2013**, *27*, 841–853. [CrossRef]
5. Yamawo, A.; Mukai, H. Outcome of interspecific competition depends on genotype of conspecific neighbours. *Oecologia* **2020**, *193*, 415–423. [CrossRef]
6. Hierro, J.L.; Callaway, R.M. The ecological importance of allelopathy. *Annu. Rev. Ecol. Evol. Syst.* **2021**, *52*, 25–45. [CrossRef]
7. Xu, Y.; Chen, X.; Ding, L.; Kong, C.H. Allelopathy and allelochemicals in grasslands and forests. *Forests* **2023**, *14*, 562. [CrossRef]
8. Dudley, S.A. Plant cooperation. *AoB Plants* **2015**, *7*, plv113. [CrossRef]
9. Ryan, M.R. Crops better when grown together. *Nat. Sustain.* **2021**, *4*, 926–927. [CrossRef]
10. Fréville, H.; Montazeaud, G.; Forst, E.; David, J.; Papa, R.; Tenaillon, M. Shift in beneficial interactions during crop evolution. *Evol. Appl.* **2022**, *15*, 905–918. [CrossRef]
11. Biedrzycki, M.L.; Bais, H.P. Kin recognition in plants: A mysterious behaviour unsolved. *J. Exp. Bot.* **2010**, *61*, 4123–4128. [CrossRef] [PubMed]
12. Bilas, R.D.; Bretman, A.; Bennett, T. Friends, neighbours and enemies: An overview of the communal and social biology of plants. *Plant Cell Environ.* **2021**, *44*, 997–1013. [CrossRef]
13. Dudley, S.A.; Murphy, G.P.; File, A.L. Kin recognition and competition in plants. *Funct. Ecol.* **2013**, *27*, 898–906. [CrossRef]
14. Xu, Y.; Cheng, H.F.; Kong, C.H.; Meiners, S.J. Intra-specific kin recognition contributes to inter-specific allelopathy: A case study of allelopathic rice interference with paddy weeds. *Plant Cell Environ.* **2021**, *44*, 3479–3491. [CrossRef] [PubMed]
15. Nakamura, R.R. Plant kin selection. *Evol. Theory* **1980**, *5*, 113–117.
16. Biedrzycki, M.L.; Bais, H.P. Kin recognition in plants: Did we learn anything from roots? *Front. Ecol. Evol.* **2022**, *9*, 785019. [CrossRef]
17. Dudley, S.A.; File, A.L. Kin recognition in an annual plant. *Biol. Lett.* **2007**, *3*, 435–438. [CrossRef]
18. Biedrzycki, M.L.; Jilany, T.A.; Dudley, S.A.; Bais, H.P. Root exudates mediate kin recognition in plants. *Commun. Integr. Biol.* **2010**, *3*, 28–35. [CrossRef]

19. Lepik, A.; Abakumova, M.; Zobel, K.; Semchenko, M. Kin recognition is density-dependent and uncommon among temperate grassland plants. *Funct. Ecol.* **2012**, *26*, 1214–1220. [CrossRef]
20. Pereira, M.L.; Sadras, V.O.; Batista, W.; Casal, J.J.; Hall, A.J. Light mediated self-organization of sunflower stands increases oil yield in the field. *Proc. Natl. Acad. Sci. USA* **2017**, *114*, 7975–7980. [CrossRef]
21. Yang, X.F.; Li, L.L.; Xu, Y.; Kong, C.H. Kin recognition in rice (*Oryza sativa*) lines. *New Phytol.* **2018**, *220*, 567–578. [CrossRef] [PubMed]
22. Pezzola, E.; Pandolfi, C.; Mancuso, S. Resource availability affects kin selection in two cultivars of *Pisum sativum*. *Plant Growth Regul.* **2020**, *90*, 321–329. [CrossRef]
23. Bhatt, V.M.; Khandelwal, A.; Dudley, S.A. Kin recognition, not competitive interactions predicts root allocation in young *Cakile edentula* seeding pairs. *New Phytol.* **2011**, *189*, 1135–1142. [CrossRef]
24. Torices, R.; Gómez, J.M.; Pannell, J.R. Kin discrimination allows plants to modify investment towards pollinator attraction. *Nat. Commun.* **2018**, *9*, 2018. [CrossRef] [PubMed]
25. Malik, M.S.; Burgos, N.R.; Talbert, R.E. Confirmation and control of propanil-resistant and quinclorac-resistant barnyardgrass (*Echinochloa crus-galli*) in rice. *Weed Technol.* **2010**, *24*, 226–233. [CrossRef]
26. Norsworthy, J.K.; Wilson, M.J.; Scott, R.C.; Gbur, E.E. Herbicidal activity on acetolactate synthase-resistant barnyardgrass (*Echinochloa crus-galli*) in Arkansas, USA. *Weed Biol. Manag.* **2014**, *14*, 50–58. [CrossRef]
27. Goh, S.S.; Vila-Aiub, M.M.; Busi, R.; Powles, S.B. Glyphosate resistance in *Echinochloa colona*: Phenotypic characterisation and quantification of selection intensity. *Pest Manag. Sci.* **2016**, *72*, 67–73. [CrossRef]
28. Yang, X.F.; Kong, C.H.; Yang, X.; Li, Y.F. Interference of allelopathic rice with penoxsulam-resistant barnyardgrass. *Pest Manag. Sci.* **2017**, *73*, 2310–2317. [CrossRef]
29. Gealy, D.R.; Anders, M.; Watkins, B.; Duke, S. Crop performance and weed suppression by weed-suppressive rice cultivars in furrow- and flood-irrigated systems under reduced herbicide inputs. *Weed Sci.* **2014**, *62*, 303–320. [CrossRef]
30. Li, L.L.; Zhao, H.H.; Kong, C.H. (−)-Loliolide, the most ubiquitous lactone, is involved in barnyardgrass-induced rice allelopathy. *J. Exp. Bot.* **2020**, *71*, 1540–1550. [CrossRef] [PubMed]
31. Sun, B.; Wang, P.; Kong, C.H. Plant-soil feedback in the interference of allelopathic rice with barnyardgrass. *Plant Soil* **2014**, *377*, 309–321. [CrossRef]
32. Yang, X.F.; Kong, C.H. Interference of allelopathic rice with paddy weeds at the root level. *Plant Biol.* **2017**, *19*, 584–591. [CrossRef] [PubMed]
33. Kong, C.H.; Zhang, S.Z.; Li, Y.H.; Xia, Z.C.; Yang, X.F.; Meiners, S.J.; Wang, P. Plant neighbor detection and allelochemical response are driven by root-secreted signaling chemicals. *Nat. Commun.* **2018**, *9*, 3867. [CrossRef] [PubMed]
34. Li, L.L.; Li, Z.; Lou, Y.G.; Meiners, S.J.; Kong, C.H. (−)-Loliolide is a general signal of plant stress that activates jasmonate-related responses. *New Phytol.* **2022**. [CrossRef] [PubMed]
35. Anten, N.P.R.; Chen, B.J.W. Detect thy family; Mechanisms, ecology, and agricultural aspects of kin recognition. *Plant Cell Environ.* **2021**, *44*, 1059–1071. [CrossRef]
36. Anten, N.P.R.; Chen, B.J.W. Kin discrimination in allelopathy and consequences for agricultural weed control. *Plant Cell Environ.* **2021**, *44*, 3705–3708. [CrossRef]
37. Baucom, R.S. Evolutionary and ecological insights from herbicide-resistant weeds: What have we learned about plant adaptation, and what is left to uncover? *New Phytol.* **2019**, *223*, 68–82. [CrossRef]
38. Belter, P.R.; Cahill Jr, J.F. Disentangling root system responses to neighbours: Identification of novel root behavioural strategies. *AoB Plants* **2015**, *7*, plv059. [CrossRef]
39. Goddard, E.L.; Varga, S.; John, E.A.; Soulsbury, C.D. Evidence for root kin recognition in the clonal plant species *Glechoma hederacea*. *Front. Ecol. Evol.* **2020**, *8*, 578141. [CrossRef]
40. Falik, O.; Hoffmann, I.; Novoplansky, A. Say it with flowers: Flowering acceleration by root communication. *Plant Signal. Behav.* **2014**, *9*, e28258. [CrossRef]
41. Li, F.L.; Chen, X.; Luo, H.M.; Meiners, S.J.; Kong, C.H. Root-secreted (−)-loliolide modulates both belowground defense and aboveground flowering in Arabidopsis and tobacco. *J. Exp. Bot.* **2023**, *74*, 964–975. [CrossRef] [PubMed]
42. Bawa, K.S. Kin selection and the evolution of plant reproductive traits. *Proc. Roy. Soc. B* **2016**, *283*, 20160789. [CrossRef] [PubMed]
43. Shivaprakash, K.N.; Bawa, K.S. The evolution of placentation in flowering plants: A possible role for kin selection. *Front. Ecol. Evol.* **2022**, *10*, 784077.
44. Munguía-Rosas, M.A.; Ollerton, J.; Parra-Tabla, V.; De-Nova, J.A. Meta-analysis of phenotypic selection on flowering phenology suggests that early flowering plants are favoured. *Ecol. Lett.* **2011**, *14*, 511–521. [CrossRef]
45. Crepy, M.A.; Casal, J.J. Photoreceptor-mediated kin recognition in plants. *New Phytol.* **2015**, *205*, 329–338. [CrossRef]
46. Wang, N.Q.; Kong, C.H.; Wang, P.; Meiners, S.J. Root exudate signals in plant–plant interactions. *Plant Cell Environ.* **2021**, *44*, 1044–1058. [CrossRef]
47. Semchenko, M.; Saar, S.; Lepik, A. Plant root exudates mediate neighbour recognition and trigger complex behavioural changes. *New Phytol.* **2014**, *204*, 631–637.
48. Bateson, P. Kin recognition. *Science* **1984**, *224*, 446. [CrossRef]
49. Lehmann, L.; Perrin, N. Altruism, dispersal, and phenotype-matching kin recognition. *Am. Nat.* **2002**, *159*, 451–468. [CrossRef]

50. Karban, R.; Shiojiri, K.; Ishizaki, S.; Wetzel, W.C.; Evans, R.Y. Kin recognition affects plant communication and defence. *Proc. Roy. Soc. B* **2013**, *280*, 20123062.
51. Yamawo, A. Relatedness of neighboring plants alters the expression of indirect defense traits in an extrafloral nectary-bearing plant. *Evol. Biol.* **2015**, *42*, 12–19. [CrossRef]
52. Kalske, A.; Shiojiri, K.; Uesugi, A.; Sakata, Y.; Morrell, K.; Kessler, A. Insect herbivory selects for volatile-mediated plant-plant communication. *Cur. Biol.* **2019**, *29*, 3128–3133. [CrossRef] [PubMed]
53. Kong, C.H.; Chen, X.H.; Hu, F.; Zhang, S.Z. Breeding of commercially acceptable allelopathic rice cultivars in China. *Pest Manag. Sci.* **2011**, *67*, 1100–1106. [CrossRef] [PubMed]
54. Wang, C.Y.; Li, L.L.; Meiners, S.J.; Kong, C.H. Root placement patterns in allelopathic plant -plant interactions. *New Phytol.* **2023**, *237*, 563–575. [CrossRef] [PubMed]

Article

Interspecific Drought Cuing in Plants

Omer Falik [1,2] and Ariel Novoplansky [2,*]

[1] Achva Academic College, Arugot 7980400, Israel
[2] Mitrani Department of Desert Ecology, Blaustein Institutes for Desert Research, Ben-Gurion University of the Negev, Sede Boqer Campus, Midreshet Ben-Gurion, 8499000, Israel
* Correspondence: anovopla@bgu.ac.il; Tel.: +972-52-5793045

Abstract: Plants readily communicate with their pollinators, herbivores, symbionts, and the predators and pathogens of their herbivores. We previously demonstrated that plants could exchange, relay, and adaptively utilize drought cues from their conspecific neighbors. Here, we studied the hypothesis that plants can exchange drought cues with their interspecific neighbors. Triplets of various combinations of split-root *Stenotaphrum secundatum* and *Cynodon dactylon* plants were planted in rows of four pots. One root of the first plant was subjected to drought while its other root shared its pot with one of the roots of an unstressed target neighbor, which, in turn, shared its other pot with an additional unstressed target neighbor. Drought cuing and relayed cuing were observed in all intra- and interspecific neighbor combinations, but its strength depended on plant identity and position. Although both species initiated similar stomatal closure in both immediate and relayed intraspecific neighbors, interspecific cuing between stressed plants and their immediate unstressed neighbors depended on neighbor identity. Combined with previous findings, the results suggest that stress cuing and relay cuing could affect the magnitude and fate of interspecific interactions, and the ability of whole communities to endure abiotic stresses. The findings call for further investigation into the mechanisms and ecological implications of interplant stress cuing at the population and community levels.

Keywords: *Cynodon dactylon*; drought stress; phenotypic plasticity; interspecific plant communication; root communication; *Stenotaphrum secundatum*; stomata; stress cues

Citation: Falik, O.; Novoplansky, A. Interspecific Drought Cuing in Plants. *Plants* **2023**, *12*, 1200. https://doi.org/10.3390/plants12051200

Academic Editors: Frantisek Baluska and Gustavo Maia Souza

Received: 7 October 2022
Revised: 17 February 2023
Accepted: 28 February 2023
Published: 6 March 2023

1. Introduction

Coping with environmental variation is one of the most prominent and ubiquitous challenges of biological existence. At the population level, it is one of the major drivers of Darwinian evolution and genetic diversity [1]. However, at spatiotemporal scales relevant to individual organisms, behavioral responses and phenotypic plasticity have clear adaptive advantages [2–10]. In contrast to natural selection that does not require organismal awareness or involvement, adaptive behavior and phenotypic plasticity are based on the ability of individuals to perceive and integrate accurate information relevant to challenges and opportunities in their immediate environments [11,12]. Unlike rapid biochemical, physiological, and some behavioral responses, developmental plasticity could require substantial time, and, thus, relevant information must pertain to anticipated rather than to prevalent conditions. Accordingly, given sufficiently tight corrections between predictive cues and signals, and ensuing conditions, natural selection is expected to favor preemptive responses to forthcoming rather than to current conditions [8,13–16]. Highly relevant and reliable information is often available from conspecific neighbors that already experience environmental changes and challenges, such as in the case of bacterial quorum sensing [17] or interplant communication of warning cues related to herbivory (e.g., refs. [18–20]), salinity [21], or pathogen attack [22–24].

Useful information could also be perceived from other species. For example, many animals readily respond to allospecific alarm calls related to the presence of predators [25–28], (but see [29]). As most plants require the same resources and are susceptible to similar stresses and enemies, it is not surprising that many plants are able to take advantage of cues emitted from taxonomically remote neighbors. Besides their ability to communicate with their pollinators, herbivores, predators, and pathogens of their herbivores, and myriad symbionts, many plants can emit, eavesdrop on, and respond to a large variety of cues released from plants belonging to other species [30]. In one of the most studied systems, wild *Nicotiana attenuate* plants have been demonstrated to incur significantly lower herbivore damage when receiving volatile cues from neighboring damaged *Artemisia tridentata* shrubs [31]. Similarly, UV-C-stressed *Arabidopsis thaliana* and *Nicotiana tabacum* plants readily exchange volatile cues with their neighbors regardless of their taxonomic identity [32]. Interestingly, in some cases, interspecific communication is facilitated by mycorrhizal networks, indicating that environmental information can be readily transmitted and relayed across kingdom barriers, e.g., refs. [33,34].

We previously demonstrated that unstressed *Pisum sativum* plants rapidly close their stomata in response to interplant cuing from drought-stressed conspecific neighbors, and that 'relay cuing' can elicit stomatal closure in multiple increasingly distant unstressed plants [35]. Interplant drought cuing and relay cuing were only observed between plants that shared their rooting media, implying reliance on root–root communication rather than on aboveground volatile cuing [35]. The involvement of ABA in interplant drought cuing has been demonstrated from experiments in which interplant drought cuing was drastically reduced in plants with diminished ABA synthesis [16] and additional analyses showing elevated ABA levels in the rhizosphere of both drought-stressed plants and their unstressed neighbors [36].

An additional study demonstrated both direct and relayed interplant drought cuing in the wild plants *Cynodon dactylon, Digitaria sanguinalis,* and *Stenotaphrum secundatum* [37]. In a recent study, we have shown that cuing from drought-stressed plants significantly increased the survival of both directly and relayed-cued target plants under drought [16].

Here, we tested the hypotheses that plants are able to perceive and respond to both direct and relayed interspecific drought cuing and that responsiveness to drought cuing relies on the identity of the emitting plant and its inherent drought tolerance. Responsiveness to drought cuing was studied by following stomatal aperture in unstressed relatively xeric *C. dactylon* and relatively mesic *S. secundatum* plants that were subjected to either direct or relayed drought cues from intra- or interspecific neighbors.

2. Results

2.1. Intraspecific Drought Cuing

As expected, both *S. secundatum* and *C. dactylon* demonstrate interplant drought cuing and relayed cuing. Subjecting one of the roots of the IND plant to 60 min of drought (pot 1, Figure 1) causes 28–45% and 30–39% decreases in stomatal aperture, in drought-treated or cued *S. secundatum,* and *C. dactylon* triplets, respectively, compared to their unstressed controls (Figure 2a,d). Stomatal closure in response to drought was non-significantly different in the two species (Student's *t*-test: $t = 0.22$, $p = 0.413$), or in plants located at different positions in the triplet (stressed IND, directly cued T1, and relayed T2 neighbors) in either species (one-way ANOVA: *S. secundatum*: $F = 2.13$, $p = 0.134$; *C. dactylon*: $F = 1.03$, $p = 0.367$).

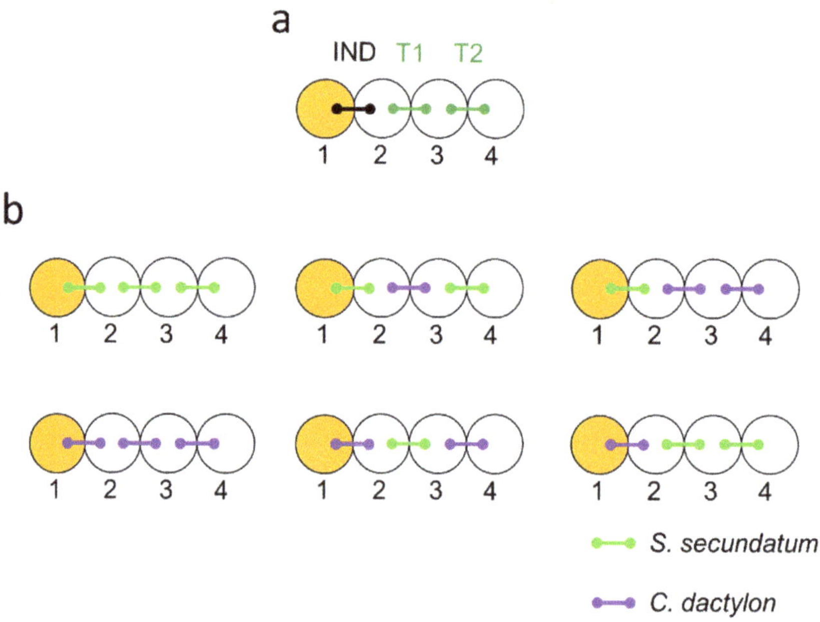

Figure 1. Testing for interspecific drought cuing—the experimental setup. Circles represent pots and connector lines represent split-root plants. Plants neighboring an externally stressed plant (IND) shared their pots with their immediate unstressed (T1) neighbors and target plants either shared their other pots with another target plant (T1) or only with their immediate neighbors (T2) (**a**). Drought or control treatments were imposed by replacing the water in pot 1 (orange) with either dry (drought) or wet (benign control) vermiculite–bentonite (VB) mixture for 60 min. Stomatal width was measured in paired triplet sets. interplant drought cuing was tested and compared between treatments in which the identity of the IND and target plants varied (**b**), to reveal the ability of both species to emit and respond to both direct (T1) and relayed (T2) drought cues from either intra- and interspecific neighbors.

2.2. Interspecific Drought Cuing

Interplant drought cuing was observed in all interspecific treatment combinations. Subjecting one of the roots of an IND plant of either *S. secundatum* or *C. dactylon* to drought caused significant stomatal closure in both directly cued T1 and relayed cued T2 plants, regardless of species combination, but its strength varied with plant identity and position (Figure 2b,c,e,f).

Drought treatment causes similar 34–40% decreases in stomatal aperture in stressed (IND) *C. dactylon* plants and their unstressed (T1) *S. secundatum* neighbors (Figure 2e,f; non-significant difference in stomatal closure between IND and T1 plants in both C.d.–S.s.–C.d, and C.d.–S.s.–S.s. treatments; Table 1). Drought-stressed (IND) *S. secundatum* plants demonstrate 51–67% decreases in relative stomatal aperture when immediately neighboring *C. dactylon* T1 plants (Figure 2b,c; S.s.–C.d.–S.s.— Student's *t*-test: *t* = 1.46, *p* = 0.079; S.s.–C.d.–C.d.— Student's *t*-test: *t* = 3.00, *p* = 0.003); however, these decreases mostly resulted from increased stomatal aperture in the control S.s. IND plants when neighboring C.d. T1 plants (Figure 2b,c).

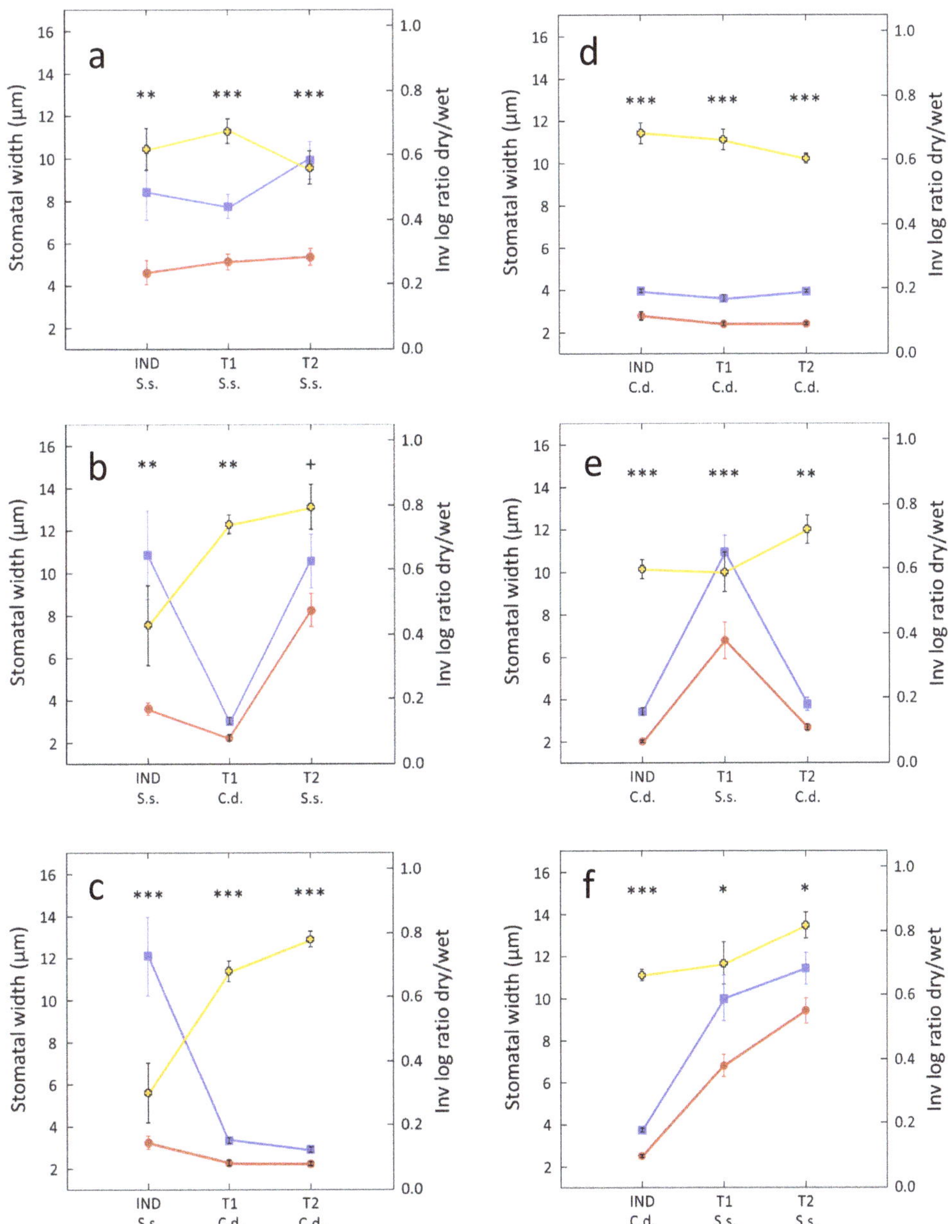

Figure 2. Testing for interplant drought cuing in *S. secundatum* (S.s.) and *C. dactylon* (C.d.). Data (mean ± 1 SEM; *n* = 12), are for stomatal width of treated IND plants and their untreated (T1, T2) target neighbors in benign control triplets (blue lines) and drought-cuing triplets (red lines). IND-T1-T2 triplets comprised different combinations of S.s and C.d.: S.s-S.s-S.s (**a**); S.s-C.d.-S.s (**b**); S.s-C.d.-C.d. (**c**); C.d.-C.d.-C.d. (**d**); C.d-S.s.-C.d. (**e**); C.d-S.s.-S.s. (**f**). To properly compare the relative effects of drought (IND plants), drought cuing (T1 plants), and relayed cuing (T2 plants) on stomatal aperture regardless of absolute stomatal width, a second y-axis (right) presents the inverse of the log of (treated/control) ratio of each data pair (yellow lines). Paired *t*-test, + 0.1 < *p* > 0.05; * *p* < 0.05; ** *p* < 0.01; *** *p* < 0.001.

Table 1. Comparing stomatal closure between drought-treated (IND) plants and their cued (T1) and relayed cued (T2) neighbors in different interspecific triplet sets.

Triplet Combination IND–T1–T2	Difference in Relative Stomatal Width between T1 or T2 and IND	Student's *t*-Test
T1 vs. IND		
S.s.–C.d.–S.s.	+30%	$p = 0.075$
S.s–C.d.–C.d.	+78%	$p = 0.005$
C.d.–S.s.–C.d.	-6%	$p = 0.338$
C.d.–S.s.–S.s.	+19%	$p = 0.177$
T2 vs. IND		
S.s.–C.d.–S.s.	+57%	$p = 0.028$
S.s–C.d.–C.d.	+100%	$p < 0.001$
C.d.–S.s.–C.d.	+24%	$p = 0.071$
C.d.–S.s.–S.s.	+28%	$p = 0.027$

In contrast to monospecific triplet combinations (Figure 2a,d), both species demonstrate significant but weaker stomatal closure in response to relayed cuing (T2 plants) than to direct interspecific drought cuing (T1 plants), regardless of triplet combination (Figure 2b,c,e,f), with significant or marginally significant differences in stomatal closure between IND and T2 plants in all interspecific triplet combinations (Table 1).

3. Discussion

Adaptive phenotypic plasticity relays on the perception and integration of information pertaining to anticipated internal physiological states, growth conditions, challenges, and opportunities [8,13,15]. We previously demonstrated that certain plants can anticipatorily adapt to impending drought by perceiving root cuing from their stressed conspecific neighbors [16,35,37]. Although we cannot rule out the possibility that the observed interplant cuing involves volatile organic compounds, e.g., refs. [38,39], our previous studies demonstrated that, at least in *Pisum sativum*, interplant drought cuing is mostly, if not solely, based on inter-root cuing as it could be only observed between plants that shared their rooting media [16].

Here we demonstrate, for the first time, the existence of interspecific drought cuing and relayed cuing in plants. As expected (see [37]), both drought-stressed *S. secundatum* and *C. dactylon* plants, and their unstressed conspecific neighbors closed their stomata to similar extents (Figure 2a,d). In contrast to our expectations, we could not find a significant greater effect of drought cuing of the more xeric *C. dactylon* in comparison to its more mesic *S. secundatum* counterpart.

The greater relative stomatal width in stressed IND *S. secundatum* plants when neighboring unstressed *C. daclylon* (Figure 2b,c) could be only partially attributed to a decreased stomatal aperture in these plants but mostly to increases in stomatal aperture in the unstressed IND S. secundatum plants, suggesting a strong dependance of the responses of drought-stressed *S. secundatum* on the identity of its immediate (T1) neighbors.

As each plant is both perceiving and emitting stress cues, stress cuing and relay cuing could be expected to elicit cuing amplification by self-propagation (see [40] for a similar phenomenon related to volatile defense cuing), with an increasingly stronger response of plants to their own (echoed) cues. Such increased responses are reminiscent of amplified responsiveness of previously primed plants to later challenges such as insect herbivory [41], salt stress [42], or pathogen attack [43]. However, for such a self-amplified cuing system to be reliable, it is essential that plants do not engage in a runaway overly escalating state of alert [44]. Accordingly, it is expected that the level and effectiveness of ongoing stress cuing would strongly depend on eventual materialization of the anticipated stressful conditions, without which they are expected to rapidly habituate and drastically decrease their stress responsiveness over time [37].

Leakiness of honest cues or signals from stressed plants can be only expected where the average fitness benefits of information sharing outweigh their costs [16,37,45]. Although sharing information with potential competitors would be typically selected against, leaking drought cues could be beneficial due to the following reasons:

Neighbor identity: both *S. secundatum* and *C. dactylon* are clonal plants capable of creating large patches where most interactions and information sharing are between clonemates [46,47].

Plant size and integration: the considerable absolute size and longevity of clonal plants imply that the distance between different organs on the same clone could be substantial and the physiological integration of the clone typically deteriorates over time due to disturbances, trampling, or grazing (e.g., ref. [48]). Under such circumstances, exogenous signaling between different parts of the same clone could be more rapid and efficient than endogenous signaling [49,50].

Facilitation: if and to the extent that drought cuing can induce increased water use efficiency and decreased water uptake in receiver plants, drought cuing could alleviate drought and increase survival and performance of larger patches of neighboring plants, regardless of their genetic identity [45]. Such circumstances can be particularly emphasized in extreme arid environments, where the importance and prevalence of facilitation could be greater than those of competitive interactions [51–53].

Diversity and stress tolerance: the possibility that information regarding impending stresses could be exchanged between different community members may not only significantly affect the magnitude and fate of interspecific interactions, but also the ability of whole communities to tolerate or resist abiotic stresses [54,55]. Recent studies have shown that increasing species richness could enhance drought tolerance and resistance (e.g., refs. [56–58]. In the context of our findings, and to the extent that they are indicatory of fitness-related implications [16], the potential advantages of interspecific drought cuing could further outweigh the possible costs of sharing viable information with potential genetically alien competitors.

Our findings suggest that the effectiveness of interplant drought cuing could depend on the identities of the emitter and the receiver plants. Previous studies demonstrate that some plants are able to detect and adaptively respond according to the identity of their neighbors [59–65]. A recent study found that the composition of VOCs emitted from focal plants following herbivory stress was affected by the identity of their neighbors [66]. Our results are consistent with the speculation that responses to stress cues could rely on the identity of the stress cue emitters, though further work is required to study the hypothesis that responsiveness to specific stress cues could depend on the ability of the responding plants to not only perceive stress cues but also to differentially respond according to the abilities of the emitters to tolerate and resist the perceived stress.

Our findings call for further investigation into the mechanisms of intra- and interspecific stress cuing and relayed cuing, in the inherent (G), environmental (E), and interactive G X E contexts of their stress tolerance and resistance. For example, it could be expected that the ability of plants to effectively exchange stress cues and signals depends on the history of their cohabitation in the same ecosystems and geographical ranges (*sensu* [67]).

4. Materials and Methods

4.1. Plant Material

C. dactylon and *S. secundatum* were chosen for their ease of handling, propagation, growth, and their xeric evolutionary backgrounds. We previously demonstrated that both species are able to communicate drought cues with their conspecific neighbors [37]. *C. dactylon* (Bermuda grass) is a prostrate perennial grass, which spreads by means of both stolons and rhizomes [68]. It is common to warm ecosystems in most continents, where it occurs in diverse-types disturbed habitats and desert washes [69–71]. *C. dactylon* cultivars are commonly used as sturdy turf and lawn grasses [72]. *S. secundatum* (buffalo grass) is a perennial stoloniferous grass native to the Caribbean region, South America, and parts of North America and Africa, and it has been introduced to many other geographical regions [73]. *S. secundatum* is a strong competitor and is commonly used as a lawn grass [74]. A few studies demonstrate that *C. dactylon* is more drought-resistant than *S. secundatum* [75–77]. *C dactylon* was collected from natural populations near the Sede Boqer campus, Israel, and *S. secundatum* was acquired from a commercial nursery (Deshe-Itzhar, Kfar Monash, Israel) as sod.

4.2. Experiment Design

Testing for drought cuing required that specific stress-induced plants (IND) would experience a drought event or benign conditions while their neighboring target plants (T1, T2) would only experience cuing from the IND plants (Figure 1). This was achieved by using triplets of various combinations of split-root *S. secundatum* and *C. dactylon* plants planted in rows of four pots (Figure 1). One of the roots of the IND plant was subjected to either drought or benign conditions while its other root shared a pot with one of the roots of its nearest unstressed neighbor (T1). The other root of T1 shared its pot with one of the roots of an additional unstressed target plant (T2). This configuration permitted T1 to exchange stress cues with both IND and T2, while preventing direct root cuing between IND and T2 and, thus, allowing us to separately study the effects of direct and relayed drought cuing on T1 and T2, respectively ([35]; Figure 1a).

Drought cuing was tested in and compared between the following plant triplet combinations (Figure 1b):

Intraspecific cuing: S.s.–S.s.–S.s., C.d.–C.d.–C.d.;
Interspecific cuing (*S. secundatum* stressed-induced): S.s.–C.s.–S.s., S.s–C.d.–C.d.;
Interspecific cuing (*C. dactylon* stressed-induced): C.d–S.s.–C.d., C.d.–S.s.–S.s.

allowing us to compare both direct and relayed interplant drought cuing between intra- and interspecific neighboring configurations.

4.3. Growth Conditions and Experimental Setup

The plants were grown in a naturally lit greenhouse, partially controlled by an automated pad-and-fan system (Termotecnica pericoli, Albenga, Italy), under 30% sunlight at the Sede Boqer campus, Israel (30°52′ N, 34°47′ E). Plants were vegetatively propagated from 10 *C. dactylon* clones, and an unknown number of *S. secundatum* mother plants. Two-ramet cuttings were planted in moist no. 2 vermiculite and grown in the greenhouse (see above) for 14–21 d, during which each ramet regenerated 3–5 leaves and 4–6 cm long roots.

Triplets of similarly sized two-ramet plants were planted in rows of four 0.2 L, 7 cm diameter, 9 cm high pots (Miniplast, Ein Shemer, Israel). In stoloniferous plants such as *C. dactylon* and *S. secundatum*, resource translocation is commonly acropetal (e.g., ref. [78]) and in response to herbivory stress, systemic warning signals were shown to travel more rapidly acropetally than basipetally [79], implying that planting orientation might affect the rate and effectiveness of signal transmission within and among plants. To increase uniformity and the probability of finding communicative cuing, potential differential effects of axis polarity were avoided by directing the plants so their proximal ramets were rooted in (IND) or nearer (T1–T2) the induction pot (pot 1, Figure 1a).

Upon transplantation into the experimental pots, all roots were trimmed to 3 cm to encourage root regeneration and intermingling in the shared target pots. Plants were allowed to regenerate and habituate to the experimental systems for 14 days before the onset of the experiment, during which time they were individually irrigated to field capacity with 100 mL nutrient solution (Ecogan, Caesarea, Israel) every 3–4 days. Pots were individually wrapped with aluminum foil to block light from reaching the roots. Pots were individually drained into separate drip trays to prevent the seepage and capillary migration of root exudates between the pots.

To allow rapid and non-destructive initiation of drought conditions, the induction pot (pot 1, Figure 1) was filled with tap water and the other pots were filled with a commercial soil mixture (Deshanit, Be'er Yaakov, Israel).

The experiment was conducted in the greenhouse starting on April 29, 2012. Drought stress was inflicted to the proximal root of the IND plant as described in [35], by carefully pumping the water from pot 1 (orange; Figure 1a) using a flexible-tip syringe and filling it with 8 g of either dry or wet mixture of 4:1 mixture of no. 1 vermiculite (Agrekal, Habonim, Israel) and bentonite (Minerco, Netanya, Israel) (VB) for 60 min [35,37]. To account for potential handing effects, control (benign) sets were induced by filling pot 1 with a mixture of wet VB (5.5 g VB and 45 mL distilled water), reflecting the effects of drought cuing rather than potential responses to the physical handing of the plants or the chemical components of VB. Accordingly, stomatal aperture in the IND plant reflected the direct effects of partial (only to one of the two roots) drought, and stomatal aperture in the T1 and T2 plants reflected the effects of direct and relayed drought cuing, respectively.

4.4. Stomata Measurements

Stomatal aperture was measured for its highly sensitive responsiveness to various environmental stresses, especially drought, e.g., ref. [80]. Stomatal aperture was estimated from epidermal impressions following [35]: negative impressions of the lower surfaces of 1–2 fully unfurled 20–30 mm^2 leaves of each sampled plant were obtained using a fresh mixture of vinyl polysiloxane dental impression silicone elastomer (Elite HD+, Badia Polesine, Rovigo, Italy). Following hardening, a positive impression of the leaf surface was obtained from the silicone molds using clear nail polish, which resulted in transparent preparations suitable for microscopic examination [35]. Stomatal aperture was estimated using AxioVision software (Carl Zeiss MicroImaging, Thornwood, NY, USA) on digital images of the nail-polish microscopic preparations. Average stomatal width was calculated from the data obtained from at least 10 stomata per plant, selected haphazardly from 2–5 0.02 mm^2 areas in the center of each microscopic preparation. To avoid observer bias, all samples were handled and analyzed using a single-blind protocol, whereby the observer was unaware of the treatment identity of the samples.

4.5. Data Analyses

Stomata size greatly differed between the studied species, with *S. secundatum* having ca. double stomatal aperture than *C. dactylon* (Figure 2a,d). The main studied treatment effects were analyzed by pairwise comparisons of stomatal aperture between plant triplets in which one of the roots of the IND plant (pot 1) was treated by drought and a control triplet in which all plants were kept under well-hydrated benign conditions (Figure 1). To easily visualize and properly compare the treatment effects on the two species, we also calculated the inverse logged ratios between stomatal aperture of the treated and the control plants to provide equal weights to cases in which either the treated or the control plants in each replication pair had a larger average stomatal aperture then its counterpart [81]. Differences between treated (drought, drought cuing) and control (benign conditions) groups were tested using paired *t*-tests. Comparisons between non-paired treatment groups, such as between relative stomatal width (inv LOG ratio dry/wet; Figure 2) were carried out using either Student's t-tests, where comparing two treatment groups or one-way ANOVAs when comparing more than two treatment groups [82]. All statistical analyses were conducted using SYSTAT 13 (SPSS).

Author Contributions: O.F. conducted the experiments, A.N. conceived the project, designed the experiment, analyzed the results, and wrote the paper. All authors have read and agreed to the published version of the manuscript.

Funding: The study was partially supported by a research grant from the Israel Science Foundation to A.N.

Institutional Review Board Statement: Not applicable.

Informed Consent Statement: Not applicable.

Data Availability Statement: Not applicable.

Acknowledgments: We thank Ishai Hoffman, Miri Vanunu, and Oron Goldstein for the technical assistance.

Conflicts of Interest: The authors declare no conflict of interest.

References

1. Messer, P.W.; Ellner, S.P.; Hairston, N.G. Can Population Genetics Adapt to Rapid Evolution. *Trends Genet.* **2016**, *32*, 408–418. [CrossRef] [PubMed]
2. Bradshaw, A.D. Evolutionary Significance of Phenotypic Plasticity in Plants. *Adv. Gen.* **1965**, *13*, 115–155. [CrossRef]
3. Levins, R. *Evolution in Changing Environments*; Princeton University Press: Princeton, NJ, USA, 1968.
4. Schlichting, C.D. The Evolution of Phenotypic Plasticity in Plants. *Annu. Rev. Ecol. Syst.* **1986**, *17*, 667–693. [CrossRef]
5. Gavrilets, S.; Scheiner, S.M. The Genetics of Phenotypic Plasticity. VI. Theoretical Predictions for Directional Selection. *J. Evol. Biol.* **1993**, *6*, 49–68. [CrossRef]
6. Alpert, P.; Simms, E. The Relative Advantages of Plasticity and Fixity in Different Environments: When Is It Good for a Plant to Adjust. *Evol. Ecol.* **2002**, *16*, 285–297. [CrossRef]
7. Lande, R. Adaptation to an Extraordinary Environment by Evolution of Phenotypic Plasticity and Genetic Assimilation. *J. Evol. Biol.* **2009**, *22*, 1435–1446. [CrossRef]
8. Novoplansky, A. Picking Battles Wisely: Plant Behaviour under Competition. *Plant. Cell Environ.* **2009**, *32*, 726–741. [CrossRef]
9. Botero, C.A.; Weissing, F.J.; Wright, J.; Rubenstein, D.R. Evolutionary Tipping Points in the Capacity to Adapt to Environmental Change. *Proc. Natl. Acad. Sci. USA* **2015**, *112*, 184–189. [CrossRef]
10. Tufto, J. Genetic Evolution, Plasticity, and Bet-Hedging as Adaptive Responses to Temporally Autocorrelated Fluctuating Selection: A Quantitative Genetic Model. *Evolution* **2015**, *69*, 2034–2049. [CrossRef]
11. Pigliucci, M. *Phenotypic Plasticity: Beyond Nature and Nurture*; Johns Hopkins University Press: Baltimore, MD, USA, 2001.
12. Sultan, S.E. Plasticity as an Intrinsic Property of Organisms. In *Phenotypic Plasticity and Evolution: Causes, Consequences, Controversies*; Pfennig, D., Ed.; CRC Press: Boca Raton, FL, USA, 2021; pp. 3–24. [CrossRef]
13. Aphalo, P.J.; Ballare, C.L. On the Importance of Information-Acquiring Systems in Plant-Plant Interactions. *Funct. Ecol.* **1995**, *9*, 5–14. [CrossRef]
14. Shemesh, H.; Rosen, R.; Eshel, G.; Novoplansky, A.; Ovadia, O. The Effect of Steepness of Temporal Resource Gradients on Spatial Root Allocation. *Plant Signal. Behav.* **2011**, *6*, 1356–1360. [CrossRef]
15. Novoplansky, A. Future Perception in Plants. In *Anticipation Across Disciplines*; Nadin, M., Ed.; Springer International Publishing: Berlin/Heidelberg, Germany, 2016; pp. 57–70. [CrossRef]
16. Falik, O.; Mauda, S.; Novoplansky, A. The Ecological Implications of Interplant Drought Cuing. *J. Ecol.* **2022**, *111*, 23–32. [CrossRef]
17. Miller, M.B.; Bassler, B.L. Quorum Sensing in Bacteria. *Annu. Rev. Microbiol.* **2001**, *55*, 165–199. [CrossRef]
18. Bruin, J.; Dicke, M.; Sabelis, M. Plants Are Better Protected against Spider-Mites after Exposure to Volatiles from Infested Conspecifics. *Experientia* **1992**, *48*, 525–529. [CrossRef]
19. Arimura, G.; Matsui, K.; Takabayashi, J. Chemical and Molecular Ecology of Herbivore-Induced Plant Volatiles: Proximate Factors and Their Ultimate Functions. *Plant Cell Physiol.* **2009**, *50*, 911–923. [CrossRef]
20. Song, Y.Y.; Zeng, R.S.; Xu, J.F.; Li, J.; Shen, X.; Yihdego, W.G. Interplant Communication of Tomato Plants through Underground Common Mycorrhizal Networks. *PLoS ONE* **2010**, *5*, e13324. [CrossRef]
21. Caparrotta, S.; Boni, S.; Taiti, C.; Palm, E.; Mancuso, S.; Pandolfi, C. Induction of Priming by Salt Stress in Neighboring Plants. *Environ. Exp. Bot.* **2017**, *147*, 261–270. [CrossRef]
22. Shulaev, V.; Silverman, P.; Raskin, I. Airborne Signalling by Methyl Salicylate in Plant Pathogen Resistance. *Nature* **1997**, *385*, 718–721. [CrossRef]
23. Yi, H.S.; Heil, M.; Adame-Álvarez, R.M.; Ballhorn, D.J.; Ryu, C.M. Airborne Induction and Priming of Plant Defenses against a Bacterial Pathogen. *Plant Physiol.* **2009**, *151*, 2152–2161. [CrossRef]
24. Gilbert, L.; Johnson, D. Plant–Plant Communication Through Common Mycorrhizal Networks. *Adv. Bot. Res.* **2017**, *82*, 83–97. [CrossRef]
25. Oda, R.; Masataka, N. Interspecific Responses of Ringtailed Lemurs to Playback of Antipredator Alarm Calls Given by Verreaux's Sifakas. *Ethology* **1996**, *102*, 441–453. [CrossRef]
26. Zuberbühler, K. Interspecies Semantic Communication in Two Forest Primates. *Proc. Biol. Sci.* **2000**, *267*, 713–718. [CrossRef] [PubMed]
27. Dawson Pell, F.S.E.; Potvin, D.A.; Ratnayake, C.P.; Fernández-Juricic, E.; Magrath, R.D.; Radford, A.N. Birds Orient Their Heads Appropriately in Response to Functionally Referential Alarm Calls of Heterospecifics. *Anim. Behav.* **2018**, *140*, 109–118. [CrossRef]
28. Carlson, N.; Greene, E.; Templeton, C. Nuthatches Vary Their Alarm Calls Based upon the Source of the Eavesdropped Signals. *Nat. Commun.* **2020**, *11*, 526. [CrossRef] [PubMed]

29. Ton, J.; D'Alessandro, M.; Jourdie, V.; Jakab, G.; Karlen, D.; Held, M.; Mauch-Mani, B.; Turlings, T.C.J. Priming by Airborne Signals Boosts Direct and Indirect Resistance in Maize. *Plant J.* **2007**, *49*, 16–26. [CrossRef]
30. Karban, R. Plant Communication. *Annu. Rev. Ecol. Evol. Syst.* **2021**, *52*, 1–24. [CrossRef]
31. Karban, R.; Baldwin, I.; Baxter, K.J.; Laue, G.; Felton, G. Communication between Plants: Induced Resistance in Wild Tobacco Plants Following Clipping of Neighboring Sagebrush. *Oecologia* **2000**, *125*, 66–71. [CrossRef]
32. Yao, Y.; Danna, C.H.; Zemp, F.J.; Titov, V.; Ciftci, O.N.; Przybylski, R.; Ausubel, F.M.; Kovalchuk, I. UV-C–Irradiated Arabidopsis and Tobacco Emit Volatiles That Trigger Genomic Instability in Neighboring Plants. *Plant Cell* **2011**, *23*, 3842–3852. [CrossRef]
33. Pearse, I.S.; Porensky, L.M.; Yang, L.H.; Stanton, M.L.; Karban, R.; Bhattacharyya, L.; Cox, R.; Dove, K.; Higgins, A.; Kamoroff, C.; et al. Complex Consequences of Herbivory and Interplant Cues in Three Annual Plants. *PLoS ONE* **2012**, *7*, e38105. [CrossRef]
34. Gorzelak, M.A.; Asay, A.K.; Pickles, B.J.; Simard, S.W. Inter-Plant Communication through Mycorrhizal Networks Mediates Complex Adaptive Behaviour in Plant Communities. *AoB Plants* **2015**, *7*, plv050. [CrossRef]
35. Falik, O.; Mordoch, Y.; Quansah, L.; Fait, A.; Novoplansky, A. Rumor Has It: Relay Communication of Stress Cues in Plants. *PLoS ONE* **2011**, *6*, e23625. [CrossRef]
36. Falik, O.; Novoplansky, A. Is ABA the exogenous vector of interplant drought cuing. *Plant Signal. Behav.* **2022**, *17*, 2129295. [CrossRef]
37. Falik, O.; Mordoch, Y.; Ben-Natan, D.; Vanunu, M.; Goldstein, O.; Novoplansky, A. Plant Responsiveness to Root-Root Communication of Stress Cues. *Ann. Bot.* **2012**, *110*, 271–280. [CrossRef]
38. Balacey, S.L. Investigating the Role of Volatile Signalling In Plant Responses To Drought. Ph.D Thesis, University of Adelaide, Adelaide, Australia, 2021.
39. Jiang, Y.; Ye, J.; Niinemets, Ü. Dose-dependent methyl jasmonate effects on photosynthetic traits and volatile emissions: Biphasic kinetics and stomatal regulation. *Plant Signal. Behav.* **2021**, *16*, 1917169. [CrossRef]
40. Wenig, M.; Ghirardo, A.; Sales, J.H.; Pabst, E.S.; Breitenbach, H.H.; Antritter, F.; Weber, B.; Lange, B.; Lenk, M.; Cameron, R.K.; et al. Systemic Acquired Resistance Networks Amplify Airborne Defense Cues. *Nat. Commun.* **2019**, *10*, 3813. [CrossRef]
41. Wu, J.; Baldwin, I.T. New Insights into Plant Responses to the Attack from Insect Herbivores. *Annu. Rev. Genet.* **2010**, *44*, 1–24. [CrossRef]
42. Johnson, R.; Puthur, J.T. Seed Priming as a Cost Effective Technique for Developing Plants with Cross Tolerance to Salinity Stress. *Plant Physiol. Biochem.* **2021**, *162*, 247–257. [CrossRef]
43. Bernsdorff, F.; Döring, A.C.; Gruner, K.; Schuck, S.; Bräutigam, A.; Zeier, J. Pipecolic Acid Orchestrates Plant Systemic Acquired Resistance and Defense Priming via Salicylic Acid-Dependent and -Independent Pathways. *Plant Cell* **2016**, *28*, 102–129. [CrossRef]
44. Farmer, E.E. Jasmonate Perception Machines. *Nature* **2007**, *448*, 659–660. [CrossRef]
45. Dicke, M.; Bruin, J. Chemical Information Transfer between Plants: Back to the Future. *Biochem. Syst. Ecol.* **2001**, *29*, 981–994. [CrossRef]
46. Cheplick, G.P. Sibling Competition Is a Consequence of Restricted Dispersal in an Annual Cleistogamous Grass. *Ecology* **1993**, *74*, 2161–2164. [CrossRef]
47. Herben, T.; Novoplansky, A. Implications of Self/Non-Self Discrimination for Spatial Patterning of Clonal Plants. *Evol. Ecol.* **2008**, *22*, 337–350. [CrossRef]
48. Whigham, D.; Chapa, A. Timing and Intensity of Herbivory: Its Influence on the Performance of Clonal Woodland Herbs. *Plant Species Biol.* **1999**, *14*, 29–37. [CrossRef]
49. Karban, R.; Shiojiri, K.; Huntzinger, M.; McCall, A.C. Damage-induced Resistance in Sagebrush: Volatiles are key to understand Intra- and Interplant Communication. *Ecology* **2006**, *87*, 922–930. [CrossRef] [PubMed]
50. Rodriguez-Saona, C.R.; Rodriguez-Saona, L.E.; Frost, C.J. Herbivore-Induced Volatiles in the Perennial Shrub, Vaccinium Corymbosum, and Their Role in Inter-Branch Signaling. *J. Chem. Ecol.* **2009**, *35*, 163–175. [CrossRef]
51. Bertness, M.D.; Callaway, R. Positive Interactions in Communities. *Trends Ecol. Evol.* **1994**, *9*, 191–193. [CrossRef]
52. Butterfield, B.J. Effects of Facilitation on Community Stability and Dynamics: Synthesis and Future Directions. *J. Ecol.* **2009**, *97*, 1192–1201. [CrossRef]
53. Ploughe, L.W.; Jacobs, E.M.; Frank, G.S.; Greenler, S.M.; Smith, M.D.; Dukes, J.S. Community Response to Extreme Drought (CRED): A Framework for Drought-Induced Shifts in Plant–Plant Interactions. *New Phytol.* **2019**, *222*, 52–69. [CrossRef]
54. Simard, S.W.; Beiler, K.J.; Bingham, M.A.; Deslippe, J.R.; Philip, L.J.; Teste, F.P. Mycorrhizal Networks: Mechanisms, Ecology and Modelling. *Fungal Biol. Rev.* **2012**, *26*, 39–60. [CrossRef]
55. Remke, M.J.; Johnson, N.C.; Wright, J.; Williamson, M.; Bowker, M.A. Sympatric Pairings of Dryland Grass Populations, Mycorrhizal Fungi and Associated Soil Biota Enhance Mutualism and Ameliorate Drought Stress. *J. Ecol.* **2021**, *109*, 1210–1223. [CrossRef]
56. Kotlarz, J.; Nasiłowska, S.; Rotchimmel, K.; Kubiak, K.; Kacprzak, M. Species Diversity of Oak Stands and Its Significance for Drought Resistance. *Forests* **2018**, *9*, 126. [CrossRef]
57. Aguirre, B.A.; Hsieh, B.; Watson, S.J.; Wright, A.J. The Experimental Manipulation of Atmospheric Drought: Teasing out the Role of Microclimate in Biodiversity Experiments. *J. Ecol.* **2021**, *109*, 1986–1999. [CrossRef]

58. Liu, D.; Wang, T.; Peñuelas, J.; Piao, S. Drought Resistance Enhanced by Tree Species Diversity in Global Forests. *Nat. Geosci.* **2022**, *15*, 800–804. [CrossRef]

59. Mahall, B.E.; Callaway, R.M. Root Communication among Desert Shrubs. *Proc. Natl. Acad. Sci. USA* **1991**, *88*, 874–876. [CrossRef]

60. Callaway, R. The Detection of Neighbors by Plants. *Trends Ecol. Evol.* **2002**, *17*, 104–105. [CrossRef]

61. Gruntman, M.; Novoplansky, A. Physiologically Mediated Self/Non-Self Discrimination in Roots. *Proc. Natl. Acad. Sci. USA* **2004**, *101*, 3863–3867. [CrossRef]

62. Mescher, M.; Runyon, J.; De Moraes, C. Plant Host Finding by Parasitic Plants. *Plant Signal. Behav.* **2006**, *1*, 284–286. [CrossRef]

63. Semchenko, M.; Saar, S.; Lepik, A. Plant Root Exudates Mediate Neighbour Recognition and Trigger Complex Behavioural Changes. *New Phytol.* **2014**, *204*, 631–637. [CrossRef]

64. Ninkovic, V.; Markovic, D.; Dahlin, I. Decoding Neighbour Volatiles in Preparation for Future Competition and Implications for Tritrophic Interactions. *Perspect. Plant Ecol. Evol. Syst.* **2016**, *23*, 11–17. [CrossRef]

65. Falik, O.; Reides, P.; Gersani, M.; Novoplansky, A. Self/Non-Self Discrimination in Roots. *J. Ecol.* **2003**, *91*, 525–531. [CrossRef]

66. Kigathi, R.N.; Weisser, W.W.; Reichelt, M.; Gershenzon, J.; Unsicker, S.B. Plant Volatile Emission Depends on the Species Composition of the Neighboring Plant Community. *BMC Plant Biol.* **2019**, *19*, 58. [CrossRef] [PubMed]

67. Mahall, B.E.; Callaway, R.M. Effects of Regional Origin and Genotype on Intraspecific Root Communication in the Desert Shrub *Ambrosia Dumosa* (Asteraceae). *Am. J. Bot.* **1996**, *83*, 93–98. [CrossRef]

68. Fernandez, O.N. Establishment of Cynodon Dactylon from Stolon and Rhizome Fragments. *Weed Res.* **2003**, *43*, 130–138. [CrossRef]

69. Holm, L.G.; Plunknett, D.L.; Pancho, J.V.; Herberger, J.P. *The World's Worst Weeds: Distribution and Biology*; Krieger Publishing Company: Malabar, FL, USA, 1991.

70. Gould, F.W. *Grasses of the Southwestern United States*; The University of Arizona Press: Tucson, AZ, USA, 1951.

71. Parker, K.F. *An Illustrated Guide to Arizona Weeds*; The University of Arizona Press: Tucson, AZ, USA, 1972.

72. Horowitz, M. Bermudagrass (Cynodon Dactylon): A History of the Weed and Its Control in Israel. *Phytoparasitica* **1996**, *24*, 305–320. [CrossRef]

73. CABI. Stenotaphrum Secundatum. Invasive Species Compendium. Available online: www.cabi.org/isc (accessed on 1 October 2022).

74. Busey, P. St. Augustine Grass, Stenotaphrum Secundatum (Walt.) Kuntze. In *Biology, Breeding, and Genetics of Turfgrasses*; Casler, M.D., Duncan, R.R., Eds.; John and Wiley and Sons: Hoboken, NJ, USA, 2003; pp. 309–330.

75. Geren, H.; Avcioglu, R.; Curaoglu, M. Performances of Some Warm-Season Turfgrasses under Mediterranean Conditions. *African J. Biotechnol.* **2009**, *8*, 4469–4474.

76. Fuentealba, M.P.; Zhang, J.; Kenworthy, K.; Erickson, J.; Kruse, J.; Trenholm, L. Transpiration Responses of Warm-Season Turfgrass in Relation to Progressive Soil Drying. *Sci. Hortic.* **2016**, *198*, 249–253. [CrossRef]

77. Kim, K.S.; Beard, J.B. Comparative Drought Resistances Among Eleven Warm-Season Turfgrasses and Associated Plant Parameters. *Weed Turfgrass Sci.* **2018**, *7*, 239–245. [CrossRef]

78. Price, E.A.C.; Hutchings, M.J. The Causes and Developmental Effects of Integration and Independence between Different Parts of Glechoma Hederacea Clones. *Oikos* **1992**, *63*, 376–386. [CrossRef]

79. Gutbrodt, B.; Mody, K.; Wittwer, R.; Dorn, S. Within-Plant Distribution of Induced Resistance in Apple Seedlings: Rapid Acropetal and Delayed Basipetal Responses. *Planta* **2011**, *233*, 1199–1207. [CrossRef]

80. Miura, K.; Okamoto, H.; Okuma, E.; Shiba, H.; Kamada, H.; Hasegawa, P.M.; Murata, Y. SIZ1 Deficiency Causes Reduced Stomatal Aperture and Enhanced Drought Tolerance via Controlling Salicylic Acid-Induced Accumulation of Reactive Oxygen Species in Arabidopsis. *Plant J.* **2013**, *73*, 91–104. [CrossRef]

81. Novoplansky, A. Hierarchy Establishment among Potentially Similar Buds. *Plant. Cell Environ.* **1996**, *19*, 781–786. [CrossRef]

82. Sokal, R.R.; Rohlf, F.J. *Biometry*, 4th ed.; Freeman: New York, NY, USA, 2012.

Article

Classifying Circumnutation in Pea Plants via Supervised Machine Learning

Qiuran Wang [1,*], Tommaso Barbariol [2], Gian Antonio Susto [2], Bianca Bonato [1], Silvia Guerra [1] and Umberto Castiello [1]

1 Department of General Psychology, University of Padova, 35132 Padova, Italy
2 Department of Information Engineering, University of Padova, 35131 Padova, Italy
* Correspondence: qiuran.wang@gmail.com or qiuran.wang@phd.unipd.it

Abstract: Climbing plants require an external support to grow vertically and enhance light acquisition. Climbers that find a suitable support demonstrate greater performance and fitness than those that remain prostrate. Support search is characterized by oscillatory movements (i.e., circumnutation), in which plants rotate around a central axis during their growth. Numerous studies have elucidated the mechanistic details of circumnutation, but how this phenomenon is controlled during support searching remains unclear. To fill this gap, here we tested whether simulation-based machine learning methods can capture differences in movement patterns nested in actual kinematical data. We compared machine learning classifiers with the aim of generating models that learn to discriminate between circumnutation patterns related to the presence/absence of a support in the environment. Results indicate that there is a difference in the pattern of circumnutation, depending on the presence of a support, that can be learned and classified rather accurately. We also identify distinctive kinematic features at the level of the junction underneath the tendrils that seems to be a superior indicator for discerning the presence/absence of the support by the plant. Overall, machine learning approaches appear to be powerful tools for understanding the movement of plants.

Keywords: plant movement; circumnutation; machine learning; classification; kinematics

Citation: Wang, Q.; Barbariol, T.; Susto, G.A.; Bonato, B.; Guerra, S.; Castiello, U. Classifying Circumnutation in Pea Plants via Supervised Machine Learning. *Plants* **2023**, *12*, 965. https://doi.org/10.3390/plants12040965

Academic Editor: Frantisek Baluska

Received: 6 January 2023
Revised: 13 February 2023
Accepted: 16 February 2023
Published: 20 February 2023

1. Introduction

When observing plants, they seem relatively immobile, stuck to the ground in rigid structures. But for careful observers, as Darwin was in the 19th century, it is quite clear that plants do produce movement. Darwin was fascinated by the graceful movements of twining plants revolving in large arcs, winding around a support, and forming a helical tube of tissue. He described this movement as "a continuous self-bowing of the whole shoot, successively directed to all points of the compass" [1] and later named this movement circumnutation [2].

Circumnutation is a common phenomenon in plants but is exaggerated in twining stems. By circumnutating, twiners increase the probability of encountering a support to grow vertically and enhance light acquisition. Vines that find a suitable support have greater performance and fitness than those that remain prostrate [3,4]. Therefore, locating a suitable support is a key process in the life history of climbing plants. Numerous studies on climbing plant behavior have elucidated mechanistic details of support searching and attachment [e.g., 3]. This body of research relies chiefly on field observations reporting on morphological or physiological responses [4], as well as on laboratory studies focused on the characterization of kinematical patterning through the use of time-lapse photography [5,6]. Although this body of research provides some quantitative data, the process is admittedly subjective and rather preliminary. In other words, it does not offer a clear explanation of what happens in the pattern of circumnutation when climbers perceive a potential support and decide to orient their movement towards it.

Machine learning approaches might be an alternative method of addressing this issue and enabling accurate phenotyping. The application of machine learning to questions in plant biology is still in its infancy, yet the applicability of these methods to a broad range of problems is clear. These technologies have recently achieved impressive performance on a variety of predictive tasks, such as species identification [7,8], plant species distribution modeling [9,10], weed detection [11], and mercury damage to herbarium specimens [12]. They are also being applied to questions of comparative genomics [13], gene expression [14], and to conducting high-throughput phenotyping [15,16] for agricultural and ecological research. Machine learning methods, however, have never been used for modeling or predicting the movement of plants. Predicting plant behavior through their movement is important for several reasons. Realistic predictions could aid in the formation of conservation strategies to combat the decline in biodiversity. For example, predicting movement might be important in the context of understanding the spread of infectious diseases through plant species. Many diseases are spread through different means of communication between individuals. Realistic predictions of the movement of infected individuals can suggest interventions that will optimally alleviate the further spread of diseases.

In this connection, here we use machine learning methods to classify plant movement behavior, and to predict movement patterns which will enable us to build stochastic movement generators, useful in scenarios where collecting actual movement data is laborious. Given that predicting plant movement is important when building simulators, we tested whether simulation-based machine learning methods can capture the movement patterns nested in actual kinematical data. We compared several machine learning classifiers to model plants' movement with the goal of generating models that, on the basis of a binary labeled dataset, learn to discriminate between the presence/absence of a support in the environment so as to formulate precise predictions based on an unlabeled dataset. We found that there is a difference in the pattern of circumnutation that can be learned and classified rather accurately depending on the presence or absence of the support. Furthermore, we identified the most distinctive kinematic features that contribute to the classification tasks and provide additional information for driving future circumnutation studies. Overall, machine learning appears to be a valid tool for studying the movement of plants.

2. Results

2.1. Classifiers Are Able to Perform Accurate Predictions Depending on the Presence/Absence of the Support

To test whether the pattern of circumnutation differed depending on the presence/absence of a support in the environment, we exposed pea plants to a condition in which a support was not present in the environment ("no support" condition; Figure 1a) and a condition in which a support was present in the environment ("support" condition; Figure 1b). The plants that grew in the presence of the support oriented their movement towards it and prepared for grasping. The plants that grew in the absence of the support continued to circumnutate toward the light source and then fell down. From the 3D reconstruction of movement trajectories for the considered anatomical landmarks (i.e., the tendrils and the point where the tendrils tie, from now on "junction"; Figure 1c), we extracted a set of kinematic features that were used for machine learning classification (see details in Section 4 Material and Methods; also see Supplementary Materials S1–S3). Three classifiers, namely random forest (RF), logistic regression (LR), and support vector classifier (SVC), were used as a cross-model validation [17]. These approaches have been optimized and validated in a variety of fields [18,19]. The classifiers generated models based on a binary-labeled training set, learned to discriminate between the presence/absence of the support, and formulated precise predictions based on an unlabeled test set. The performance corresponds to the accuracy of classification (i.e., the rate of discriminating plants growing in the presence/absence of the support on the test set correctly). When considering the totality of the circumnutations performed by the plant (i.e., "overall movement classification"), the classifiers were able to distinguish between the "support" and the "no support" conditions with a mean accuracy across all classifiers and all features of 66.80 % (SD 15.39; Table 1). When considering circumnutations singularly (i.e., "circumnutation classification"), the mean accuracy was 68.52% (SD 12.63; Table 2). These results demonstrate that the classifiers were capable of differentiating the pattern of circumnutation depending on the presence/absence of the support rather accurately above the chance level (>50.00%).

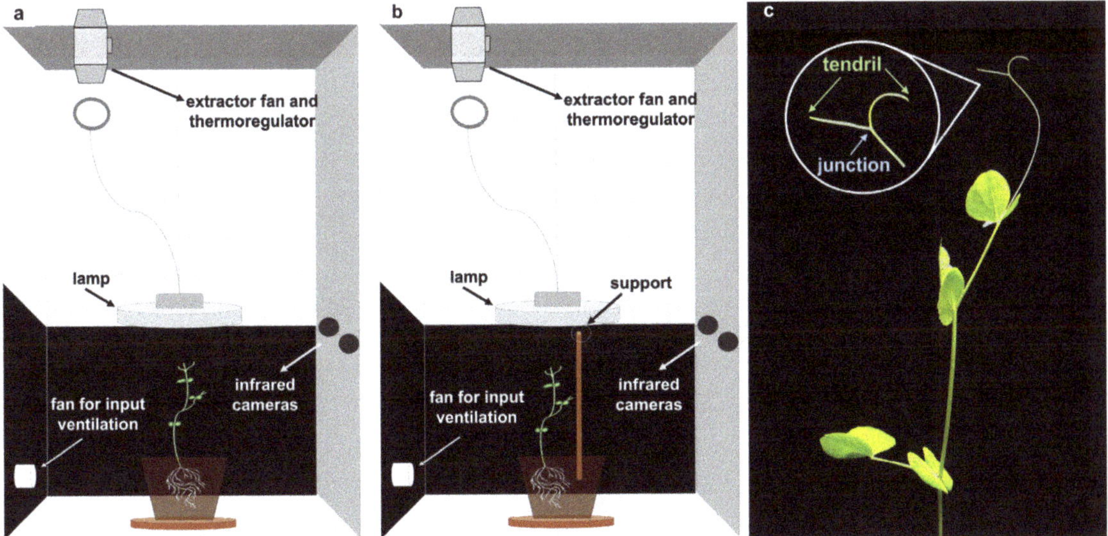

Figure 1. Experimental conditions and anatomical landmarks. Experimental setup, experimental conditions and anatomical landmarks considered. Each chamber is equipped with two infrared cameras on one side, a thermoregulator for controlling the temperature at 26 °C, two fans for input and output ventilation, and a lamp on top of the plant. (**a**) "No support" condition, n = 19. (**b**) "Support" condition, n = 13. (**c**) The anatomical landmarks of interest were the "tendrils" and the "junction" developing from the considered leaf. The "tendrils" refers to the tip of the shoot, and the "junction" refers to the point where the tendrils tie together.

Table 1. Accuracy in "overall movement classification" task. This table shows the mean and standard deviation of the accuracy for each classifier.

	Accuracy % Mean (Standard Deviation)			
	Random Forest	**Logistic Regression**	**SVC**	**Feature Mean Accuracy**
Junction trajectory	71.00 (18.30)	80.50 (13.54)	71.50 (9.89)	74.30 (14.80)
Junction velocity	78.50 (12.24)	78.00 (9.04)	75.50 (12.23)	77.30 (11.19)
Junction acceleration	66.50 (11.81)	72.00 (12.12)	71.00 (11.81)	69.80 (11.99)
Tendril trajectory	67.00 (16.49)	56.50 (14.93)	66.00 (11.13)	63.2 (14.95)
Tendril velocity	75.50 (10.51)	68.00 (15.34)	72.50 (10.21)	72.00 (12.47)
Tendril acceleration	51.00 (11.92)	57.00 (10.87)	63.50 (10.16)	57.20 (12.01)
Tendril aperture	62.50 (15.73)	49.50 (12.23)	60.00 (6.25)	57.30 (13.17)
Movement duration	48.50 (17.43)	65.00 (16.54)	56.50 (10.90)	56.70 (16.48)
All features	76.50 (12.14)	71.00 (13.84)	72.00 (10.38)	73.20 (12.27)
Classifier's mean accuracy	66.30 (17.36)	66.40 (16.37)	67.60 (11.94)	66.80 (15.39)

Note. A string of accuracy for each classifier and feature is obtained after repeating the classification task 25 times.

Table 2. Accuracy in "circumnutation movement". This table shows the mean and standard deviation for accuracy for each classifier.

	Accuracy % Mean (Standard Deviation)			
	Random Forest	**Logistic Regression**	**SVC**	**Feature Mean Accuracy**
Junction trajectory	71.84 (10.71)	74.87 (12.14)	71.54 (14.03)	72.75 (12.29)
Junction velocity	65.09 (11.09)	71.01 (15.23)	70.42 (14.44)	68.84 (13.78)
Junction acceleration	67.12 (9.50)	70.27 (10.44)	69.33 (12.22)	68.91 (10.72)
Tendril trajectory	59.49 (9.10)	68.65 (14.56)	67.38 (12.01)	65.17 (12.61)
Tendril velocity	67.35 (11.39)	70.84 (15.23)	70.37 (14.28)	69.52 (13.63)
Tendril acceleration	62.87 (10.42)	65.62 (12.31)	66.20 (11.23)	64.90 (11.29)
Tendril aperture	64.82 (11.28)	65.60 (11.80)	64.67 (12.79)	65.03 (11.82)
Circumnutation movement duration	63.24 (12.18)	72.98 (12.82)	69.92 (12.58)	68.71 (13.02)
All features	73.74 (12.91)	73.37 (10.35)	72.14 (11.54)	73.08 (11.51)
Classifier's mean accuracy	66.20 (11.60)	70.29 (12.98)	69.07 (12.96)	68.52 (12.63)

Note. A string of accuracy for each classifier and feature is obtained after repeating the classification task 25 times.

2.2. Specific Contribution of the Considered Features across Classifiers for the Overall Movement Classification

As shown in Table 1 (also see Supplementary Materials Figure S1), the SVC performs with a slightly higher average accuracy (mean 67.60%, SD 11.94) compared to the RF (mean 66.30%, SD 17.36) and LR (mean 66.40%, SD 16.37) classifiers. Regarding those features that contributed to the successful classification, the "junction velocity" (mean 77.30%, SD 11.99), the "junction trajectory" (mean 74.30%, SD 14.80), and "all features" (mean 73.20%, SD 12.27) show generally better performance compared with the "tendril aperture" (mean 57.30%, SD 13.17), the "tendril acceleration" (mean 57.2%, SD 12.01), and "movement duration" (mean 56.70%, SD 16.48). With a mean accuracy of 80.50% (SD 13.54) obtained with the LR classifier, "junction trajectory" seems to be the best indicator for distinguishing between the "support" and "no support" conditions. Overall, this suggests that the plants exhibit differences in behavioral patterns depending on the presence/absence of the support.

2.3. Specific Contribution of the Considered Features When Considering Single Circumnutations

On the basis of the features derived from a single circumnutation, the classifiers can reliably predict whether the plants are moving in the presence/absence of a potential support (Table 2; also see Supplementary Materials Figure S2). In comparison to the RF (mean 66.20%, SD 11.60) and the SVC (mean 70.29%, SD 12.98), the LR has a slightly greater average accuracy (mean 69.07%, SD 12.96). As for the contribution of the different features, "all features" (mean 73.08%, SD 11.51), "junction trajectory" (mean 72.75%, SD 12.29), and "tendril velocity" (mean 69.52%, SD 13.63) exhibit better performance compared with "tendril trajectory" (mean 65.17%, SD 12.61), "tendril aperture" (mean 65.03%, SD 11.82), and "tendril acceleration" (mean 64.90%, SD 11.29). With a mean accuracy of 74.87% (SD 12.14) obtained with the LR classifier, "junction trajectory" seems to be the best indicator for distinguishing between the "support" and the "no support" conditions. This is in accordance with the findings for the "overall movement classification." Again, this demonstrates that the classifiers are able to extract from the kinematics of circumnutation whether the plant is moving in the presence/absence of a potential support.

2.4. The Accuracy of the Classification Depends on Organs and Features

When looking more deeply into the contributory role played by the features considered for classification, we found that kinematic features for the tendrils appear to be less relevant with respect to junction-related features for both classification tasks (Tables 1 and 2). When considering movement duration, this feature appears to be less informative when the overall movement classification is considered. However, this very same feature appears to be a reliable indicator when single circumnutations are considered (68.71%, SD 13.02).

2.5. A Full Kinematic Profile Favors Classification

When we combined all the extracted features, we achieved a high level of accuracy across all classifiers (overall movement classification: mean 73.20%, SD 12.27; Table 1; circumnutation classification: mean 73.08%, SD 11.51; Table 2). After the models had been fitted, the importance of kinematic features was determined by applying permutation importance (Figure 2a,b). Different feature importance is detected among classifiers when considering overall movement and single circumnutations separately. For instance, when the overall movement is considered, "junction velocity", "junction trajectory", and "junction acceleration" appear to be the most crucial classification features, whereas "tendril acceleration", "tendril aperture", "tendril trajectory", and "movement duration" appear to be less essential. The negative value (<0.00) for the less important features mentioned above indicates that predictions based on shuffled data typically turn out to be more accurate than real data. "Junction trajectory" and "junction acceleration" appear to be more important than "tendril acceleration" and "tendril aperture" for classification when single circumnutations are considered. Movement duration is an important feature for distinguishing between the presence/absence of the support when it is referred to single circumnutation, but not when "overall movement duration" is considered.

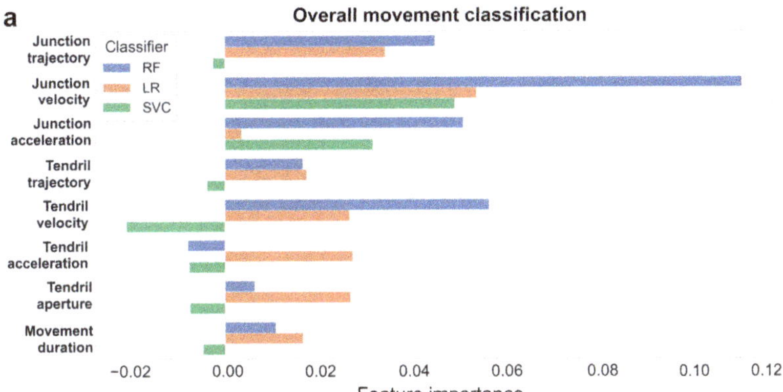

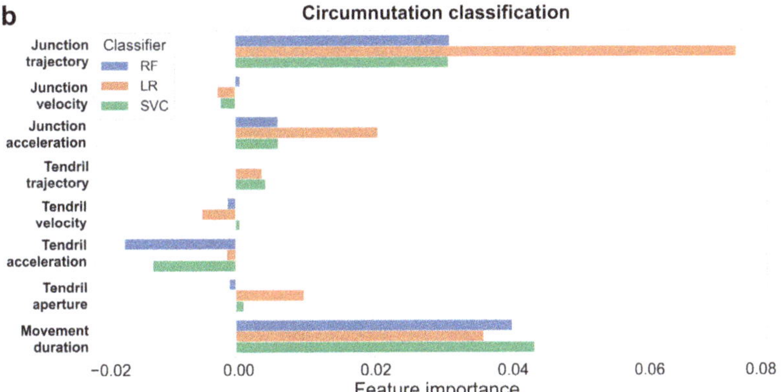

Figure 2. Feature importance in "all features". Kinematic feature importance of three classifiers random forest (RF, blue), logistic regression (LR, orange), support vector classifier (SVC, green). (**a**) Feature importance in "overall movement classification" task. (**b**) Feature importance in "circumnutation classification" task.

3. Discussion

In this study, we propose a general framework to classify pea plants' circumnutation movement. We applied this framework using various machine learning models 'fed' with kinematic data. Our findings show that machine learning techniques have the ability to unveil how kinematic patterning is modulated in key organs when pea plants 'hunt' for a support.

Nutation kinematics of different organs has served to lay a foundation of several mechanisms postulated as responsible for the movement in question with tendrils being amongst the most investigated [3,20–22]. Tendrils serve climbing plants by providing a parasitic alternative to building independently stable structural supports, allowing the plant to wend its way to sunlight and numerous ecological niches [23]. Accordingly, previous evidence provides a degree of support that some climbing plants can modify their *circumnutation* patterns to a greater or lesser extent depending on the presence/absence of a potential support in the environment [24,25]. Experimental evidence demonstrating that this is the case has been forthcoming from recent studies that used kinematic analysis to characterize the movements of the tendrils of pea plants [6,26–28]. Guerra and colleagues [29], for example, demonstrated that pea plants (*Pisum sativum* L.) are able to perceive a stimulus and modulate the kinematics of the tendrils according to the features of a potential support. Therefore, it seems that the tendrils of climbing plants reaching to grasp a stimulus plays a pivotal role as far as support detection is concerned.

The findings of the present study, however, seem to suggest that, rather than the tendrils, the junction underneath them is a superior indicator for discerning the presence/absence of the support. The fact that the kinematics of the junction is a stronger predictor than the kinematics of the tendrils for the presence of the support points to this organ as a navigator guiding the tendrils towards the support. Indeed, if we look carefully at how circumnutation unfolds once the support has been somewhat detected, it is evident that the junction of the tendrils modifies its velocity and timing to launch the tendrils toward the support. In addition, once informed that the 'take-off' is approaching, the tendrils open and assume a choreography so as to accommodate the thickness and the shape of the support [29]. All of this corroborates the idea that plant movements are adaptive, flexible, anticipatory, and goal-directed. Put simply, they are somewhat organized and structured, with different organs 'cross-talking' to accomplish a crucial endeavor for the plant's survival. Our study using machine learning techniques illuminates and quantifies this proposal.

Another novel observation that stems from our study is the classifiers being able to extract a tremendous amount of information from a single circumnutation, which represents the smallest unit of the entire movement. The very fact that the classifiers can make accurate predictions from the emergence of the very first circumnutation reveals that the plants, at the time they initiated to circumnutate, were already well-aware of their surroundings.

At this stage, a central question that could be asked is whether climbers actually make 'decisions' when it comes to support searching and selection. In this respect, our study supports the notion that climbers do not find support/hosts merely by chance. Apart from the evidence of oriented growth towards experimental stakes as discussed above, the methodology used here might be useful to understand climbing plants' preferences. This was first described by Darwin for tendrils in *B. capreolata* initially seizing but then loosing sticks that were inappropriate [1]. A similar phenomenon is observed when herbaceous twining vines get in contact with a very thick trunk and wind up on themselves instead of attempting to twine around it. In the case of annual vines, Darwin remarked that, even without support diameter constraints, it would be maladaptive to twine around thick—and hence large—trees, as these vines would hardly reach high-light layers by the end of the growing season [1].

Further machine learning research should aim at characterizing how circumnutation changes as far as support characteristics are concerned. Predictions and modeling of the cost-benefit analysis of climbing plant behavior should be helpful to infer the selective pressures that have operated to shape current climber ecological communities. In addition to plant movement, as a direct reflection of plants' internal state, other physiological markers could be added to obtain a more complete, reliable, and consistent picture of how the environment shapes climbers' behavior. Such technologies will enable the investigation of unknown aspects of the helical growth performed by the tendrils and their junction on an evolutionary scale, shedding some light on the mechanisms involved in the acquisition and evolution of the climbing habits of terrestrial plants.

4. Materials and Methods

4.1. Subjects and Materials

A total of 32 snow peas (*Pisum sativum* var. Saccharum cv Carouby de Maussane) were chosen as study plants. For each pot, 6 seedlings were potted at 8 cm from the pot's border and sowed at a depth of 2.5 cm. Once germinated, one healthy-looking sprout was selected and randomly assigned to the experimental conditions: 19 plants were grown individually in chambers without the presence of a support ("no support" condition; Figure 1a), while 13 plants were grown individually in chambers where a potential support was present ("support" condition; Figure 1b). Sprouts were placed 8 cm from the pot's border and sowed at a depth of 2.5 cm. The support was a wooden pole (54 cm in height and 1.3 cm in diameter) inserted 7 cm below the soil surface and positioned 12 cm away from the plant's first unifoliate leaf.

4.2. Growth Setup

Each plant was positioned in a thermo-light-controlled growth chamber (Cultibox SG combi 80 × 80 × 160 cm; Figure 1). The temperature was set at 26 °C by means of an extractor fan equipped with a thermo-regulator (TT125 vents; 125 mm-diameter; max 280 mc/h) and an input-ventilation fan (Blauberg Tubo 100–102 m^3/h). The two-fan combination allowed for a steady air flow rate with a mean air residence time of 60 s. The fan was carefully placed so that air circulation did not affect the plants' movements.

Cylindrical pots (diameter 30 cm, depth 20 cm) were filled with river sand (type 16SS, dimension 0.8/1.2 mm, weight 1.4) and positioned at the center of the growth chamber. A cool white led lamp (V-TAC innovative LED lighting, VT-911-100W, Des Moines, IA, USA) was positioned 50 cm above each seedling, and each plant was grown under an 11:25 h light regime (5:45 a.m. to 5 p.m.). The Photosynthetic Photon Flux Density at 50 cm under the lamp in correspondence of the seedling was 350 $\mu mol_{ph}/(m^2 s)$ (quantum sensor LI-190R, Lincoln, Nebraska, USA). The plants were watered three times a week and fertilized using a half-strength nutrient solution (Murashige and Skoog Basal Salt Micronutrient Solution; see components & organics).

4.3. Data Acquisition and Data Processing

For each growth chamber, a pair of RGB-infrared cameras (IP 2.1 Mpx outdoor varifocal IR 1080P) were placed 110 cm above the ground, spaced at 45 cm to record stereo images of the plant (see Figure 1a and Supplementary Materials S1). The two cameras synchronously acquired a frame every 180 s (frequency 0.0056 Hz). RGB images were acquired during the daylight cycle and infrared images during the night cycle. The anatomical landmarks of interest were the "tendrils" and the "junction" (Figure 1c), developing from the considered leaf. The initial frame was the one corresponding to the appearance of the tendrils and the junction. The final frame was defined as either the frame in which the tendrils start to coil for the "support" condition (number of selected images: 699.62, SD 379.28), or the frame just before the plant fell on the ground for the "no support" condition (number of selected images: 1617.11, SD 1112.82). Images from both left and right cameras were used in order to reconstruct 3D trajectories. An ad hoc software (Ab.Acus s.r.l., Milan, Italy) developed in Matlab was used to identify anatomical points to be investigated by means of markers, and to track their position frame-by-frame on the images acquired by the two cameras to reconstruct the 3D trajectory of each marker. The markers on the anatomical landmarks of interest, namely the tip of the tendrils and the junction, were inserted post-hoc (Figure 1c). The tracking procedures were at first performed automatically throughout the time course of the movement sequence using the Kanade-Lucas-Tomasi (KLT) algorithm on the frames acquired by each camera, after distortion removal. The tracking was manually verified by the experimenter, who checked the position of the markers frame-by-frame. The 3-D trajectory of each tracked marker was computed by triangulating the 2-D trajectories obtained from the two cameras (Figure 1). The 3D coordinates were obtained up to 15 digits after the decimal. The frames corresponding to the time at which the plants grasped the support or touched the ground in the absence of the support were removed from the data set. This was done to prevent classifiers from using these final frames to distinguish between the two conditions. Therefore, the classifiers were only fed with helical organ movements (i.e., circumnutation). Moreover, since each plant has its own starting position, we used the coordinates for the first frame as the origin (0,0,0) for all plants in order to prevent a location bias that could affect learning by the classification models (Figure 3a–c).

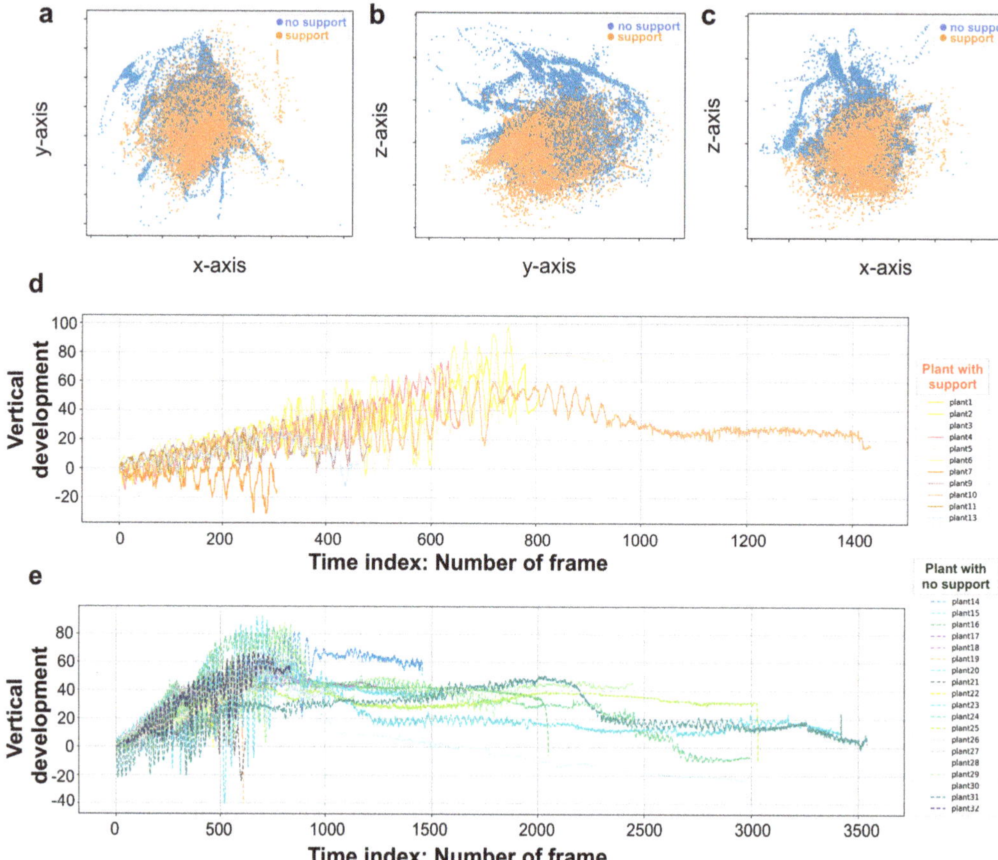

Figure 3. Data acquisition. Coordinates of junction trajectory and plant vertical development in time. (**a**) Junction trajectory for all plants in the x-y dimension for the two experimental conditions. (**b**) Junction trajectory for all plants in the y-z dimension. (**c**) Junction trajectory for all plants in the x-z dimension. (**d**) Junction vertical development in time for the "support" condition. (**e**) Junction vertical development in time for the "no support" conditions. In panels '(**d**)' and '(**e**)', different colors represent different plants. Note that for the "no support" conditions, the length of the time index which is indicated as the number of frames has a longer range than the "support" conditions.

4.4. Derived Features

Kinematic variables (hereafter, 'features') were analyzed in order to ascertain whether they differed for the "support" and the "no support" conditions. This aspect is fundamental in order to verify the ability of machine learning tools to discriminate across the conditions. To do this, we used the Wilcoxon rank-sum test in R-studio, and the size of the effect calculated as $r = z\sqrt{N}$ where z is the z-score and N is the total number of observations was also considered. In line with previous studies [29,30], we found statistically significant differences between the "support" and the "no support" conditions (Table 3; see also Figure 4 for a graphical example). On the basis of the obtained results, the features considered for model classifications were: (a) junction trajectory; (b) tendril trajectory; (c) junction velocity; (d) tendril velocity; (e) junction acceleration; (f) tendril acceleration; (g) tendrils aperture; (h) overall movement duration; (i) movement duration for each circumnutation; (j) all features (i.e., full kinematics). Please refer to Supplementary Materials S3 for details regarding feature extraction.

Table 3. Kinematic data for the "support" and the "no support" conditions. Statistical values obtained when comparing the two conditions are also reported.

	Median		**W**	*p*	**r**
	No Support	**Support**			
Junction velocity (mm/min)	1.7488	3.5035	166	0.007	0.299
Junction acceleration (mm/min)	0.0006	−0.0001	51	0.021	0.257
Tendril velocity (mm/min)	2.5289	4.4670	1242	0.000	0.510
Tendril acceleration (mm/min)	0.0008	−0.0001	361	0.000	0.439
Tendrils aperture (mm)	25.2039	14.7132	245	0.000	0.394
Overall movement duration (min)	3744	1683	59	0.013	0.545
Circumnutation movement duration (min)	201.0857	217.000	143	0.103	0.181

Note. mm = millimeters; mm/min = millimeters by minutes.

4.5. Data Pre-Processing

We adopted a z-score as a data normalization method (standard scaling), by using the formula $Z = (x-\mu)/\sigma$, where μ stands for the mean value of the feature and σ for the standard deviation of the features. A value equal to the mean of all the features was normalized to 0 and the standard deviation to 1. To avoid biases toward features of the dataset and, at the same time, to prevent the classifiers from learning information from the test dataset, we utilized the transform method to keep the same features from the training data to transform the test data.

To split the training and test sets, each derived feature was labeled with two different conditions, "support" and "no support", as a binary labeled dataset. The stratified shuffle split cross-validator was applied to the dataset, which is a merge of StratifiedKFold and SuffleSplit to return stratified randomized folds [17]. The set number of re-shuffling and splitting iterations equals 25, test size as 0.25, default random state.

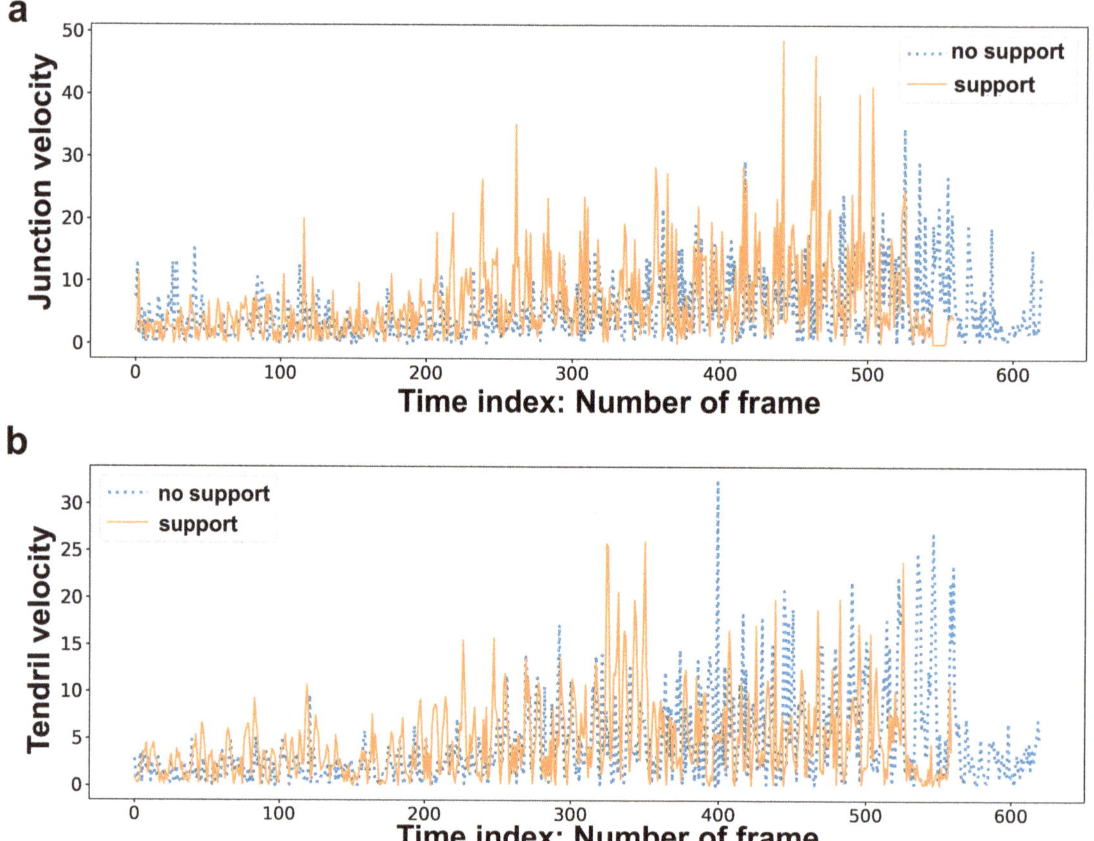

Figure 4. Graphical examples for: (**a**) junction velocity and (**b**) tendril velocity for the "no support" condition and the "support" condition.

4.6. Models' Classifications Tasks

The modeling process was carried out with Python. We performed modelling of pea plants' behavior based on supervised classification frameworks. The purpose of a Machine Learning Classifier is to produce models that, on the basis of a binary-labelled training set, learn to discriminate between different growth circumstances and to provide exact predictions on the basis of an unlabeled test set. Random decision forests (RF), logistic regression (LR), and support vector classifier (SVC) are the classifiers that were applied and compared through cross-validation (see Supplementary Materials S4) [17]. These approaches are optimized and validated in a wide variety of fields [18,19]. The percentage of test data that were successfully classified for the two conditions is counted under the accuracy of classification. The classification task employed each of the generated kinematic features separately, and the classification accuracy for each feature was evaluated. We also assessed the accuracy of "all features", where permutation importance was computed following the fitting of the classifiers [31]; we analyzed feature importance for all the derived features. The "overall movement classification" and the "circumnutation classification" are the two broad categories that constitute the model classification task. Each classification task consists of 25 trials, which include 25 iterations of the training and test. The absolute movement duration was typically longer for the plants growing in the presence of a support (Figure 3d) than for the plants growing in the absence of a support. For the "overall movement classification" task, we considered the features extracted from the whole movement for each individual plant (Figure 3e). For the "circumnutation classification task", we partitioned the data into circumnutations, smoothing the data set by generating an approximation function that captured the key patterns, namely the waves of the movement in coordinates (i.e., circumnutation). Then, by cutting between peaks, we split between the waves. The features that were extracted from each circumnutation helped in compensating the dataset for this task. For classifying which condition a single circumnutation corresponded to, the classifiers were fitted to the dataset.

Supplementary Materials: The following supporting information can be downloaded at: https://www.mdpi.com/article/10.3390/plants12040965/s1, Figure S1: Specific contribution of the considered features across classifiers for the overall movement classification; Figure S2: Specific contribution of the considered features when considering single circumnutation. References [32–34] are cited in the Supplementary Materials

Author Contributions: Conceptualization, Q.W., T.B. and U.C.; methodology, Q.W. and T.B.; software, Q.W. and T.B.; validation, Q.W.; formal analysis, Q.W.; investigation, Q.W., B.B. and S.G.; resources, G.A.S.; data curation, Q.W.; writing—original draft preparation, Q.W. and U.C.; writing—review and editing, U.C., Q.W., G.A.S. and T.B.; visualization, Q.W. and T.B; supervision, U.C.; project administration, Q.W. All authors have read and agreed to the published version of the manuscript.

Funding: This research is partially funded by the SEED BADA^3 project, the Department of Information Engineering, University of Padova.

Data Availability Statement: The data is available online: https://doi.org/10.6084/m9.figshare.21215711 (accessed on 17 February 2023). The classifiers used for plant-growing environment identifications and classifications based on pea plants' movement reported in this paper are available online (https://github.com/qiuranwang/Pea_movement.git, accessed on 17 February 2023).

Acknowledgments: Qiuran Wang is funded by the China Scholarship Council (CSC). We thank Andrè Geremia Parise for his invaluable comments on a previous version of this manuscript.

Conflicts of Interest: The authors declare no conflict of interest.

References

1. Darwin, C. *The Movements and Habits of Climbing Plants*; John Murray: London, UK, 1875.
2. Darwin, C. *The Power of Movement in Plants*; D. Appleton: Boston, MA, USA, 1897.
3. Gianoli, E. The behavioural ecology of climbing plants. *AoB Plants* **2015**, *7*, plv013. [CrossRef] [PubMed]
4. Putz, F.E.; Holbrook, N.M. Biomechanical studies of vines. In *The Biology of Vines*; Cambridge University Press: Cambridge, UK, 1992; pp. 73–98. [CrossRef]
5. Stolarz, M.; Dziubińska, H. Spontaneous action potentials and circumnutation in Helianthus annuus. *Acta Physiol. Plant.* **2017**, *39*, 234. [CrossRef]
6. Guerra, S.; Bonato, B.; Wang, Q.; Peressotti, A.; Peressotti, F.; Baccinelli, W.; Bulgheroni, M.; Castiello, U. Kinematic Evidence of Root-to-Shoot Signaling for the Coding of Support Thickness in Pea Plants. *Biology* **2022**, *11*, 405. [CrossRef] [PubMed]
7. Unger, E.K.; Keller, J.P.; Altermatt, M.; Liang, R.; Matsui, A.; Dong, C.; Hon, O.J.; Yao, Z.; Sun, J.; Banala, S. Directed evolution of a selective and sensitive serotonin sensor via machine learning. *Cell* **2020**, *183*, 1986–2002.e1926. [CrossRef]
8. Carranza-Rojas, J.; Goeau, H.; Bonnet, P.; Mata-Montero, E.; Joly, A. Going deeper in the automated identification of Herbarium specimens. *BMC Evol. Biol.* **2017**, *17*, 181. [CrossRef]
9. Zhang, J.; Li, S. A review of machine learning based species' distribution modelling. In Proceedings of the 2017 International Conference on Industrial Informatics-Computing Technology, Intelligent Technology, Industrial Information Integration (ICIICII), Wuhan, China, 2–3 December 2017; pp. 199–206.
10. Botella, C.; Joly, A.; Bonnet, P.; Monestiez, P.; Munoz, F. Species distribution modeling based on the automated identification of citizen observations. *Appl. Plant Sci.* **2018**, *6*, e1029. [CrossRef]
11. Yu, J.; Sharpe, S.M.; Schumann, A.W.; Boyd, N.S. Deep learning for image-based weed detection in turfgrass. *Eur. J. Agron.* **2019**, *104*, 78–84. [CrossRef]
12. Schuettpelz, E.; Frandsen, P.B.; Dikow, R.B.; Brown, A.; Orli, S.; Peters, M.; Metallo, A.; Funk, V.A.; Dorr, L.J. Applications of deep convolutional neural networks to digitized natural history collections. *Biodivers. Data J.* **2017**, *5*, e21139. [CrossRef]
13. Xu, C.; Jackson, S.A. Machine learning and complex biological data. *Genome Biol.* **2019**, *20*, 76. [CrossRef]
14. Mochida, K.; Koda, S.; Inoue, K.; Nishii, R. Statistical and machine learning approaches to predict gene regulatory networks from transcriptome datasets. *Front. Plant Sci.* **2018**, *9*, 1770. [CrossRef]

15. Singh, A.; Ganapathysubramanian, B.; Singh, A.K.; Sarkar, S. Machine learning for high-throughput stress phenotyping in plants. *Trends Plant Sci.* **2016**, *21*, 110–124. [CrossRef] [PubMed]

16. Ubbens, J.R.; Stavness, I. Deep plant phenomics: A deep learning platform for complex plant phenotyping tasks. *Front. Plant Sci.* **2017**, *8*, 1190. [CrossRef] [PubMed]

17. Pedregosa, F.; Varoquaux, G.; Gramfort, A.; Michel, V.; Thirion, B.; Grisel, O.; Blondel, M.; Prettenhofer, P.; Weiss, R.; Dubourg, V. Scikit-learn: Machine learning in Python. *J. Mach. Learn. Res.* **2011**, *12*, 2825–2830.

18. Salvatore, C.; Cerasa, A.; Castiglioni, I.; Gallivanone, F.; Augimeri, A.; Lopez, M.; Arabia, G.; Morelli, M.; Gilardi, M.; Quattrone, A. Machine learning on brain MRI data for differential diagnosis of Parkinson's disease and Progressive Supranuclear Palsy. *J. Neurosci. Methods* **2014**, *222*, 230–237. [CrossRef] [PubMed]

19. Wijeyakulasuriya, D.A.; Eisenhauer, E.W.; Shaby, B.A.; Hanks, E.M. Machine learning for modeling animal movement. *PLoS ONE* **2020**, *15*, e0235750. [CrossRef]

20. Isnard, S.; Silk, W.K. Moving with climbing plants from Charles Darwin's time into the 21st century. *Am. J. Bot.* **2009**, *96*, 1205–1221. [CrossRef]

21. Raja, V.; Silva, P.L.; Holghoomi, R.; Calvo, P. The dynamics of plant nutation. *Sci. Rep.* **2020**, *10*, 19465. [CrossRef]

22. Simonetti, V.; Bulgheroni, M.; Guerra, S.; Peressotti, A.; Peressotti, F.; Baccinelli, W.; Ceccarini, F.; Bonato, B.; Wang, Q.; Castiello, U. Can Plants Move Like Animals? A Three-Dimensional Stereovision Analysis of Movement in Plants. *Animals* **2021**, *11*, 1854. [CrossRef]

23. Gerbode, S.J.; Puzey, J.R.; McCormick, A.G.; Mahadevan, L. How the cucumber tendril coils and overwinds. *Science* **2012**, *337*, 1087–1091. [CrossRef]

24. Tronchet, A. Suite de nos observations sur le comportement des vrilles en présence de tuteurs. *Bull. Société Bot. Fr.* **1946**, *93*, 13–18. [CrossRef]

25. Tronchet, A. *La Sensibilité des Plantes*; Masson Paris: Paris, France, 1977.

26. Ceccarini, F.; Guerra, S.; Peressotti, A.; Peressotti, F.; Bulgheroni, M.; Baccinelli, W.; Bonato, B.; Castiello, U. On-line control of movement in plants. *Biochem. Biophys. Res. Commun.* **2020**, *564*, 86–91. [CrossRef]

27. Ceccarini, F.; Guerra, S.; Peressotti, A.; Peressotti, F.; Bulgheroni, M.; Baccinelli, W.; Bonato, B.; Castiello, U. Speed—accuracy trade-off in plants. *Psychon. Bull. Rev.* **2020**, *27*, 966–973. [CrossRef] [PubMed]

28. Castiello, U. (Re) claiming plants in comparative psychology. *J. Comp. Psychol.* **2020**, *135*, 127–141. [CrossRef] [PubMed]

29. Guerra, S.; Peressotti, A.; Peressotti, F.; Bulgheroni, M.; Baccinelli, W.; D'Amico, E.; Gomez, A.; Massaccesi, S.; Ceccarini, F.; Castiello, U. Flexible control of movement in plants. *Sci. Rep.* **2019**, *9*, 16570. [CrossRef]

30. Guerra, S.; Bonato, B.; Wang, Q.; Ceccarini, F.; Peressotti, A.; Peressotti, F.; Baccinelli, W.; Bulgheroni, M.; Castiello, U. The coding of object thickness in plants: When roots matter. *J. Comp. Psychol.* **2021**, *135*, 495. [CrossRef]

31. Breiman, L. Random forests. *Mach. Learn.* **2001**, *45*, 5–32. [CrossRef]

32. Liaw, A.; Wiener, M. Classification and regression by randomForest. *R News* **2002**, *2*, 18–22. [CrossRef]

33. Dreiseitl, S.; Ohno-Machado, L. Logistic regression and artificial neural network classification models: A method-ology review. *J. Biomed. Inform.* **2002**, *35*, 352–359. [CrossRef]

34. Manevitz, L.M.; Yousef, M. One-class SVMs for document classification. *J. Mach. Learn. Res.* **2001**, *2*, 139–154. [CrossRef]

Article

Systemic Signals Induced by Single and Combined Abiotic Stimuli in Common Bean Plants

Ádrya Vanessa Lira Costa [1], Thiago Francisco de Carvalho Oliveira [1], Douglas Antônio Posso [1], Gabriela Niemeyer Reissig [1], André Geremia Parise [2], Willian Silva Barros [1] and Gustavo Maia Souza [1,*]

1   Laboratory of Plant Cognition and Electrophysiology, Department of Botany, Institute of Biology, Federal University of Pelotas, Capão do Leão CEP 96160-000, Rio Grande do Sul, Brazil
2   School of Biological Sciences, University of Reading, Reading RG6 6AH, UK
*   Correspondence: gmsouza@ufpel.edu.br

Abstract: To survive in a dynamic environment growing fixed to the ground, plants have developed mechanisms for monitoring and perceiving the environment. When a stimulus is perceived, a series of signals are induced and can propagate away from the stimulated site. Three distinct types of systemic signaling exist, i.e., (i) electrical, (ii) hydraulic, and (iii) chemical, which differ not only in their nature but also in their propagation speed. Naturally, plants suffer influences from two or more stimuli (biotic and/or abiotic). Stimuli combination can promote the activation of new signaling mechanisms that are explicitly activated, as well as the emergence of a new response. This study evaluated the behavior of electrical (electrome) and hydraulic signals after applying simple and combined stimuli in common bean plants. We used simple and mixed stimuli applications to identify biochemical responses and extract information from the electrical and hydraulic patterns. Time series analysis, comparing the conditions before and after the stimuli and the oxidative responses at local and systemic levels, detected changes in electrome and hydraulic signal profiles. Changes in electrome are different between types of stimulation, including their combination, and systemic changes in hydraulic and oxidative dynamics accompany these electrical signals.

Keywords: electrome; heat shock; time dispersion analysis; turgor pressure; wound; plant attention

Citation: Costa, Á.V.L.; Oliveira, T.F.d.C.; Posso, D.A.; Reissig, G.N.; Parise, A.G.; Barros, W.S.; Souza, G.M. Systemic Signals Induced by Single and Combined Abiotic Stimuli in Common Bean Plants. *Plants* **2023**, *12*, 924. https://doi.org/10.3390/plants12040924

Academic Editor: Elena Loreti

Received: 22 December 2022
Revised: 10 January 2023
Accepted: 10 February 2023
Published: 17 February 2023

1. Introduction

As sessile organisms, plants cannot move away from threats. To circumvent these conditions, they have developed monitoring systems, signaling mechanisms, and responses to various stimuli enabling their survival, development, and reproduction [1].

Sometimes, a stimulus on the plant can occur at a local level (e.g., cell or tissue) and trigger responses in regions far away from the affected site (i.e., systemic response). The local and systemic level of responses can be induced by different signals that propagate over long distances in plants [2,3]. This systemic signaling aspect relies on different traveling molecules, such as phytohormones, small peptides, micro RNAs, calcium waves, reactive oxygen species (ROS), volatile organic compounds, and hydraulic and electrical signals (see reviews [4,5]). These signals can travel from one cell to another via apoplast or symplast pathways and by vascular tissues throughout the plant [5,6].

Among all the long-distance signals, hydraulic and electrical ones have higher propagation speeds [5,7], enabling, in certain cases, prompt decoding of the stimulus and assisting in the coordination of appropriate responses [8].

Plants exhibit electrical signals and spontaneous, non-evoked electrical activity associated with basic physiological processes [9] that can be altered as a result of internal or external stimulation. This stimulation leads to ion imbalances across plasma membranes, leading to transient voltage variations promoted by ion flux through ion channels and electrogenic pumps [5,7].

Plant electrophysiological studies have described four main types of electrical signals: action potentials (AP), variation potentials (VP), local potentials (LP), and systemic potentials (SP) (see review by [10]). However, most investigations focus on a single signal type in isolation, mostly AP or VP, considering parameters such as frequency, amplitude, the distance of propagation, and time frame. However, quite often, mixed electrical potential waves are recorded, for instance, as a result of overlapping APs and VPs (among other voltage variations), creating a complex bioelectrical information network in which several electrical signals may be superposed [4]. As a result, the term "electrome" was coined to designate the totality of ionic currents of any living entity, evoked naturally or spontaneously, recorded by non-polarized Ag/AgCl electrodes [11], and more recently, the term "plant electrome" was employed to refer to the emergent complexity of bioelectrical activity in plants [12,13].

Recent studies analyzing plant electrome [1,13–17] revealed its potential as a tool for early stress diagnosis in plants since the electrome exhibits specific patterns of responses that function as a classifiable electrical signature by standard time series analysis and machine learning techniques [15–19].

Hydraulics also play an important role in the network of systemic signaling in plants. The hydraulic signal results from changes in water potential (Ψw), which in turn is propagated through the hydraulic connections of the xylem, enabling the hydraulic signal to transmit over long distances in the form of pressure waves [20]. The hydraulic wave can be fast, traveling nearly at the speed of sound [21], but its relevance relies on how significant the change in turgor pressure within the cell is. The pressure changes originate in the xylem vessels and, due to low axial resistance, can be propagated rapidly to surrounding cells [22] and throughout the plant. However, dead xylem cells did not perceive pressure changes, so they must be decoded by adjacent parenchyma cells, containing a series of sensors capable of detecting turgor-induced mechanical forces [20]. Among the turgor sensors, there are the mechanosensitive channels such as MscS and MscL (small/large conductance mechanosensitive channels), TPK, MCAs [23], as well as kinase receptors such as AHK1 [23] involved in the cell wall integrity (CWI) pathway, critical during pollen germination [24]. The perception of the hydraulic signal by putative sensors leads to the conversion of the physical signal into a chemical one, which will mediate various adaptive responses [20]. Although it is not known how hydraulic waves are linked to electrical, Ca^{2+}, and ROS signals, it has been proposed that mechanosensitive channels permeable to Ca^{2+} and other cations and anions could detect systemic hydraulic waves in tissues distant from the stress site and convert them into Ca^{2+} signals [6,10].

Even though much of the research evaluates the signals separately, studies point out that the integration of these signals is necessary for a more efficient response to stimuli [5,10,24,25].

Signal integration is based on the activation of various membrane proteins that mediate different signaling pathways. Electrical signals and Ca^{2+} waves, which occur during systemic acquired resistance (SAR), depend on Ca^{2+}-permeable glutamate receptor (GLR) type channels, and Ca^{2+} and ROS waves that occur during systemic acquired acclimation (SAA) are associated with the function of the respiratory burst oxidase homologous protein D (RBOHD), involved in ROS generation [26].

Although our knowledge of how stresses affect plants when applied individually is vast, in nature, plants often encounter more than one environmental stimulation/stress condition at a time, resulting in a condition termed "stress combination" [27], and little is known about how plants acclimate to a combination of different stresses, let alone how the different signals interact. The occurrence of two or more different stresses (either biotic and/or abiotic) can be challenging for plants. This challenge can be solved by additive, subtractive, and/or combinatorial effects of different pathways, networks, and mechanisms that are activated by each of the different stresses, or by triggering new responses that are specifically activated during stress combination, promoting emergent responses [28].

Evaluating how the integration of these signals in crop plants behaves when single and combined stimuli are applied may enable a more realistic representation of the response of these plants. Therefore, considering that local stimulation can induce systemic signaling integrating different types of signals and stress-specific responses, we investigated whether local application of single and combined abiotic stimuli induces specific changes in plant electrome and hydraulic dynamics that are able to propagate over long distances in common bean (*Phaseolus vulgaris* L.) plants.

To test this hypothesis, we (1) evaluated the dynamics of systemic signals induced by simple stimuli, followed by (2) the evaluation of the dynamics of systemic signals induced by combined stimuli. Changes in the profile of bioelectrical (electrome) and hydraulic (turgor pressure variation) signals were analyzed with techniques for time series analysis, comparing conditions before and after different stimulation conditions, as well as oxidative responses at a local and systemic level.

2. Results

2.1. Local and Systemic ROS Responses

The results of hydrogen peroxide responses (Figure 1A) showed an interaction between the analyzed factors (tissue and treatment), meaning that the effect on tissues was dependent on the treatments. The highest values, compared with the control plants, were found in the wound and thermal shock treatments, both in local and systemic tissues. However, it was verified that with the combined stresses (W+HS), hydrogen peroxide content was similar to the non-stimulated control plants.

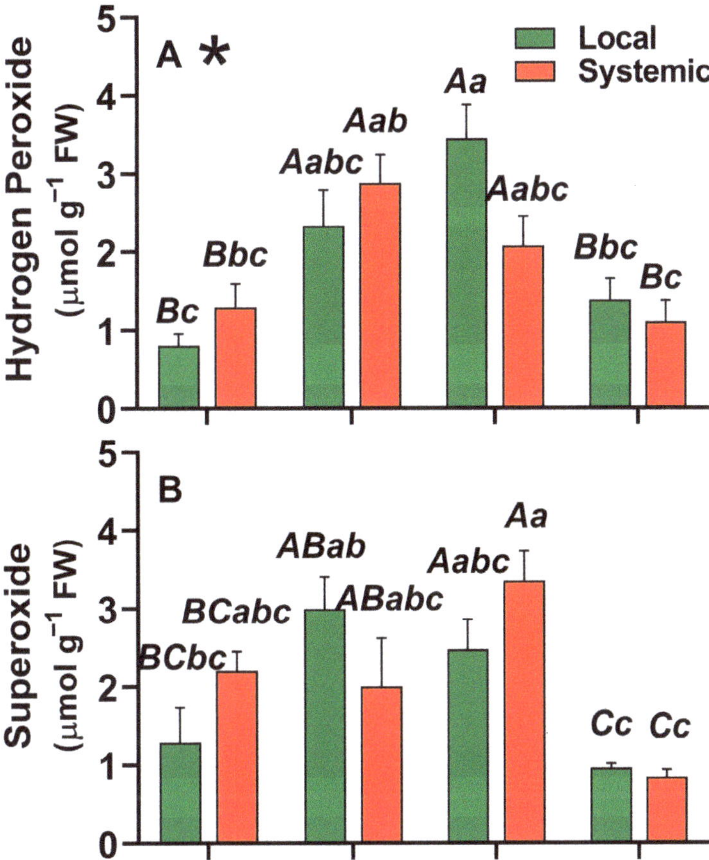

Figure 1. *Cont.*

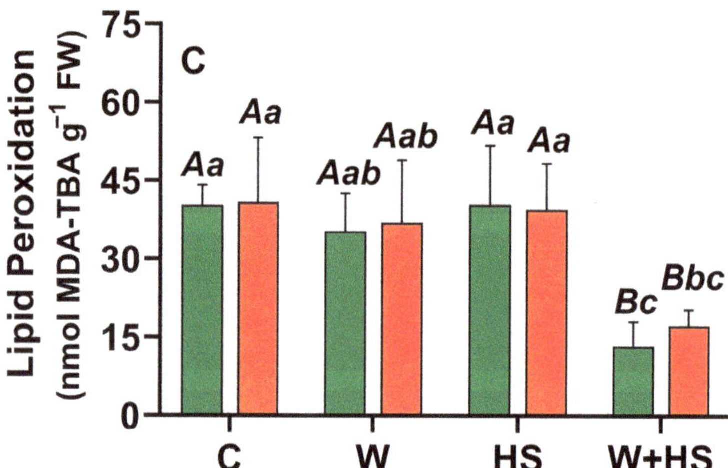

Figure 1. Hydrogen peroxide (**A**), superoxide (**B**), and lipid peroxidation (**C**) of local and systemic leaves submitted to different treatments. Capital letters represent differences between the treatment factor by the post hoc Tukey test, an asterisk next to the indicating letter in the figure indicates interaction between the factors through two-way ANOVA, and lowercase letters indicate the differences found by the Tukey test for the interaction between the factors of treatments and fabrics. C = control; W = wounding; HS = heat shock; W+HS = wounding + heat shock. $p \leq 0.05$; n = 4.

Regarding the superoxide anion content (Figure 1B), there was no interaction between the factors, with the highest values observed in the thermal shock and wound treatments, followed by control plants, and the lowest values were found in the combined treatment (W+HS). Concerning the comparison between local and systemic tissues, there was no significant difference in any treatment.

Lipid peroxidation content (Figure 1C) did not differ between local and systemic tissues, regardless of the treatments. However, the lowest lipid peroxidation was observed in the combined treatment (W+HS) ($p < 0.05$).

2.2. Electrome Dynamics

2.2.1. Visual Analysis

Through the visual analysis of the original time series, it was possible to detect qualitative changes in the electrome when comparing the measurements before and after the stimuli. In the series after the application of W, HS, and W + HS stimuli, the spikes (higher amplitude voltage variations, typically above 50 μV) exhibited evident changes after the stimuli in local and systemic leaves (Figure 2). In systemic leaves, the amplitude of the spikes was smaller than in local leaves. Moreover, the observations indicated that in the first minutes after stimulation, the electrome was more affected than the rest of the series.

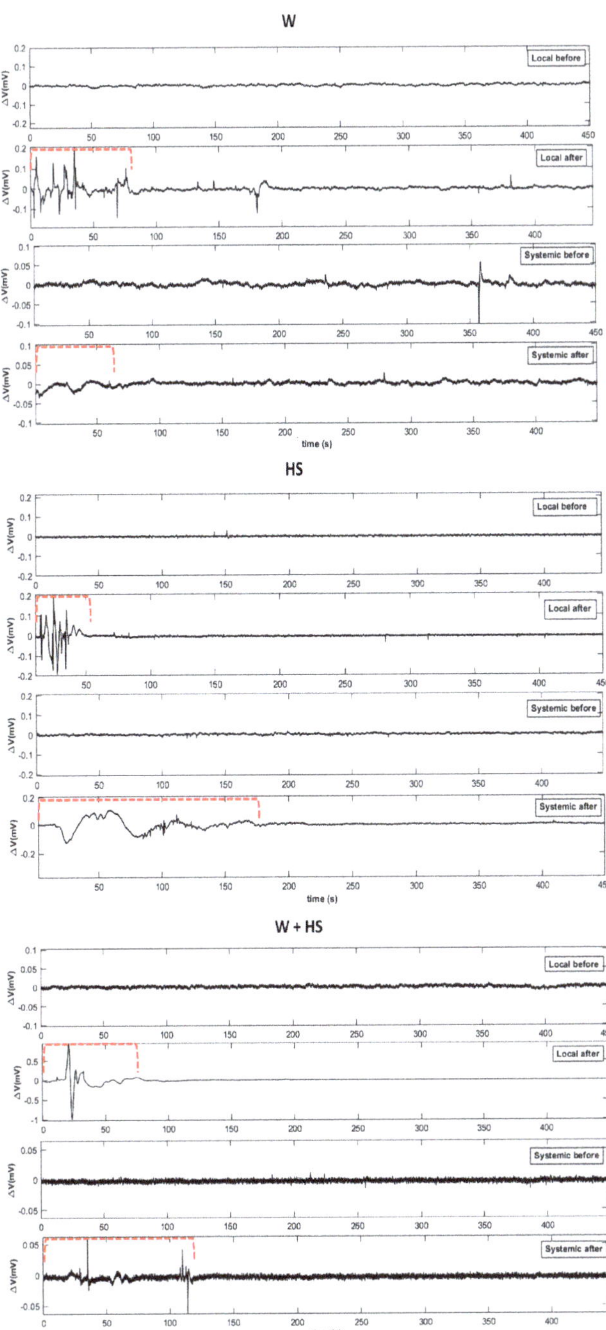

Figure 2. Original representation of the time series before and after the stimuli for wounding (W), heat shock (HS), and combined stimuli (W+HS) in local and systemic leaves. To facilitate comparison, the total length of the series (x = 1 h) was reduced to 7.5 min of measurement. The y-axis (ΔV) was adjusted for each situation. The red dotted line indicates the moment of greatest electrome change after stimulation.

2.2.2. ApEn

Figure 3 shows the results from ApEn analysis for the W, HS, and W+HS essays comparing before and after stimulation in the local leaf. It is possible to observe that in the first moments (typically within the first 10 min), there is a significant change in the ApEn values. On the other hand, as expected, no changes were found in the control plants. On average, there was a decrease of 0.3 in the ApEn values in all treatments. However, after approximately 10 min, all values tended to match the range observed in non-stimulated control plants.

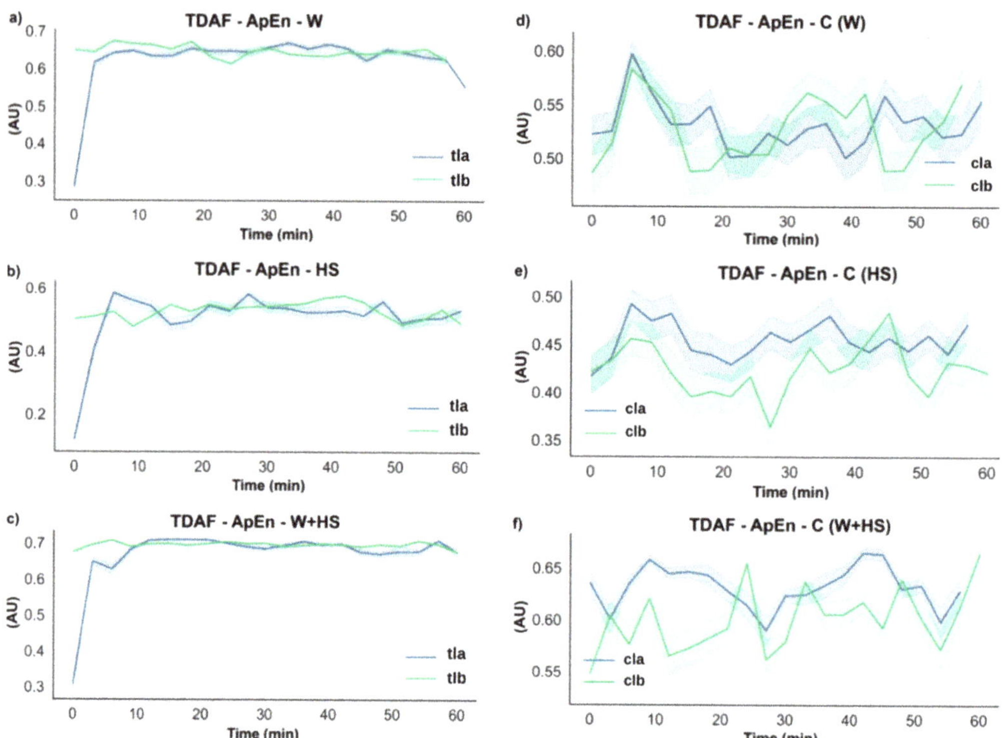

Figure 3. TDAF plots for ApEn analysis. Results for (**a**) the local wound (W), (**b**) heat shock (HS), and (**c**) wound + heat shock (W+HS). Results of the control plants for (**d–f**) wound C (W), heat shock C (HS), and wound + heat shock C (W+HS), respectively. tlb: treated local before stimulus; tla: treated local after stimulus; clb: control local before stimulus; cla: control local after stimulus. AU: arbitrary unit.

Interestingly, although a significant difference before and after stimulation in the non-stimulated plants was not observed, the control for W+HS indicated a difference of around 0.1 in complexity in the first minutes.

The ApEn results for the systemic leaves are shown in Figure 4, and as observed in the local ones, the stimuli change the entropy of the signal, with the difference being that these changes are presented in a smaller amplitude than that observed for the local leaves. A difference of 0.1 in complexity was found in the plants subjected to wounding. However, after the first few minutes, the complexity increased and remained higher than before. Plants stimulated with heat shock showed a 0.3 decrease in complexity at the beginning of the runs and returned to pre-stimulation values after around 10 min. There was a 0.2 decrease in complexity after the application of the combined stimulus, but unlike the isolated stimuli, a biphasic behavior was observed, where an increase in complexity in the first minutes was noted, followed by a reduction around 10 min later. In the second phase, the complexity increases again until it returns to the values found before the stimulus around 15 min later.

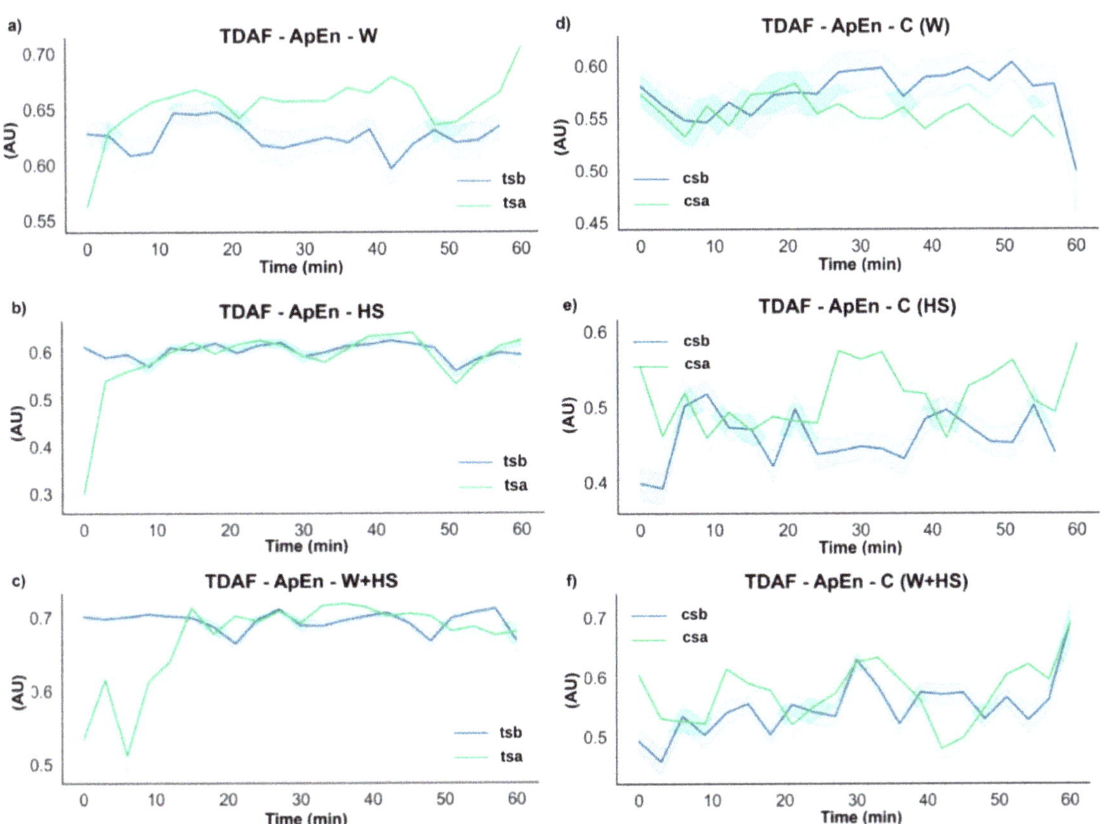

Figure 4. TDAF plots for ApEn analysis. Results for (**a**) systemic wounding (W), (**b**) heat shock (HS), and (**c**) wounding + heat shock (W+HS). Results of the control plants for (**d**–**f**) wounding C (W), heat shock C (HS), and wounding + heat shock C (W+HS), respectively. tsb: treated systemic before stimulus; tsa: treated systemic after stimulus; csb: control systemic before stimulus; csa: control systemic after stimulus. AU: arbitrary unit.

Unexpectedly, an increase in the complexity (higher ApEn values) during the first minutes was noticed in the control plants (not stimulated) of the HS and W+HS essays.

2.2.3. DFA

The results from the DFA analysis showed an increase in the correlation right after the stimuli (Figure 5). However, this value returns to a pre-stimulated condition before the first 10 min. As expected, no remarkable difference was observed in non-simulated plants through time.

The results of the DFA analysis for the systemic leaves found in Figure 6 supported the changes found by the ApEn analysis (Figure 4). However, differences in the duration of the changes observed in the electrome after combined stimulation indicated that at the systemic level, this stimulus lasted longer. No specific changes were observed in non-stimulated plants.

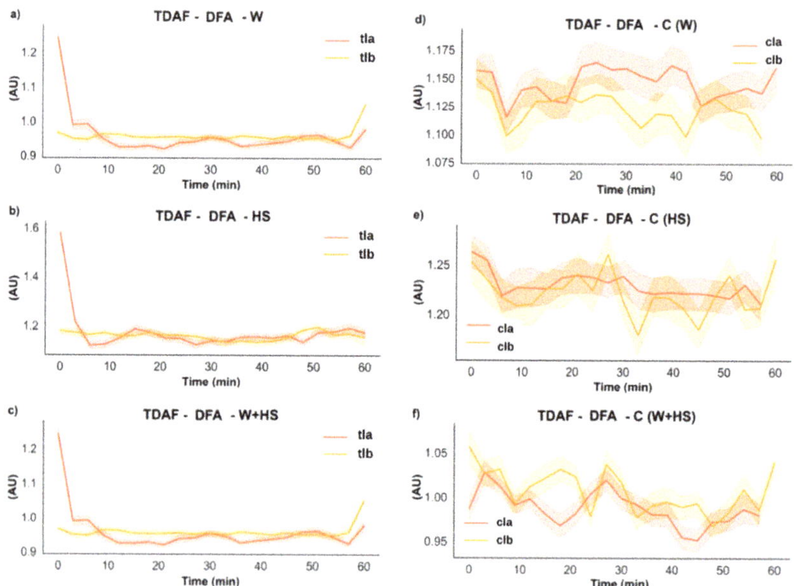

Figure 5. TDAF plots for DFA analysis. Results for (**a**) the local wound (W), (**b**) heat shock (HS), and (**c**) wound + heat shock (W+HS). Results of the control plants for (**d–f**) wound C (W), heat shock C (HS), and wound + heat shock C (W+HS), respectively. tlb: treated local before stimulus; tla: treated local after stimulus; clb: control local before stimulus; cla: control local after stimulus. AU: arbitrary unit.

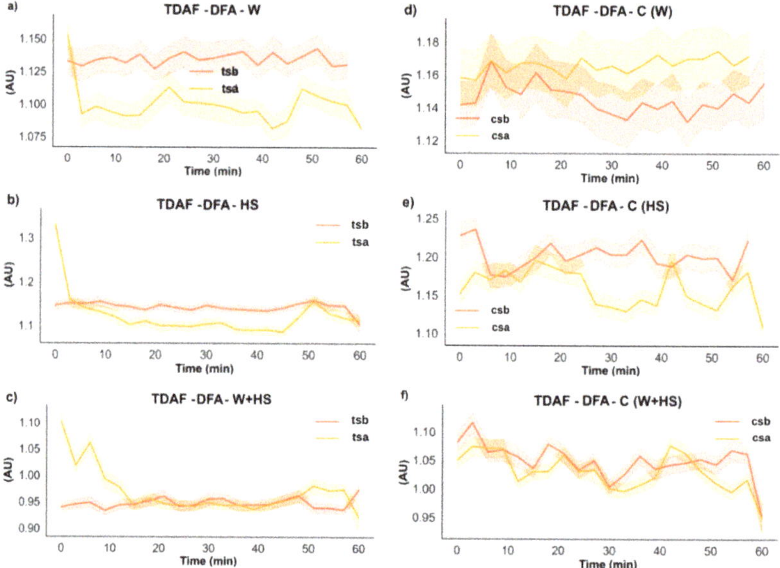

Figure 6. TDAF plots for DFA analysis. Results for (**a**) the systemic wound (W), (**b**) heat shock (HS), and (**c**) wound + heat shock (W+HS). Results of the control plants for (**d–f**) wound C (W), heat shock C (HS), and wound + heat shock C (W+HS), respectively. tsb: treated systemic before stimulus; tsa: treated systemic after stimulus; csb: control systemic before stimulus; csa: control systemic after stimulus. AU: arbitrary unit.

2.2.4. Average Band Power (ABP)

The stimulus induced an increase in the energy of the low frequencies (0–0.5 and delta 0.5–4 Hz), but this increase returned to the previous values after the first 6 min (Figure 7). However, for W+HS (Figure 7c), changes in frequency energy were observed in two moments: first, an increase in the energy of the frequency was observed at the beginning of the runs, which was maintained for up to 10 min, followed by a reduction in the energy; second, the energy increased again around 50 min later and then decreased once more. The differences in the controls found in the first moments are smaller than the global average; therefore, we do not consider that there is a remarkable difference in these cases. However, the control for W+HS (Figure 7f) maintained a steady increase for more than 16 min.

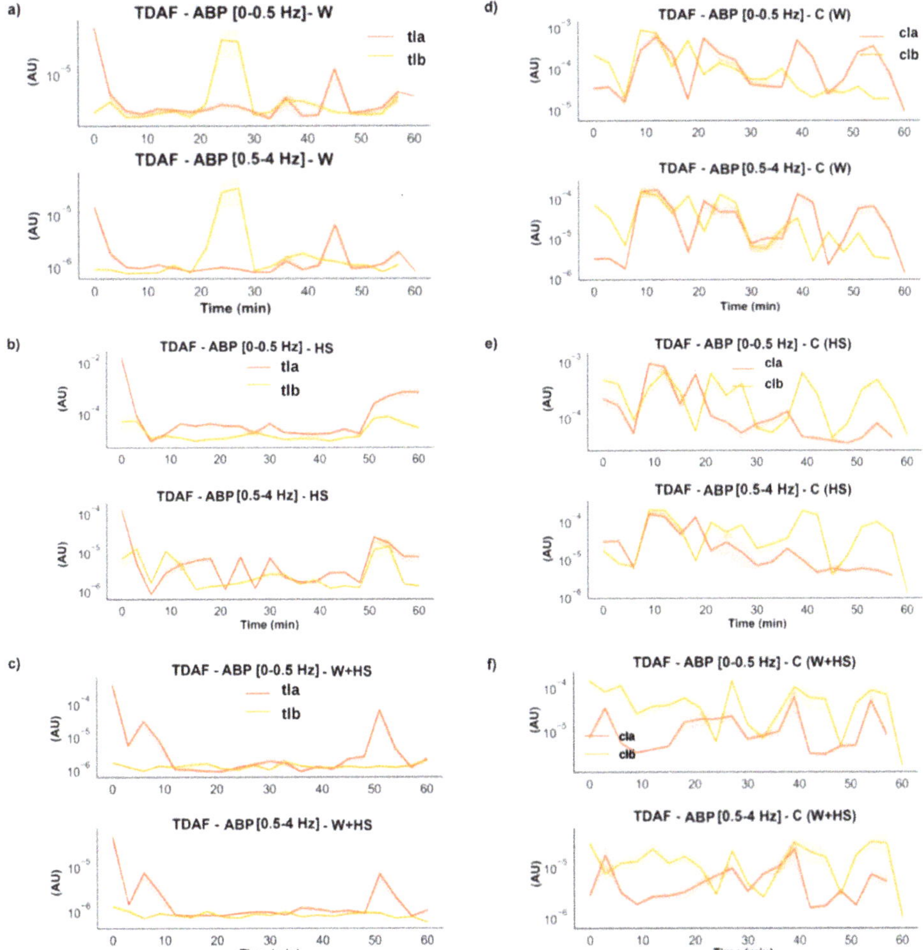

Figure 7. TDAF graphs for ABP analysis for low and Delta frequencies. Results for (**a**) the local wound (W), (**b**) heat shock (HS), and (**c**) wound + heat shock (W+HS). Results of the control plants for (**d–f**) wound C (W), heat shock C (HS), and wound + heat shock C (W+HS), respectively. tlb: treated local before stimulus; tla: treated local after stimulus; clb: control local before stimulus; cla: control local after stimulus. AU: arbitrary unit.

In Figure 8, the results for low and delta ABP for the systemic leaves indicate that there was a change for the plants that suffer the injuries, but no substantial change was found for the control plants.

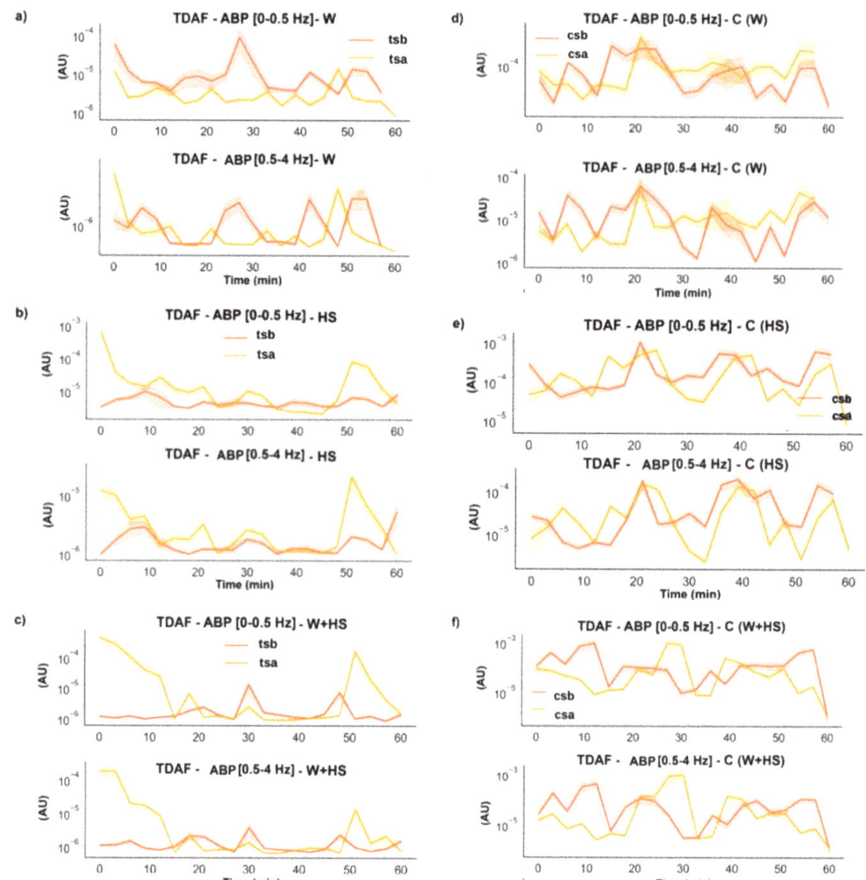

Figure 8. TDAF graphs for ABP analysis for low and delta frequencies. Results for (**a**) the systemic wound (W), (**b**) heat shock (HS), and (**c**) wound + heat shock (W+HS). Results of the control plants for (**d**–**f**) wound C (W), heat shock C (HS), and wound + heat shock C (W+HS), respectively. tsb: treated systemic before stimulus; tsa: treated systemic after stimulus; csb: control systemic before stimulus; csa: control systemic after stimulus. AU: arbitrary unit.

2.3. Measures of Turgor Pressure Variation

Measurements of turgor pressure variation indicated that after applying stimuli to local leaves, slight changes at a systemic level can be observed. In each stimulus, it was possible to observe some specific changes in the turgor pressure variation (Figure 9). Just after heat shock only, a sharper oscillation can be observed, followed by an average increase in the coefficient of variation (18.84 to 27.06%), indicating higher irregularity of the turgor pressure dynamics (Figure 9b). After the wound alone, a specific change was not clear, except for a damping in the turgor pressure oscillations approx. 15 min after the stimulus (Figure 9c). The main alterations were observed in combined stimuli (W+HS), which induced a quick and sharp increase in the turgor pressure after approx. 4 min. Moreover, the average coefficient of variation increased from 9.89% to 28.27%.

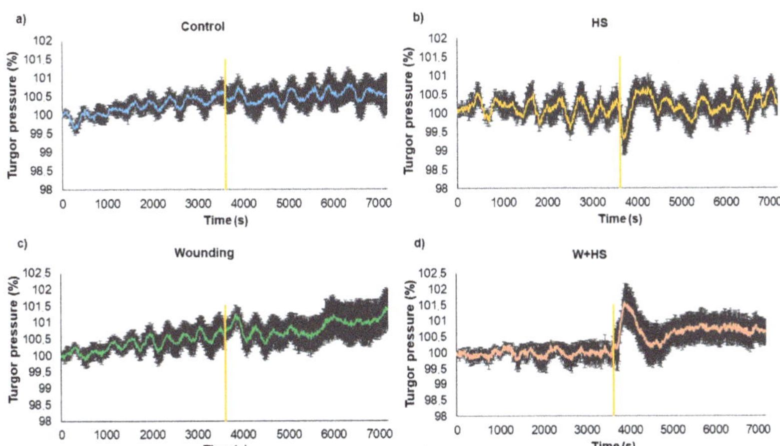

Figure 9. Measurements of the turgor pressure variation of systemic leaves over 2 h. Hydraulic pressure is represented as a percentage of the initial turgor pressure at 0 s. The stimuli (**a**–**d**) were applied after 1 h of measurement (represented by the vertical orange line). Each graph represents the average of 6 biological replicates (n = 6) and the standard deviation in each sampled point (time step = 10 s). Heat shock (HS), wound (W), and combined stimuli (W+HS). The average coefficients of variation are shown for the data before (C.V.B) and after (C.V.A) stimulation.

Interestingly, although no specific changes were observed in control plants as expected, the average CV% decreased (21.5–14%) after the moment when the other plants were stimulated (Figure 9a). Considering that both plants (control and stimulated) were close to each other in the Faraday cage during the electrome measurements, it would be plausible to consider a likely secondary effect caused by emitted VOCs (volatile organic compounds) from stimulated to non-stimulated plants [29].

3. Discussion

Environmental factors can affect plants in a spatially heterogeneous way; thus, long-distance signaling systems play an important role in the emergence of plant adaptations, allowing local responses to propagate throughout all parts of the plant body [1,30,31].

In addition, each stimulus can affect the plant in a specific way and trigger different responses in each module [3]. Therefore, the focus of recent research has been understanding how plants respond to different stimuli, which signals are activated, and if there is a pattern that can be identified [15–18,27].

Stress combination is a term used to describe the situation in which a plant is subjected to two or more abiotic stresses simultaneously. Although the combination of stress has been recognized as one of the main causes of crop loss worldwide [28,32], only more recently has this been addressed in laboratory studies at the molecular level [33,34], while there are few studies that evaluate the dynamics of signals after/during the occurrence of combined stimuli [35].

Given the complexity of the systemic signals involved, particularly the plant electrome, in this work, by applying time series analysis methods for the electrome of common bean leaves, we identified that wound stimuli (W), heat shock (HS), and the combined stimuli (W+HS) caused local and rapid systemic reactions and that the pattern of dynamics observed for single stimuli is different from those applied in combination.

Frequencies with higher amplitude and more energy, as shown by the results of ABP and DFA exponent (Figures 5–8), were found in the first stretches of the time series, followed by a reduction in ApEn values that indicates greater regularity, which may be associated with stress effects on dynamical aspects of plant physiology [14]. The studies by Souza et al. [36] and Saraiva et al. [14] evaluated the β exponent (an energy indicator of frequencies) after submitting soybean plants to different stress conditions increasing the exponent concerning the control treatment. Stressful conditions demand more energy from the system, and this energy may be behind the triggering of the systemic change in the electrome, considering that after the cut, the frequencies of local and systemic electromes showed an increase in the amplitude for W and W+HS, but at a lower intensity than that observed in local and systemic HS. Therefore, the need for frequencies with sufficient energy to trigger a consistent systemic change in electrome is apparent.

Differently from the isolated stimuli, the combination of stimuli (W+HS) promoted a biphasic behavior of the electrome, indicating a possible integration of the signals triggered by these stimuli. Similar behavior was observed when a combination of heat and re-irrigation was applied to maize plants, generating a heat-induced variation potential, followed by a transient depolarization induced by re-irrigation [37].

The results of turgor pressure variation in the systemic leaves indicated that a hydraulic signal accompanies the electrical signal in inducing a specific systemic response to stimuli, as well as changes in ROS levels. Among the changes in hydraulic dynamics, the most evident was observed after the HS and W+HS stimuli. In the study by Vuralhan-Eckert et al. [35], in which stomatal conductivity and photosynthesis were evaluated, the application of both stimuli at the same time showed that maize plants respond first to injury events and then to processes induced by re-irrigation. This observation is consistent with our results for turgor pressure variation, in which the hydraulic dynamics after W+HS present a pattern similar to W, but with greater amplitude and duration, possibly associated with the additive effect of the hydraulic waves induced by the combined stimulus.

When leaflets are damaged by burning or wound, there is an immediate loss of water content in the plants, and on many occasions, these injuries can disturb the plant's vascular system, which has a direct effect on the turgor pressure of epidermal cells [37,38]. According to Johns et al. [5], physical damage that disrupts the integrity of the xylem should release tension in these vessels, and due to the relatively incompressible nature of water, this pressure change will be transmitted almost instantly through the vascular tissues. For this reason, it is possible to observe changes in hydraulic dynamics immediately after the stimuli.

Recently, a strong connection between REDOX metabolism and systemic signaling in plants during stress has been uncovered [39]. Studies have shown that ROS levels, the expression of different ROS scavenging enzymes, and the level of different antioxidants exhibit a unique pattern during combination stress that is different from that found to be induced by each of the different stresses applied separately [40]. Herein, an increase in hydrogen peroxide and superoxide content after W and HS were observed in both local and systemic leaves (Figure 1). Therefore, the stimuli triggered an increase in the production of ROS in the local leaf which, in turn, signaled the production of ROS in the systemic leaves.

Studies evaluating systemic stomatal responses in Arabidopsis have shown dependence on systemic reactive oxygen species (ROS) signals and include systemic stomatal closing responses to light or injury, as well as systemic opening responses to heat [1,26,40,41]. In soybean plants, systemic stomatal responses were also associated with a dependence on systemic ROS signaling [6,26]. It is possible to associate the results mentioned above with those observed in our study. The results showed an increase in ROS after W and an increase in turgor pressure which is related to stomatal closure, while for HS, it is possible to see a significant increase in ROS and a decrease in turgor pressure, often related to opening stomata. In effect, stomatal opening increases leaf transpiration rate and, consequently, decreases water pressure inside leaves, reducing the whole turgor pressure [42].

The reduction in the level of lipid peroxidation observed only for the combination of stimuli suggests a rapid activation of defense mechanisms that could be better clarified through the evaluation of enzymatic and non-enzymatic defense mechanisms, not explored in this study. According to Dvořák et al. [43], in general, increased ROS production caused by a plethora of environmental stimuli rapidly triggers antioxidant defense by several mechanisms, including retrograde signaling, transcriptional control, post-transcriptional regulation, post-translational redox modifications or phosphorylation, and protein–protein interactions.

Although stimulation caused changes in hydraulic and electrical dynamics for a while, the behavior of these signals returns to pre-stimulation values minutes after the stimulus ceased (Figures 2–9). According to Kranner et al. [44], the response of a plant to stress will vary depending on the duration (short and long term) and severity of stress. "Lower stress events" can be partially compensated by acclimatization repair mechanisms, while severe or chronic stress events cause considerable damage and can lead to cell and tissue death. Herein, local stimuli for a short period (see Section 4) did not lead to a "collapse" of the plant cells, which allowed the plant to return to a non-stimulated condition minutes after the application of the stimuli.

It has not gone unnoticed that after the stimuli, the electrome behaved in a manner that previous studies suggest to be a state of attention in plants [16,45]. According to the hypothesis developed by Parise et al. [45], attention in plants is a disproportionate investment of energy in an activity or the perception of a stimulus or set of stimuli [46], and it could be observed through electromic analyses when there is a drop in the electrome complexity accompanied by an increase in the correlation of the signals and, likely, an increase in the energy of the electrome [45]. This is precisely what was observed here. After stimulation, there was a transient decrease in the ApEn in both local and systemic leaves, together with an increase in the correlation and ABP. According to the hypothesis, when a plant faces a challenge, the modules must synchronize their functioning to respond in coordination to the stimuli perceived. Since they will be working in concert, there will be a decrease in the complexity of the electrical signals, which will become more regular and predictable. At the same time, there will necessarily be an increase in the correlation of the signals and a likely increase in their energy. This is what is called plant attention [45]. This state is not expected to last long, but only until the problem is solved, or the actions needed to face it are completed. Examples of such actions would be the delivery of information to distant modules and/or the achievement of a new physiological state (acclimatization) when the signs of attention in the electrome would not be detectable anymore. The transient behavior of plant attention, emphasized by Parise et al. [45], was also observed in this study, where attention-related alterations in the electrome lasted for around 15 min. Therefore, this work corroborates the hypothesis of plant attention.

Furthermore, we have noticed some unexpected changes in bioelectrical activity in the control (non-stimulated) plants, although slight and transitory (Figures 2–8). Because the control plants were in the same Faraday cage as stimulated ones, we cannot discard the possibility that volatile organic compounds (VOCs) could have been released from the stressed plants and somehow affected the behavior of non-stimulated ones. It is known that VOCs are powerful signaling molecules allowing plant–plant communication, especially under stressful situations [33]. In this vein, it would be instructive to carry out specific studies to test such a possibility, i.e., to test the hypothesis that plant–plant communication mediated by VOCs can trigger bioelectrical changes in plants. Such a hypothesis has already been suggested by Parise et al. [16] when the electrome of *Cuscuta racemosa* showed evidence of being affected by a distant host.

4. Materials and Methods

4.1. Plant Material and Growing Conditions

Plants for the experiments were obtained by germination of common bean seeds (*Phaseolus vulgaris* L. cv. IAC Netuno) sowed in Gerbox® boxes lined with germitest paper moistened with 15 mL of distilled water. When the roots were 1 to 2 cm long, the seedlings were transplanted into 380 mL polystyrene pots (drilled in the base) containing 450 g of washed and sterilized sand and kept under a customized lighting system composed of LED lamps providing a photosynthetically active photon flux density (DFFFA) of approximately 350 μmol m^2 s^{-1}. The lighting system was connected to a timer that set the photoperiod at 14 h light and 10 h dark. Air humidity was maintained around 74% and the temperature at 25 ± 1 °C.

During the growth period, the plants were watered daily with distilled water (40 mL), and three times a week, they were supplemented with Hoagland and Arnon (20 mL) nutrient solution [47]. The plants were kept under these conditions until the third trifoliate leaf was fully expanded.

4.2. Experimental Design and Evaluation of the Dynamics of Systemic Responses Induced by Simple and Combined Stimuli

To test our hypothesis, three trials were conducted, where in each one, the stimuli (treatments) of heat shock (HS), wounding (W), and heat shock + wounding were applied locally (i.e., on single leaves). Then, the time series of the bioelectrical signals and the variation in the turgor pressure, in addition to the samples for the analysis of ROS content and lipid peroxidation, were evaluated as follows.

For the evaluation of single stimuli, the central leaflet of the second trifoliate leaf was selected for the local application of the HS and W treatments, while the first trifoliate leaf was not stimulated (systemic tissue) (Figure 10A). For the combined stimuli, the central leaflet of the third and second leaves was subjected simultaneously to heat and wounding stimuli (HS+W treatment), and the first trifoliate leave remained not stimulated (systemic tissue) (Figure 10B). The essays performed are detailed below.

- Essay I—Heat shock (HS): The heat shock stimulus (HS) was applied by placing a flame approximately 10 cm from the central leaflet for 20 s. Measurements in unstimulated (control) plants were also obtained. Leaf temperature was measured with an infrared camera (FLIR Systems) on local leaves before and after stimulation to have an idea of leaf temperature variation during the test. According to the measurements, after stimulation, the local leaf temperature increased by an average of $\pm 23\ ^{\circ}\mathrm{C}$ (data not shown).
- Essay II—Wounding (W): With calibrated scissors, 2 cm cuts in the central leaflet of the second trifoliate leaf were made to induce systemic wound signaling and response.
- Essay III—Heat shock + Wounding (HS+W): The stimuli were applied simultaneously to different leaves of the same plant to induce a signaling and systemic response to the combined stimulus. The 2 cm cut was applied to the third leaflet and the application of thermal shock was to the second leaflet for 20 s.

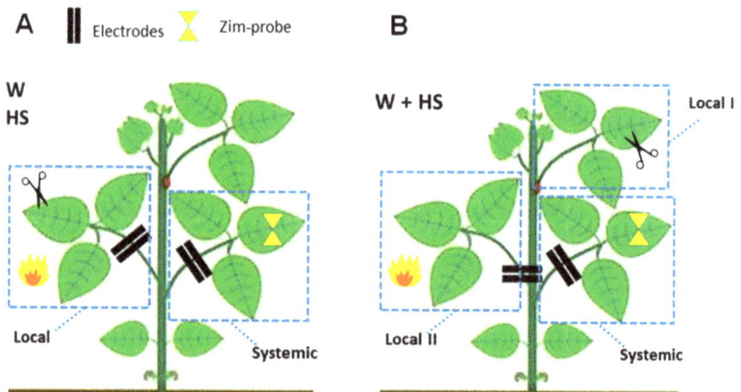

Figure 10. Representation of the experimental design indicating the place of application of the stimuli (W, HS, and W+HS) and where the systemic evaluation of the signals was performed. (**A**,**B**) The dashed rectangular border indicates the location (local tissue) where the different stimuli were applied. The yellow hourglass represents the turgor pressure probe installed on the systemic sheet. The two black lines indicate the electrode insertion region.

In each HS, W, and HS+W essay, electrome measurements, turgor pressure variation, quantification of ROs, and lipid peroxidation were obtained, in addition to measurements in unstimulated plants (control).

4.3. Quantification of ROS and Lipid Peroxidation

The leaf samples were collected a few seconds after the application of the stimuli, in different plants than those that were being used for bioelectrical analysis since leaf detachment will cause extreme bioelectrical reactions. The samples were immediately weighted, frozen with liquid nitrogen, and kept in an ultra-freezer.

The superoxide (O_2^-) content was determined according to Li et al. [48] in all experiments. For extraction, tissues (0.2 g) were ground in 65 mM phosphate buffer, pH 7.8, and centrifuged at 5000 g for 10 min. The supernatant was mixed with 65 mM phosphate buffer, pH 7.8, and 10 mM hydroxylamine hydrochloride, and placed at 25 °C for 20 min. Then, 17 mM sulfanilamide and 7 mM alfa-naphthylamine at a final concentration were added to the mixture. The absorbance of the solution at 530 nm was measured after incubation for 20 min at 25 °C. A standard curve with nitrite dioxide (NO_2^-) radical was used to calculate the rate of O_2^- generation.

The levels of hydrogen peroxide (H_2O_2) were determined according to Velikova et al. [49] in each experiment. Tissues (0.2 g) were powdered in 0.1% acid (w:v) trichloroacetic acid (TCA). The homogenate was centrifuged (12,000× g, 4 °C, 20 min), and the supernatant was added to 10 mM potassium phosphate buffer, pH 7.0, and 1 M potassium iodide. The absorbance of the reaction was measured at 390 nm. The H_2O_2 content was given on a standard curve prepared with known concentrations of H_2O_2.

To measure lipid peroxidation, the thiobarbituric acid (TBA) test was used, which determines malondialdehyde (MDA) as the end product of lipid peroxidation [49]. The material (0.1 g) was homogenized in a 0.1% (w:v) TCA solution. The homogenate was centrifuged (12,000× g, 4 °C, 20 min), and the supernatant was added to 0.5% (w:v) TBA in a 10% TCA solution. The mixture was incubated in hot water (90 °C) for 20 min, and the reaction was stopped by placing the reaction tubes in an ice bath for 10 min. Then, the samples were centrifuged at 10,000× g for 5 min, and the absorbance was read at 535 nm. The value for non-specific absorption at 600 nm was subtracted. The amount of MDA-TBA complex (red pigment) was calculated from the extinction coefficient ($\varepsilon = 155 \times 10^3$ M^{-1} cm^{-1}).

4.4. Electrome Acquisition and Analysis

4.4.1. Electrophytogram (EPG)

Bioelectrical measurements were performed using the electrophytogram (EPG) method [14]. A Biopac Student Lab System was employed for bioelectrical data acquisition system, model MP-36 (Goleta, CA, USA), with 4 channels of high impedance (10 GΩ), SSL2 cables, and 12 mm thick stainless steel needle electrodes (model EL-452). A pair of electrodes, placed 1 cm from each other, was inserted in the plants to capture the signals from the local and systemic leaves. The measurements were performed in 1 plant not stimulated and 1 plant stimulated, simultaneously inside a Faraday cage to avoid electrostatic effects from the laboratory power net. In Essay III, one of the electrodes was introduced in the internode between stimulated leaves (sites I and II) and another in the petiole of the systemic leaf (not stimulated).

The voltage variations (ΔV, measured in μV) captured by the electrodes go through a series of steps starting from the data acquisition system, amplification, filtering, and conversion of these signals. In this study, we used the electrocardiogram (ECG-AHA) function present in the MP-36's BSL-PRO software [15,16]. We used amplifiers with high input impedance (>109 Ω) to avoid signal distortions. A minimum high-pass filter of 0.5 Hz and another low-pass filter of 1.5 kHz was used to filter the signal, so that all frequencies below 0.5 Hz and above 1.5 kHz were attenuated. In addition, a 60 Hz band-stop was adopted on all channels to avoid noise from electronic components present in the laboratory. In all experiments, a data capture frequency of 62.5 Hz with a gain (amplification) of 1000 times was used.

A period of 24 h of acclimatization to the electrodes was necessary before data acquisition since the moment of insertion of the electrodes causes wound responses in the plant that normalize after a few hours [50].

Electrome measurements of each plant were obtained in the form of time series of microvolt variation as $\Delta V = \{\Delta V1, \Delta V2, \ldots, \Delta Vn\}$, where ΔV is the potential difference between the electrodes inserted in the leaf petiole and n is the time series size. The duration of the series is derived from a one-hour sample of data acquisition, with an acquisition rate of 62.5 Hz, totaling n = 225,000 points.

The measurements of the local and systemic leaves were divided into two moments: (1) one hour of measurement before stimulation, and (2) one hour of measurement after stimulation. This form was used for each of the treatments. The W, HS, and W+HS treatments were classified as follows: tlb—treated local b; tla—treated local a; tsb—treated systemic b; tsa—treated systemic a. Control groups were classified as: clb—control local b; cla—control local a; csb—control systemic b; csa—control systemic a. Ninety time series (n = 225,000 points) were obtained in each group of the different treatments.

4.4.2. Electrophysiological Analyzes

After acquiring the measurements, each time series was analyzed by the following techniques.

Visual Inspection

Visual inspection of time series allows a preliminary search for patterns or changes in the raw runs. Although descriptive and subjective, it allows a first analysis of the behavior of the time series, as well as some comparisons between them, such as the high presence of voltage variation peaks in specific stretches of the series [15,16].

Analysis of the Dispersion of Features over Time (Time Dispersion Analysis of Features—TDAF)

The TDAF method, developed by our research group, aims to demonstrate the dynamics of the electrome run characteristics (features). The latest studies published in the literature suggest that 3 min long measurements already contain enough information to identify behaviors in the plant electrome [16–18]. Thus, the TDAF assumes the use of the smallest possible unit of the analyzed series, considering the measurement equipment, and uses the dispersion theory [51] to analyze the dataset as a whole. Sometimes, analyses considering an entire time series (TS) can mask behaviors that happen within a few seconds or minutes. Thus, through several tests, we were able to observe the behavior of bioelectrical runs in different time frames.

Considering this information, the first action in the generation of the features is the division (cut) of all the TSs of each treatment in a series of less than 1 min interchangeable, with a delay of 20%, aiming to eliminate tendencies that the TS may have in the cut point. For each group of treatments (W, HS, W+HS, and Control), a total of 9300 samples were obtained. The groups were classified as follows: tlb—treated local b; tla—treated local a; tsb—treated systemic b; tsa—treated systemic a; clb—local control b; cla—local control a; csb—systemic control b; csa—control systemic a.

In summary, the TDAF divides the series at the same instant in time for all the TSs analyzed. Afterward, each piece with its respective position in time receives this index (Figure 11). This marker is used to return all results to their respective bins. Thus, in the end, the bins will contain all the time series and features; with these values, it is possible to obtain the data dispersion (maximum/minimum values, median quartiles) minute by minute.

An advantage of this analysis is the visualization of the behavior of the resources and how they differ, or not, for each class, without the need for any extra calculation, just manipulation via software. However, the major drawback is that it can only be used for time-bound event classifications, and highly diversified TS causes inaccurate and noisy analysis.

For each sample, we define the features based on different time series analysis techniques, which together help in electrome classification.

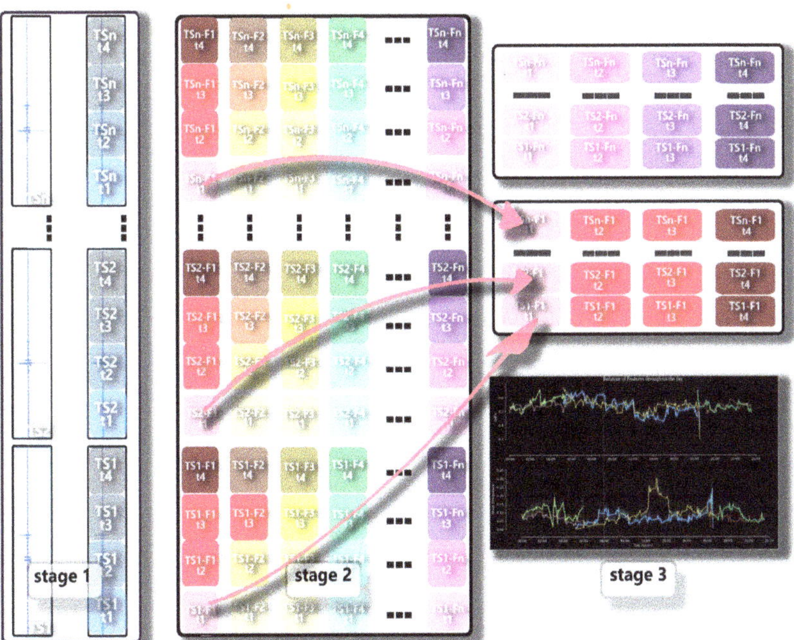

Figure 11. TDAF schematic. In stage 1, we have the time series (TS1, TS2, . . . , TSn) and the process of dividing and storing the position information referring to the time of each cut (t1, t2, t3, t4). In stage 2, the characteristics (F1, F2, F4, . . . , Fn) are calculated for each slice and together with the result, the timing information is virtually saved. In stage 3, time information is used for each generation and bin placement (data grouping). With each bin in place, the original time series is reconstructed, now with the results of each analyzed feature.

Detrended Fluctuation Analysis (DFA)

Detrended fluctuation analysis (DFA) calculates a power law scale estimate (similar to the Hurst exponent). For instance, by observing changes in the scale exponents, this method was able to identify heart diseases by analyzing the time series of an electrocardiogram [43]. Therefore, this analysis can calculate the autocorrelation of a time series and indicate its long-term and short-term memory index. The DFA calculation returns values that vary as follows: $1 - \alpha < 1$ the signal is stationary (oscillating around a constant mean) and can be modeled as fractional Gaussian noise; $2 - \alpha = 0.5$ indicates a lack of correlation or memory, indicating a noisy signal; $3 - 0.5 > \alpha < 1$ corresponds to a signal with positive correlation memory; $4 - \alpha < 0.5$ the correlation is negative; $5 - \alpha > 1$ the signal is non-stationary and can be modeled as fractional Brownian motion as $H = a - 1$ [52]. The Nolds library's DFA method was used for resource generation.

Average Band Power (Average Band Power—ABP)

The ABP was calculated with the values obtained from the PSD. Briefly, this method analyzes the signal at specific frequencies such as "delta" (0.5–4 Hz), "theta" (4–8 Hz), "alpha" (8–12 Hz), "beta" (12–30 Hz), and "gamma" (30–100 Hz). It is commonly used to generate features in neuroscience EEG analysis [53–59]. To calculate the ABP, we integrated the PSD area determined by the region of interest. Composite Simpson's rule was used, employing the Simpson method from the script library [60]. In general, Simpson's rule decomposes the area into several parabolas and then sums them up, returning the total value of the area of interest. Herein, since the sampling rate was 62.5 Hz, the ABPs of the frequencies of 0–0.5 Hz (here called low {low}), 0.5–4 Hz delta, 4–8 Hz theta, 8–12 Hz alpha, and 12–30 Hz beta were calculated.

Approximate Entropy

Approximate entropy (ApEn) provides information about the level of organization of the time series and considers the temporal order of the points in the sequence of a time series; therefore, it is preferable to measure the regularity or randomness of biological signals [61,62]. Higher ApEn values indicate the existence of more irregular dynamics (greater complexity), while lower values indicate that the dynamics are more regular and deterministic [63–65].

The development of this analysis aimed to discriminate between two time series generated by different systems, or two time series generated by the same system under different physiological conditions [63–65].

Lower ApEn values were associated with compromised and deteriorated physiological processes; that is, they present greater regularity, while healthy physiological processes are more complex [61].

The definition of approximate entropy comes from Takens' theorem [63], from which the set of M vectors $\Delta vj = (\Delta vj, \Delta vj\text{-}\tau, \Delta vj\text{-}2\tau, \ldots, \Delta vj\text{-}(m\text{-}1)\tau)$ of m embedding dimension with lag τ, para j = 1, 2, 3, $\ldots$, M and $M = N - (m - 1)\tau$, where N is the size of the one-dimensional series $(\Delta V = \Delta V1, \Delta V2, \Delta V3, \ldots, \Delta VN)$. For each vector j, the number of neighbors within a hypersphere of radius r is calculated. Subsequently, the mean of the logarithm of the number of neighbors is taken, which is given by $(r)= 1M \text{ J-}1 \text{ M} ln (N(j)viz \leq r)$. Thus, ApEn is defined in the following relation:

$$ApEn = \Phi m(r) - \Phi m + 1(r), \tag{1}$$

For the calculations, we set m = 2, τ = 1 e r = 0.02. σ, where σ is the standard deviation of the original time series. The choice of m = 2 is due to the better efficiency of the ApEn calculation for small time series [63].

In this study, the approximate entropy obtained from the time series was compared between the moments "before" and "after" each stimulus (single or combined).

4.5. Measurements of Turgor Pressure Variation

The non-invasive Yara ZIM® probe measures the pressure difference between two magnets and leaf turgor, called patch pressure (Pp), and therefore provides information on relative changes in leaf turgor in real time over long periods under laboratory or field conditions [66,67].

Here, we chose to install the probes only on the systemic leaf, due to the limited number of probes available, as well as difficulties during installation on the delicate leaves of common beans that sometimes led to tissue damage by the pressure exerted by the magnets. Furthermore, before the application of stimuli, we considered three days of observation after the probes' installation to make sure that the installation was successful. After this period, measurements were carried out one hour before and one hour after the application of W, HS, and HS+W stimuli, with a frequency of one measurement every 10 s.

4.6. Statistical Analysis

ROS quantification and lipid peroxidation data were analyzed as a completely randomized design in a double factorial scheme, where one of the factors was the tissue (2 levels—local/systemic) and the treatments were the second factor (W, HS, and W+HS). Analysis of variance (ANOVA) was performed, followed by the Tukey test ($p < 0.05$), to assess the interaction between the factors. For electrical signal data, in the ABP, ApEn, and DFA analyses, the data are presented as median $\pm$ SD (standard deviation of the mean). The results for the turgor pressure variation measurements obtained by the Zim probe were calculated as the control percentage, which is the pressure measured on the leaf 1 h before the stimulus application. Each dataset includes SE of 6 biological repeats. For statistical analysis, the statistical programs RStudio 1.2.1335 and Sigmaplot 12.0 were used.

5. Conclusions

Our results support the hypothesis that ROS signals likely act together with hydraulic and electrical signals inducing systemic responses triggered by simple and combined local stimuli. However, although the transmission of electrical signals is important in the way plants respond to their environment, it is not yet known how plants deal with a situation in which external stimuli act simultaneously, nor how the different signals behave under these conditions. The present work, therefore, tries to take a step forward in understanding this interesting gap in plant ecophysiology. Our results showed that changes in electrome were different between types of isolated stimuli, including the combination of these, and that these electrical signals are accompanied by systemic changes in hydraulic and ROS dynamics. It is possible that the combination of stimuli in this study promoted an additive effect, inducing the activation of enzymatic and non-enzymatic antioxidant mechanisms more quickly, as well as inducing a more intense hydraulic wave and an electrical signature resulting from the signals triggered by each of the stimuli.

In addition to the participation of electrical and hydraulic signals, associated with the responses in a variation of turgor pressure and ROS analyzed here, we do not rule out other systemic signals such as nitric oxide (NO), RNAs, hormones, and volatile organic compounds (VOCs) [68–71], acting together and helping to activate systemic responses. Indeed, a mix of systemic signals is supposed to coordinate a complex network of internal communication integrating stress responses throughout the whole plant [4–6], which could also be related to a state of attention in bean plants when perceiving and processing the stimuli [45,46].

Further studies are needed to address the interdependence of these signals. The combination of different stimuli, the application of inhibitors, or research with mutants indicate the influence of the electrical signal on the expression of a complex response. Likewise, the possible interference of plant–plant communication though volatiles or other means must be considered.

Author Contributions: Conceptualization, Á.V.L.C. and G.M.S.; methodology, Á.V.L.C. and G.M.S.; validation, W.S.B.; formal analysis, Á.V.L.C., T.F.d.C.O., D.A.P. and W.S.B.; investigation, Á.V.L.C., D.A.P. and G.N.R.; data curation, T.F.d.C.O.; writing—original draft preparation, Á.V.L.C., D.A.P., G.N.R. and G.M.S.; writing—review and editing, D.A.P., G.N.R., A.G.P. and G.M.S.; supervision, G.M.S.; funding acquisition, G.M.S. All authors have read and agreed to the published version of the manuscript.

Funding: This study was supported in part by the Coordenação de Aperfeiçoamento de Pessoal de Nível Superior, Brasil (CAPES) code 001. The authors are also grateful to Conselho Nacional de Desenvolvimento Científico e Tecnológico (CNPq) for the financial support provided (G.M.S. is supported by the CNPq grant 302592/2021-0).

Institutional Review Board Statement: Not applicable.

Data Availability Statement: Not applicable.

Acknowledgments: Coordenação de Aperfeiçoamento de Pessoal de Nível Superior, Brasil (CAPES), and Conselho Nacional de Desenvolvimento Científico e Tecnológico (CNPq).

Conflicts of Interest: The authors declare no conflict of interest.

References

1. Kollist, H.; Zandalinas, S.I.; Sengupta, S.; Nuhkat, M.; Kangasjärvi, J.; Mittler, R. Rapid Responses to Abiotic Stress: Priming the Landscape for the Signal Transduction Network. *Trends Plant Sci.* **2018**, *24*, 25–37. [CrossRef] [PubMed]
2. Hilleary, R.; Gilroy, S. Systemic signaling in response to wounding and pathogens. *Curr. Opin. Plant Biol.* **2018**, *43*, 57–62. [CrossRef]
3. Balfagón, D.; Sengupta, S.; Gómez-Cadenas, A.; Fritschi, F.B.; Azad, R.K.; Mittler, R.; Zandalinas, S.I. Jasmonic Acid Is Required for Plant Acclimation to a Combination of High Light and Heat Stress. *Plant Physiol.* **2019**, *181*, 1668–1682. [CrossRef] [PubMed]
4. Choi, W.-G.; Hilleary, R.; Swanson, S.J.; Kim, S.-H.; Gilroy, S. Rapid, Long-Distance Electrical and Calcium Signaling in Plants. *Annu. Rev. Plant Biol.* **2016**, *67*, 287–307. [CrossRef] [PubMed]
5. Johns, S.; Hagihara, T.; Toyota, M.; Gilroy, S. The fast and the furious: Rapid long-range signaling in plants. *Plant Physiol.* **2021**, *185*, 694–706. [CrossRef] [PubMed]

6. Fichman, Y.; Mittler, R. Rapid systemic signaling during abiotic and biotic stresses: Is the ROS wave master of all trades? *Plant J.* **2020**, *102*, 887–896. [CrossRef]

7. Huber, A.E.; Bauerle, T.L. Long-distance plant signaling pathways in response to multiple stressors: The gap in knowledge. *J. Exp. Bot.* **2016**, *67*, 2063–2079. [CrossRef]

8. Sukhov, V.; Sukhova, E.; Vodeneev, V. Long-distance electrical signals as a link between the local action of stressors and the systemic physiological responses in higher plants. *Prog. Biophys. Mol. Biol.* **2018**, *146*, 63–84. [CrossRef]

9. Fromm, J.; Hajirezaei, M.-R.; Becker, V.K.; Lautner, S. Electrical signaling along the phloem and its physiological responses in the maize leaf. *Front. Plant Sci.* **2013**, *4*, 239. [CrossRef]

10. Choi, W.-G.; Miller, G.; Wallace, I.; Harper, J.; Mittler, R.; Gilroy, S. Orchestrating rapid long-distance signaling in plants with Ca^{2+}, ROS and electrical signals. *Plant J.* **2017**, *90*, 698–707. [CrossRef]

11. De Loof, A. The cell's self-generated "electrome": The biophysical essence of the immaterial dimension of Life? *Commun. Integr. Biol.* **2016**, *9*, e1197446. [CrossRef] [PubMed]

12. De Toledo, G.R.A.; Parise, A.G.; Simmi, F.Z.; Costa, A.V.L.; Senko, L.G.S.; Debono, M.-W.; Souza, G.M. Plant electrome: The electrical dimension of plant life. *Theor. Exp. Plant Physiol.* **2019**, *31*, 21–46. [CrossRef]

13. Debono, M.-W. Electrome & Cognition Modes in Plants: A Transdisciplinary Approach to the Eco-Sensitiveness of the World. *Transdiscipl. J. Eng. Sci.* **2020**, *11*, 213–239. [CrossRef]

14. Saraiva, G.F.R.; Ferreira, A.S.; Souza, G.M. Osmotic stress decreases complexity underlying the electrophysiological dynamic in soybean. *Plant Biol.* **2017**, *19*, 702–708. [CrossRef]

15. Simmi, F.; Dallagnol, L.; Ferreira, A.; Pereira, D.; Souza, G. Electrome alterations in a plant-pathogen system: Toward early diagnosis. *Bioelectrochemistry* **2020**, *133*, 107493. [CrossRef]

16. Parise, A.G.; Reissig, G.N.; Basso, L.F.; Senko, L.G.S.; Oliveira, T.F.d.C.; de Toledo, G.R.A.; Ferreira, A.S.; Souza, G.M. Detection of Different Hosts From a Distance Alters the Behaviour and Bioelectrical Activity of Cuscuta racemosa. *Front. Plant Sci.* **2021**, *12*, 409. [CrossRef]

17. Reissig, G.N.; Oliveira, T.F.d.C.; de Oliveira, R.P.; Posso, D.A.; Parise, A.G.; Nava, D.E.; Souza, G.M. Fruit Herbivory Alters Plant Electrome: Evidence for Fruit-Shoot Long-Distance Electrical Signaling in Tomato Plants. *Front. Sustain. Food Syst.* **2021**, *5*, 244. [CrossRef]

18. Reissig, G.N.; Oliveira, T.F.d.C.; Costa, V.L.; Parise, A.G.; Pereira, D.R.; Souza, G.M. Machine Learning for Automatic Classification of Tomato Ripening Stages Using Electrophysiological Recordings. *Front. Sustain. Food Syst.* **2021**, *5*, 696829. [CrossRef]

19. Pereira, D.R.; Papa, J.P.; Saraiva, G.F.R.; Souza, G.M. Automatic classification of plant electrophysiological responses to environmental stimuli using machine learning and interval arithmetic. *Comput. Electron. Agric.* **2018**, *145*, 35–42. [CrossRef]

20. Christmann, A.; Grill, E.; Huang, J. Hydraulic signals in long-distance signaling. *Curr. Opin. Plant Biol.* **2013**, *16*, 293–300. [CrossRef]

21. Malone, M. Hydraulic signals. *Philos. Trans. R. Soc. B Biol. Sci.* **1993**, *341*, 33–39. [CrossRef]

22. Bramley, H.; Turner, D.; Tyerman, S.; Turner, N. Water Flow in the Roots of Crop Species: The Influence of Root Structure, Aquaporin Activity, and Waterlogging. *Adv. Agron.* **2007**, *96*, 133–196. [CrossRef]

23. Hamilton, E.S.; Schlegel, A.M.; Haswell, E.S. United in Diversity: Mechanosensitive Ion Channels in Plants. *Annu. Rev. Plant Biol.* **2015**, *66*, 113–137. [CrossRef] [PubMed]

24. Amien, S.; Kliwer, I.; Márton, M.L.; Debener, T.; Geiger, D.; Becker, D.; Dresselhaus, T. Defensin-Like ZmES4 Mediates Pollen Tube Burst in Maize via Opening of the Potassium Channel KZM1. *PLoS Biol.* **2010**, *8*, e1000388. [CrossRef] [PubMed]

25. Zandalinas, S.I.; Fichman, Y.; Devireddy, A.R.; Sengupta, S.; Azad, R.K.; Mittler, R. Systemic signaling during abiotic stress combination in plants. *Proc. Natl. Acad. Sci. USA* **2020**, *117*, 13810–13820. [CrossRef]

26. Fichman, Y.; Mittler, R. Integration of electric, calcium, reactive oxygen species and hydraulic signals during rapid systemic signaling in plants. *Plant J.* **2021**, *107*, 7–20. [CrossRef] [PubMed]

27. Suzuki, N.; Rivero, R.M.; Shulaev, V.; Blumwald, E.; Mittler, R. Abiotic and biotic stress combinations. *New Phytol.* **2014**, *203*, 32–43. [CrossRef]

28. Zandalinas, S.; Sengupta, S.; Burks, D.; Azad, R.K.; Mittler, R. Identification and characterization of a core set of ROS wave-associated transcripts involved in the systemic acquired acclimation response of Arabidopsis to excess light. *Plant J.* **2018**, *98*, 126–141. [CrossRef]

29. Midzi, J.; Jeffery, D.W.; Baumann, U.; Rogiers, S.; Tyerman, S.D.; Pagay, V. Stress-Induced Volatile Emissions and Signalling in Inter-Plant Communication. *Plants* **2022**, *11*, 2566. [CrossRef]

30. Balfagón, D.; Terán, F.; Oliveira, T.D.R.D.; Santa-Catarina, C.; Gómez-Cadenas, A. Citrus rootstocks modify scion antioxidant system under drought and heat stress combination. *Plant Cell Rep.* **2021**, *41*, 593–602. [CrossRef]

31. Mudrilov, M.; Ladeynova, M.; Grinberg, M.; Balalaeva, I.; Vodeneev, V. Electrical Signaling of Plants under Abiotic Stressors: Transmission of Stimulus-Specific Information. *Int. J. Mol. Sci.* **2021**, *22*, 10715. [CrossRef] [PubMed]

32. Mittler, R. Abiotic stress, the field environment and stress combination. *Trends Plant Sci.* **2006**, *11*, 15–19. [CrossRef] [PubMed]

33. Rizhsky, L.; Liang, H.; Shuman, J.; Shulaev, V.; Davletova, S.; Mittler, R. When Defense Pathways Collide. The Response of Arabidopsis to a Combination of Drought and Heat Stress. *Plant Physiol.* **2004**, *134*, 1683–1696. [CrossRef] [PubMed]

34. Suzuki, N.; Bassil, E.; Hamilton, J.S.; Inupakutika, M.A.; Zandalinas, S.I.; Tripathy, D.; Luo, Y.; Dion, E.; Fukui, G.; Kumazaki, A.; et al. ABA Is Required for Plant Acclimation to a Combination of Salt and Heat Stress. *PLoS ONE* **2016**, *11*, e0147625. [CrossRef]

35. Vuralhan-Eckert, J.; Lautner, S.; Fromm, J. Effect of simultaneously induced environmental stimuli on electrical signalling and gas exchange in maize plants. *J. Plant Physiol.* **2018**, *223*, 32–36. [CrossRef]

36. Souza, G.M.; Ferreira, A.S.; Saraiva, G.F.R.; Toledo, G.R.A. Plant "electrome" can be pushed toward a self-organized critical state by external cues: Evidences from a study with soybean seedlings subject to different environmental conditions. *Plant Signal. Behav.* **2017**, *12*, e1290040. [CrossRef]

37. Malone, M.; Stankovic, B. Surface potentials and hydraulic signals in wheat leaves following localized wounding by heat. *Plant, Cell Environ.* **1991**, *14*, 431–436. [CrossRef]

38. Malone, M. Rapid, long-distance signal transmission in higher plants. *Adv. Bot. Res.* **1996**, *22*, 163–228.

39. Peláez-Vico, M.; Fichman, Y.; Zandalinas, S.I.; Van Breusegem, F.; Karpiński, S.M.; Mittler, R. ROS and redox regulation of cell-to-cell and systemic signaling in plants during stress. *Free. Radic. Biol. Med.* **2022**, *193*, 354–362. [CrossRef]

40. Zandalinas, S.I.; Sales, C.; Beltrán, J.; Gómez-Cadenas, A.; Arbona, V. Activation of Secondary Metabolism in Citrus Plants Is Associated to Sensitivity to Combined Drought and High Temperatures. *Front. Plant Sci.* **2017**, *7*, 1954. [CrossRef]

41. Devireddy, A.R.; Zandalinas, S.I.; Gómez-Cadenas, A.; Blumwald, E.; Mittler, R. Coordinating the overall stomatal response of plants: Rapid leaf-to-leaf communication during light stress. *Sci. Signal.* **2018**, *11*, 1126. [CrossRef]

42. Peng, C.K.; Havlin, S.; Stanley, H.E.; Goldberger, A.G. Quantification of scaling exponents and crossover phenomena in nonstationary heartbeat time series. Chaos: An interdisciplinary journal of nonlinear. *Science* **1995**, *5*, 82–87.

43. Dvořák, P.; Krasylenko, Y.; Zeiner, A.; Šamaj, J.; Takáč, T. Signaling Toward Reactive Oxygen Species-Scavenging Enzymes in Plants. *Front. Plant Sci.* **2021**, *11*, 2178. [CrossRef]

44. Kranner, I.; Minibayeva, F.; Beckett, R.; Seal, C. What is stress? Concepts, definitions and applications in seed science. *New Phytol.* **2010**, *188*, 655–673. [CrossRef]

45. Parise, A.G.; de Toledo, G.R.A.; Oliveira, T.F.d.C.; Souza, G.M.; Castiello, U.; Gagliano, M.; Marder, M. Do plants pay attention? A possible phenomenological-empirical approach. *Prog. Biophys. Mol. Biol.* **2022**, *173*, 11–23. [CrossRef]

46. Marder, M. Plant intelligence and attention. *Plant Signal. Behav.* **2013**, *8*, e23902. [CrossRef] [PubMed]

47. Hoagland, D.R.; Arnon, D.I. Water-Culture Method Grow. Plants Without Soil. *Circ. Calif. Agric. Exp. Stn.* **1950**, *347*, 1–39.

48. Li, C.; Bai, T.; Ma, F.; Han, M. Hypoxia tolerance and adaptation of anaerobic respiration to hypoxia stress in two Malus species. *Sci. Hortic.* **2010**, *124*, 274–279. [CrossRef]

49. Velikova, V.; Yordanov, I.; Edreva, A. Oxidative stress and some antioxidant systems in acid rain-treated bean plants: Protective role of exogenous polyamines. *Plant Sci.* **2000**, *151*, 59–66. [CrossRef]

50. Volkov, A.; Haack, R. Insect-induced biolectrochemical signals in potato plants. *Bioelectrochemistry Bioenerg.* **1995**, *37*, 55–60. [CrossRef]

51. Jorgensen, B. *The Theory of Dispersion Models*; CRC Press: Boca Raton, FL, USA, 1997.

52. Hardstone, R.; Poil, S.-S.; Schiavone, G.; Jansen, R.; Nikulin, V.V.; Mansvelder, H.D.; Linkenkaer-Hansen, K. Detrended Fluctuation Analysis: A Scale-Free View on Neuronal Oscillations. *Front. Physiol.* **2012**, *3*, 450. [CrossRef] [PubMed]

53. Abhang, P.A.; Gawali, B.W.; Mehrotra, S.C. Technological Basics of EEG Recording and Operation of Apparatus. In *Introduction to EEG- and Speech-Based Emotion Recognition*; Abhang, P.A., Gawali, B., Mehrotra, S.C., Eds.; Academic Press: Cambridge, MA, USA, 2016; pp. 19–50.

54. Delimayanti, M.K.; Purnama, B.; Nguyen, N.G.; Faisal, M.R.; Mahmudah, K.R.; Indriani, F.; Kubo, M.; Satou, K. Classification of Brainwaves for Sleep Stages by High-Dimensional FFT Features from EEG Signals. *Appl. Sci.* **2020**, *10*, 1797. [CrossRef]

55. Evans, J.R. Neurofeedback. In *Encyclopedia of the Human Brain*; Elsevier: Amsterdam, The Netherlands, 2002; pp. 465–477.

56. Heraz, A.; Razaki, R.; Frasson, C. Using machine learning to predict learner emotional state from brainwaves. In Proceedings of the Seventh IEEE International Conference on Advanced Learning Technologies (ICALT 2007), Niigata, Japan, 18–20 July 2007.

57. Kora, P.; Meenakshi, K.; Swaraja, K.; Rajani, A.; Raju, M.S. EEG based interpretation of human brain activity during yoga and meditation using machine learning: A systematic review. *Complement. Ther. Clin. Pr.* **2021**, *43*, 101329. [CrossRef] [PubMed]

58. Savadkoohi, M.; Oladunni, T.; Thompson, L. A machine learning approach to epileptic seizure prediction using Electroencephalogram (EEG) Signal. *Biocybern. Biomed. Eng.* **2020**, *40*, 1328–1341. [CrossRef] [PubMed]

59. White, N.E.; Richards, L.M. Alpha–theta neurotherapy and the neurobehavioral treatment of addictions, mood disorders and trauma. In *Introduction to Quantitative EEG and Neurofeedback*; Elsevier: Amsterdam, The Netherlands, 2009; pp. 143–166.

60. Press, W.H.; Flannery, B.P.; Teukolsky, S.A.; Vetterling, W.T. *Numerical Recipes in Pascal (First Edition): The Art of Scientific Computing*; Cambridge University Press: Cambridge, UK, 1989.

61. Costa, M.; Goldberger, A.L.; Peng, C.-K. Multiscale entropy analysis of biological signals. *Phys. Rev. E* **2005**, *71*, 021906. [CrossRef]

62. Ocak, H. Automatic detection of epileptic seizures in EEG using discrete wavelet transform and approximate entropy. *Expert Syst. Appl.* **2009**, *36*, 2027–2036. [CrossRef]

63. Pincus, S.M. Approximate entropy as a measure of system complexity. *Proc. Natl. Acad. Sci. USA* **1991**, *88*, 2297–2301. [CrossRef]

64. Pincus, S.M. Approximate entropy (ApEn) as a complexity measure. *Chaos Interdiscip. J. Nonlinear Sci.* **1995**, *5*, 110–117. [CrossRef]

65. Pincus, S.M.; Goldberger, A.L. Physiological time-series analysis: What does regularity quantify? *Am. J. Physiol. Circ. Physiol.* **1994**, *266*, H1643–H16561994. [CrossRef]

66. Zimmermann, M.R.; Maischak, H.; Mithofer, A.; Boland, W.; Felle, H.H. System Potentials, a Novel Electrical Long-Distance Apoplastic Signal in Plants, Induced by Wounding. *Plant Physiol.* **2009**, *149*, 1593–1600. [CrossRef]

67. Zimmermann, U.; Bitter, R.; Marchiori, P.E.R.; Rüger, S.; Ehrenberger, W.; Sukhorukov, V.L.; Schüttler, A.; Ribeiro, R.V. A non-invasive plant-based probe for continuous monitoring of water stress in real time: A new tool for irrigation scheduling and deeper insight into drought and salinity stress physiology. *Theor. Exp. Plant Physiol.* **2013**, *25*, 2–11. [CrossRef]
68. Baudouin, E.; Hancock, J.T. Nitric oxide signaling in plants. *Front. Plant Sci.* **2014**, *4*, 553. [CrossRef] [PubMed]
69. Winter, N.; Kragler, F. Conceptual and Methodological Considerations on mRNA and Proteins as Intercellular and Long-Distance Signals. *Plant Cell Physiol.* **2018**, *59*, 1700–1713. [CrossRef] [PubMed]
70. Barbier, F.F.; Dun, E.A.; Kerr, S.C.; Chabikwa, T.G.; Beveridge, C.A. An Update on the Signals Controlling Shoot Branching. *Trends Plant Sci.* **2019**, *24*, 220–236. [CrossRef]
71. Zebelo, S.A.; Matsui, K.; Ozawa, R.; Maffei, M.E. Plasma membrane potential depolarization and cytosolic calcium flux are early events involved in tomato (Solanum lycopersicon) plant-to-plant communication. *Plant Sci.* **2012**, *196*, 93–100. [CrossRef]

Article

Development of Two-Dimensional Model of Photosynthesis in Plant Leaves and Analysis of Induction of Spatial Heterogeneity of CO_2 Assimilation Rate under Action of Excess Light and Drought

Ekaterina Sukhova, Daria Ratnitsyna, Ekaterina Gromova and Vladimir Sukhov *

Department of Biophysics, N.I. Lobachevsky State University of Nizhny Novgorod, 603950 Nizhny Novgorod, Russia
* Correspondence: vssuh@mail.ru; Tel.: +7-909-292-8643

Abstract: Photosynthesis is a key process in plants that can be strongly affected by the actions of environmental stressors. The stressor-induced photosynthetic responses are based on numerous and interacted processes that can restrict their experimental investigation. The development of mathematical models of photosynthetic processes is an important way of investigating these responses. Our work was devoted to the development of a two-dimensional model of photosynthesis in plant leaves that was based on the Farquhar–von Caemmerer–Berry model of CO_2 assimilation and descriptions of other processes including the stomatal and transmembrane CO_2 fluxes, lateral CO_2 and HCO_3^- fluxes, transmembrane and lateral transport of H^+ and K^+, interaction of these ions with buffers in the apoplast and cytoplasm, light-dependent regulation of H^+-ATPase in the plasma membrane, etc. Verification of the model showed that the simulated light dependences of the CO_2 assimilation rate were similar to the experimental ones and dependences of the CO_2 assimilation rate of an average leaf CO_2 conductance were also similar to the experimental dependences. An analysis of the model showed that a spatial heterogeneity of the CO_2 assimilation rate on a leaf surface should be stimulated under an increase in light intensity and a decrease in the stomatal CO_2 conductance or quantity of the open stomata; this prediction was supported by the experimental verification. Results of the work can be the basis of the development of new methods of the remote sensing of the influence of abiotic stressors (at least, excess light and drought) on plants.

Keywords: CO_2 assimilation; excess light; spatial heterogeneity; leaf CO_2 conductance; two-dimensional photosynthetic model; drought

Citation: Sukhova, E.; Ratnitsyna, D.; Gromova, E.; Sukhov, V. Development of Two-Dimensional Model of Photosynthesis in Plant Leaves and Analysis of Induction of Spatial Heterogeneity of CO_2 Assimilation Rate under Action of Excess Light and Drought. *Plants* **2022**, *11*, 3285. https://doi.org/10.3390/plants11233285

Academic Editors: Frantisek Baluska and Gustavo Maia Souza

Received: 10 October 2022
Accepted: 23 November 2022
Published: 29 November 2022

Publisher's Note: MDPI stays neutral with regard to jurisdictional claims in published maps and institutional affiliations.

1. Introduction

Photosynthesis is a key process in the life of green plants and the basis of their productivity. It is a complex process [1,2] that can be strongly affected by numerous abiotic stressors, including excess light [3–5] and fluctuations in light intensity [6–9], drought [10–12], decrease [13] and increase [14–16] in temperatures, and others.

Changes in the photosynthetic processes induced by the action of stressors include both the damage of photosynthetic machinery and numerous protective responses. The stressor-induced damages include photodamage under excess light [3–5], increase in proton leakage across the thylakoid membrane under heating [14], damage of photosynthetic complexes through the stimulation of the production of reactive oxygen species induced by the decrease in photosynthetic dark reactions under the action of various stressors [17], and others. The protective responses include the induction of a non-photochemical quenching [3,4,18,19], stimulation of a cyclic electron flow around photosystem I [7,19,20], translocation of Ferredoxin-NADP$^+$ Reductase [21,22], activation of photorespiration [23], changes in the positions of chloroplasts [24–26], and others. These processes can strongly interact, e.g., the stimulation of the cyclic electron flow increases the acidification of the lumen in chloroplasts and can increase an energy-dependent component of the non-photochemical quenching caused by this acidification [19,27,28] through the interacted protonation of PsbS proteins [3,4,29] and the synthesis of zeaxanthin and antheraxanthin from violaxanthin in the xanthophyll cycle [30].

The complexity of the photosynthetic stress responses is a reason for the active development of mathematical models of photosynthetic processes [31], because these models can be effective tools for the prediction of changes in photosynthesis under the action of adverse factors. There are models simulating processes on different levels of the organization of photosynthesis [31]: models of the ways of energy utilization in the reaction centers of photosystem II [32–34], models focusing on the description of photosynthetic light reactions and their regulation by stressors [5,35–40], models focusing on the description of photosynthetic dark reactions and CO_2 fluxes [41–44], complex models of plant productivity [45,46], and global photosynthetic models [47,48].

The photosynthetic model by Farquhar, von Caemmerer, and Berry (FvCB model) [42,49–51] is a widely-used model of C_3 photosynthesis that can describe the photosynthetic processes in mesophyll cells, leaves, plant canopies, and ecosystems [31]. This model is based on a stationary description of a photosynthetic CO_2 assimilation rate (A_{hv}) that is dependent on the slowest process of three processes that can limit the dark reactions of photosynthesis [50]: CO_2 fixation by Rubisco, linear electron flow (LEF) in the electron transport chain of thylakoids, and triose flux from the stroma of chloroplasts. Particularly, the FvCB model can be used for the description of the heterogeneity of the photosynthetic processes in the leaves and canopies of plants [52–56]; analysis of this heterogeneity has great importance for revealing new factors that can regulate photosynthetic processes (e.g., the influence of changes in the intensity and spectrum of light caused by an increase in the distance from the leaf surface during photosynthetic processes or the influence of 3-D microstructures of leaf tissues and chloroplast movements on photosynthesis).

However, the simulation of photosynthetic processes in the scale of a leaf surface that can also be based on the FvCB model is weakly developed. A model of photosynthetic processes in the scale of a leaf surface is a potential tool for the theoretical investigation of the spatial heterogeneity of photosynthetic parameters on this surface, including revealing possible modifications of the heterogeneity under the action of stressors. There are several reasons supporting the importance of the development of the leaf photosynthesis model and its theoretical analysis.

First, revealing stressor-induced changes in the photosynthetic heterogeneity can provide an additional indicator of the action of adverse factors on plants. It can be used for the development of new methods for remote sensing plant stress changes. Particularly, these methods can be based on the measurements of the spatial heterogeneity of the distribution of a photochemical reflectance index (PRI), which is calculated based on reflectance at 531 and 570 nm [57–60] and is strongly related to photosynthetic parameters (the non-photochemical quenching of fluorescence, effective quantum yield of photosystem II, light-use efficiency, and photosynthetic CO_2 assimilation rate) [61–67].

Second, the development of the leaf photosynthesis model and revealing stressor-induced changes in the spatial photosynthetic heterogeneity can be an important step for further investigation into new mechanisms influencing plant tolerance to stressors. Particularly, it was theoretically shown that the spatial heterogeneity in the physiological parameters of two-dimensional models of living cells can modify their responses to the actions of external factors through a diversity-induced resonance [68–70], e.g., this effect was shown for excitable plant cells under cooling [70,71]. It cannot be excluded that the spatial heterogeneity in photosynthetic processes can also influence the plant response to stressors. Potentially, the leaf photosynthesis model can also be used as an analysis tool for this influence.

Thus, there were three main purposes of our work: (i) The development and verification of the two-dimensional model of C_3 photosynthesis in the plant leaf, which was based on the FvCB model. (ii) The model-based analysis of the induction of the spatial heterogeneity of the CO_2 assimilation rate under excess light conditions and a decrease in leaf CO_2 conductance (g_S) (this g_S decrease imitated the action of a short-term drought). (iii) Additional experimental verification of the results of this analysis.

2. Description of the Two-Dimensional Model of C_3 Photosynthesis in Plant Leaves

The two-dimensional model of C_3 photosynthesis in the plant leaf was based on the round system of elements (Figure 1a). Each element included descriptions of the photosynthetic cell and the apoplast; some elements (central elements in 3×3 elements squares or in 5×5 elements squares) additionally included stomata. Figure 1b shows the main processes considered in the model. Equations and parameters of the two-dimensional model of C_3 photosynthesis in the plant leaf are described in File S1 "Equations and parameters of the two-dimensional photosynthetic model" in detail.

Briefly, the simplified FvCB model, which described only two limiting stages (the CO_2 fixation by Rubisco and the linear electron flow in the electron transport chain of thylakoids in accordance with [51]), was used as the basis for the simulation of the photosynthetic CO_2 assimilation in mesophyll cells (in accordance with standard Equation (1) [50,51]):

$$A_{hv} = \min(W_c, W_j) \frac{[CO_2]_{str} - \Gamma^*}{[CO_2]_{str}} \tag{1}$$

where W_c and W_j are carboxylation rates at the Rubisco-limited CO_2 assimilation and electron transport-limited CO_2 assimilation conditions, respectively (both values were calculated based on standard Equations (S2) and (S3) in accordance with [50]), $[CO_2]_{str}$ is the concentration of CO_2 in the stroma of chloroplasts, Γ^* is the photosynthetic CO_2 compensation point in the absence of mitochondrial respiration. It should be noted that Equation (1) was used for the estimation of the measured photosynthetic CO_2 assimilation and for comparison with the experimental results. The photosynthetic consumption of CO_2 in the stroma was described as $\min(W_c, W_j)$; i.e., the correction relating to photorespiration was not used in this case. Photorespiration was separately described as the CO_2 source in the cytoplasm in accordance with Equation (2) based on Equation (1):

$$V_{phr} = \frac{A_{hv}\Gamma^*}{[CO_2]_{str}} \tag{2}$$

A dark respiration was described as another CO_2 source in the cytoplasm. In accordance with von Caemmerer et al. [1], it was assumed that the rate of the dark respiration (R_d) was constant.

Carbon fluxes between cells and compartments were described based on Fick's law [72–74]. CO_2 fluxes across the stomata (j_S), plasma membrane (j_{PM}), and envelopes of chloroplasts (j_{Chl}), which depended on the CO_2 conductance [74,75], were analyzed in the model (Equations (3)–(5)):

$$j_S = g_S^0 \left([CO_2]_{out} - [CO_2]_{ap}\right) \tag{3}$$

$$j_{PM} = g_{PM} \left([CO_2]_{ap} - [CO_2]_{cyt}\right) \tag{4}$$

$$j_{Chl} = g_{Chl} \left([CO_2]_{cyt} - [CO_2]_{str}\right) \tag{5}$$

where $[CO_2]_{out}$, $[CO_2]_{ap}$, and $[CO_2]_{cyt}$ are concentrations of CO_2 in the air, apoplast and cytoplasm, respectively; g_S^0, g_{PM}, and g_{Chl} are CO_2 conductance for the stomata, plasma membrane, and chloroplast envelopes (j_{Chl}), respectively.

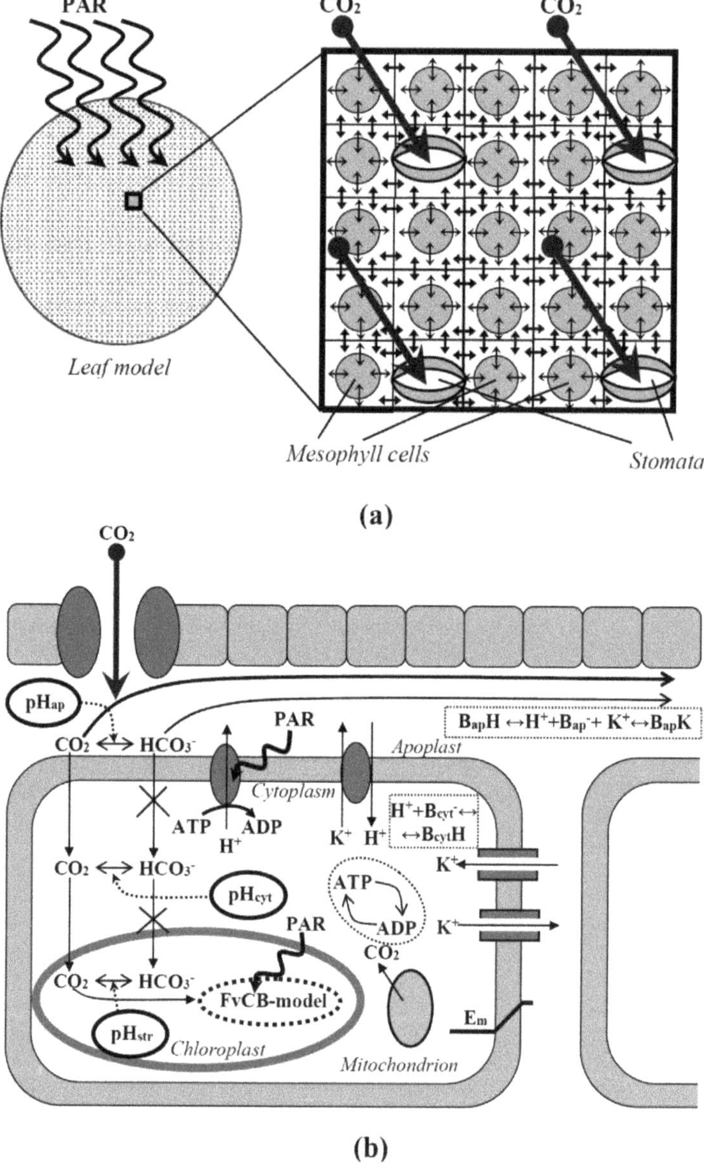

Figure 1. A general scheme of the developed two-dimensional model (**a**) and main processes described by the model on the cell level (**b**). Model elements (squares) include both mesophyll cells and stomata or only mesophyll cells (without stomata). Small arrows in the general scheme show transport of carbon dioxide, H^+, and K^+ between apoplastic volumes of neighboring cells and across the plasma membrane. PAR is the photosynthetic active radiation. pH_{ap}, pH_{cyt}, and pH_{str} are pH in the apoplast, cytoplasm, and stroma of chloroplasts, respectively. B_{cyt}^- and $B_{cyt}H$ are the free and proton-bound cytoplasmic buffers. B_{ap}^-, $B_{ap}H$, and $B_{ap}K$ are the free, proton-bound, and potassium-bound apoplastic buffers. E_m is the difference of electrical potentials across the plasma membrane. FvCB model is the Farquhar–von Caemmerer–Berry model. The main systems of ion transport at rest, including H^+-ATP-ases, H^+/K^+-antiporters, inwardly rectifying K^+ channels, and outwardly rectifying K^+ channels, are described in the two-dimensional photosynthetic model.

Similar HCO_3^- fluxes were assumed to be absent, because charged HCO_3^- should weakly diffuse across biological membranes [75].

The lateral fluxes of both neutral CO_2 and charged HCO_3^- in the apoplast were considered in the model [73] and were described by Equations (S12) and (S13). In accordance with our previous work [70], it was assumed that each cell had its section of apoplast. The lateral fluxes were described between nearest sections (four lateral fluxes for each apoplast section, Figure 1a).

The ratios between the concentrations of CO_2 and HCO_3^- were dependent on pH in the apoplast, cytoplasm, and stroma of chloroplasts [75]. It was assumed that the transitions between CO_2 and HCO_3^- were fast; this meant that the stationary distribution between these molecules could be used. Equation (6) described the portion of CO_2 in the total content of CO_2 and HCO_3^-:

$$P_{CO_2} = \frac{1}{1 + 10^{pH-pK}} \tag{6}$$

where pK is the negative logarithm of the equilibrium constant in the reaction of the transition between CO_2 and HCO_3^-.

The stromal pH was assumed to be constant; the pH in the apoplast and cytoplasm was described based on our early model of ion transport and electrogenesis in plant cells [76].

The description of H^+ and K^+ fluxes across the plasma membrane was based on our previous model [70,76], which was simplified. Only H^+-ATPase, inwardly and outwardly rectifying K^+ channels, and K^+/H^+-antiporters were described, because the interaction of these systems could support stationary H^+ concentrations in the cytoplasm and the apoplast: the H^+-ATPase provided the primary transport of H^+ across the plasma membrane; the K^+ channels provided the K^+ transport, which electrically compensated the charge transfer related to the proton transport through H^+-ATPase; the K^+/H^+-antiporter prevented the non-physiological increase in cytoplasmic pH and K^+ concentration and the decrease in apoplastic pH and K^+ concentration.

The buffer properties of the cytoplasm (for H^+) (Equation (S37)) and the apoplast (for K^+ and H^+) (Equations (S38) and (S39)) were described in accordance with Sukhova et al. [70]. H^+-ATPase was described based on the "two-state model" [77] (Equation (S18)); a regulation of its activity by blue light and ATP concentration in the cytoplasm [78] was included in the model using the Hill Equation (Equations (S19) and (S20)). We used a stationary description of this ATP concentration (Equation (S40)), which was based on the ATP synthesis dependent on the dark respiration (constant) and the CO_2 assimilation rate (the FvCB model) and the ATP hydrolysis with the assumed velocity constant.

K^+ fluxes through inwardly and outwardly rectifying K^+ channels were described based on the Goldman–Hodgkin–Katz Equation [76,79] (Equations (S21) and (S22)); the regulation of activities of these channels by the electrical potential across the plasma membrane of mesophyll cells was described based on the stationary solution of the Equation for the open probability for these channels [70] (Equations (S23) and (S24)).

H^+ and K^+ fluxes through the K^+/H^+-antiporter were described in accordance with our previous works [70,76] (Equation (S25)); this description was based on the simple Equation of the chemical kinetics. The K^+/H^+-antiporter was described as the electroneutral transporter because the transport of charges was compensated in this system.

The lateral fluxes of H^+ and K^+ were described based on Fick's law in accordance with Sukhov et al. [80], (Equations (S31) and (S32)). The electrical potential across the plasma membrane was described as the stationary value in accordance with Sukhov et al. [80], (Equation (S26)); it was assumed that the electrical conductance between cells was zero.

The developed model included numerous parameters that made it difficult for the direct experimental parameterization of a specific plant object. Considering this point, we mainly used standard parameters from the FvCB model [50] and from our previous model of ion transport and electrogenesis in plant cells [70] (Table S1 in File S1); other data from the literature were also used for the parameterization. As a result, this model could rather show the qualitive properties of forming spatial heterogeneity in the photosynthetic parameters in the leaf surface. In contrast, this model (with the current parameters) was not optimal for the quantitative predictions of the specific plant object. It should be additionally noted that using standard parameters, which provided good descriptions of investigated processes in earlier works, minimized the probability of qualitive errors in the results of the simulation. In contrast, the broad experimental search of parameters in specific species of plants could, potentially, induce these errors (strong experimental errors in the estimation of even one of numerous parameters can disrupt the results of a simulation).

The developed model was numerically analyzed using the forward Euler method. The special computer program (Microsoft Visual C++ 2019, Microsoft Corporation, Redmond, WA, USA) was developed for the numerical analysis of the model. Equation (1) was used for the calculation of the A_{hv} simulated by the developed model.

The action of excess light and drought on the spatial heterogeneity was analyzed in our work. The excess light action was provided by using the high values of the Photosynthetic Active Radiation (PAR) in Equation (S5). It was assumed that the drought action on plants was mainly related to the stomatal closure. At the model analysis, this action was imitated by using the decreased stomatal CO_2 conductance (the decreased parameter g_S^0 in Equation (3), the quantity of open stomata per leaf area was not changed) or the decreased quantity of open stomata per leaf area (from one stomata per 3×3 elements square to one stomata per 5×5 elements square, the stomatal CO_2 conductance was not changed). The average leaf CO_2 conductance (g_S) was decreased from 0.064 mol m^{-2}s^{-1} to 0.023 mol m^{-2}s^{-1} in both cases of the model analysis.

3. Results

3.1. Verification of the Developed Model on the Basis of Light Curves of Simulated and Experimental Photosynthetic CO_2 Assimilation Rate

The first question of the current analysis was: could the developed model simulate the experimental light curve of the photosynthetic CO_2 assimilation rate? We used the average photosynthetic CO_2 assimilation rates in pea plant leaves under the actinic light with different intensities and these assimilation rates at different average leaf CO_2 conductance for this verification. Experimental and simulated results were compared in a quality manner by using the standard parameters of the FvCB model [50], which were not adapted for pea plants. The details of the experimental procedures are described in Section 5 "Materials and Methods".

It is shown (Figure 2a) that the simulated dependence of average A_{hv} on the intensity of the actinic light at the basic g_S (0.064 mol m^{-2}s^{-1}) included two parts: the increase in the CO_2 assimilation rate with increasing intensity of illumination (low and moderate light intensities) and the light saturation of this assimilation rate (high light intensities). This effect was also observed at the decreased average g_S (0.023 mol m^{-2}s^{-1}), which was imitated by using the decreased stomatal CO_2 conductance; however, the maximal A_{hv} at $g_S = 0.064$ mol m^{-2}s^{-1} was higher than one at $g_S = 0.023$ mol m^{-2}s^{-1}. Additionally, the minimal light intensity for the A_{hv} saturation at $g_S = 0.064$ mol m^{-2}s^{-1} was higher than one at $g_S = 0.023$ mol m^{-2}s^{-1}.

Experimental plants were ranged in accordance with their g_S and were divided into two groups with high and low CO_2 conductance (average g_S in leaves was 0.069 ± 0.004 and 0.027 ± 0.007 mol m^{-2}s^{-1}, respectively, see Section 5.1). It is shown (Figure 2b) that experimental A_{hv} dependences on light intensity were similar to simulated ones: (i) there were stages of increase in the photosynthetic CO_2 assimilation rate and stage of A_{hv} light saturation, (ii) the maximal A_{hv} was increased with increasing g_S, and (iii) the minimal light intensity for the A_{hv} saturation was increased with increasing stomatal CO_2 conductance. It should be additionally noted that the values of the maximal A_{hv} differed in the experimental and the simulated results. This moderate quantitative difference could be caused by the used standard values of the model parameters, which were not adapted for pea seedlings (see Section 2).

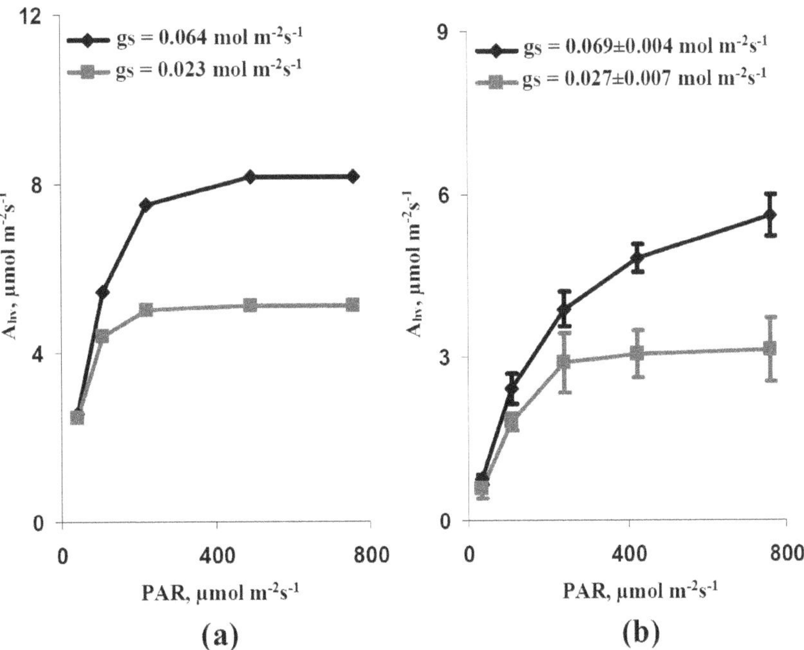

Figure 2. Simulated (**a**) and experimental (**b**) dependences of the average photosynthetic CO_2 assimilation rate (A_{hv}) on the intensity of the photosynthetic active radiation (PAR) at the varied average leaf CO_2 conductance (g_S). Simulated dependences were calculated at average $g_S = 0.064$ mol m^{-2}s^{-1} (the basic g_S) and $g_S = 0.023$ mol m^{-2}s^{-1} (the decreased g_S). Each stomata in the model was located in the center of square (3×3 elements); the average gS was calculated as the CO_2 conductance in the element with stomata divided by 9. In order to obtain experimental dependences, all experimental records in this series were ranged and divided into two groups with the low ($g_S < 0.04$ mol m^{-2}s^{-1}, $n = 5$) and high ($g_S > 0.04$ mol m^{-2}s^{-1}, $n = 9$) CO_2 conductance (see Section 5.1). A combination of Dual-PAM-300 and GFS-3000 was used in the experimental measurements of pea seedlings.

Simulated (Figure 3a) and experimental (Figure 3b) dependences of A_{hv} on g_S at the high light intensity (758 µmol m^{-2}s^{-1}) were analyzed in the next stage of our work. It is shown that both dependences were qualitatively similar and could be described by logarithmic Equations with similar coefficients. Quantitative differences between dependences were probably caused by the absence of adaptation of parameters for pea plants.

Thus, these results showed that the developed model based on the two-dimensional system of photosynthetic cells could qualitatively describe important characteristics of A_{hv}, including the shape of the light dependence of the photosynthetic CO_2 assimilation rate and changes in this shape and maximal A_{hv} during changes in the stomatal CO_2 conductance. As a result, the developed model could be used for further analysis in our current work.

3.2. Analysis of Simulated and Experimental Spatial Heterogeneities in the Photosynthetic CO_2 Assimilation Rate under Various Light Intensity and Stomatal CO_2 Conductance

The spatial heterogeneity of A_{hv} simulated by the developed model was analyzed in the next stage of investigation. First, the standard deviation of A_{hv} (SD(A_{hv})), which was calculated based on the values of A_{hv} in all elements of the two-dimensional model of the leaf, was analyzed. It is shown (Figure 4a) that SD(A_{hv}) was increased with the increase in light intensity at all variants of the average leaf CO_2 conductance. A decrease in the average g_S (from 0.064 to 0.023 mol m^{-2}s^{-1}) caused by the decrease in the stomatal CO_2 conductance strongly decreased SD(A_{hv}). In contrast, the similar decrease in the average g_S caused by the decrease in the quantity of stomata per area unit weakly influenced SD(A_{hv}).

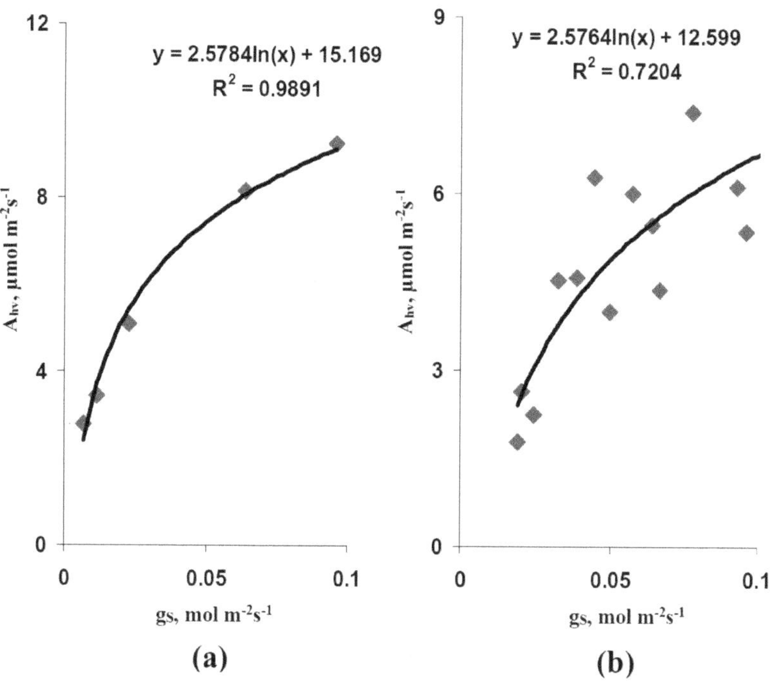

Figure 3. Simulated (**a**) and experimental (**b**) scatter plots between the average photosynthetic CO_2 assimilation rate (A_{hv}) and the average leaf CO_2 conductance (g_S) under high intensity of the photosynthetic active radiation (758 μmol m^{-2}s^{-1}). Simulated Ahv were calculated at the average g_S equaling 0.007, 0.012, 0.023, 0.064, and 0.096 mol m^{-2}s^{-1}. Each stomata in the model was located in the center of square (3×3 elements); the average g_S was calculated as the CO_2 conductance in the element with stomata divided by 9. Pea seedlings were experimentally investigated; all g_S and A_{hv} (under the 758 μmol m^{-2}s^{-1} PAR intensity) were used (n = 14). R^2 is the determination coefficient.

However, SD(A_{hv}) should be strongly related to the absolute value of A_{hv}; thus, all revealed changes could be related to changes in this value. We analyzed the coefficient of variation (CV(A_{hv})) to eliminate the influence of the absolute value of A_{hv} on the estimation of the spatial heterogeneity, because the variation coefficient was calculated as the standard deviation divided by the average value. It is shown (Figure 4b) that CV(A_{hv}) was also strongly increased with increasing light intensity in all analyzed variants. The decrease in the average g_S caused by the decrease in the quantity of stomata per area unit strongly increased CV(A_{hv}). The decrease in the average g_S caused by the decrease in the stomatal CO_2 conductance weakly influenced CV(A_{hv}); however, CV(A_{hv}) in this variant was higher than CV(A_{hv}) at the control average g_S (0.064 mol m^{-2}s^{-1}) under low and moderate light intensities.

We analyzed a ratio between SD(A_{hv}) at the control average g_S and at the decreased average g_S, which was caused by the decrease in the stomatal CO_2 conductance (with no change in the quantity of stomata), and the analogical ratio between CV(A_{hv}) to additionally estimate the last effect. It is shown (Figure 4c) that these ratios were increased under the low light intensity and the ratio of CV(A_{hv}) was also increased under the moderate light intensity.

Thus, the results of the simulation show that the increase in light intensity and the decrease in leaf CO_2 conductance could increase the spatial heterogeneity of the photosynthetic CO_2 assimilation rate. After that, we experimentally analyzed this heterogeneity to check the revealed results. The direct experimental analysis of A_{hv} was not possible. However, the FvCB model [42,49–51] predicted that the linear relation between A_{hv} and LEF could be probable at the limitation of photosynthesis by the linear electron flow. Figure 5a shows that the average A_{hv} and LEF were strongly linearly related with increasing LEF (with increasing intensity of the actinic light) to about 60 μmol m^{-2}s^{-1}; this linear relation was disrupted at higher values of LEF (75 μmol m^{-2}s^{-1} LEF at the 758 μmol m^{-2}s^{-1} light intensity). Analysis of individual A_{hv} and LEF (excluding LEF at 758 μmol m^{-2}s^{-1} light intensity) showed a similar linear relation at LEF equaling 6.5–66.2 μmol m^{-2}s^{-1} (Figure 5b). Thus, linear regression A_{hv} = 0.1 LEF was used for the calculation of A_{hv} based on the measured LEF at LEF $\leq$ 66 μmol m^{-2}s^{-1}.

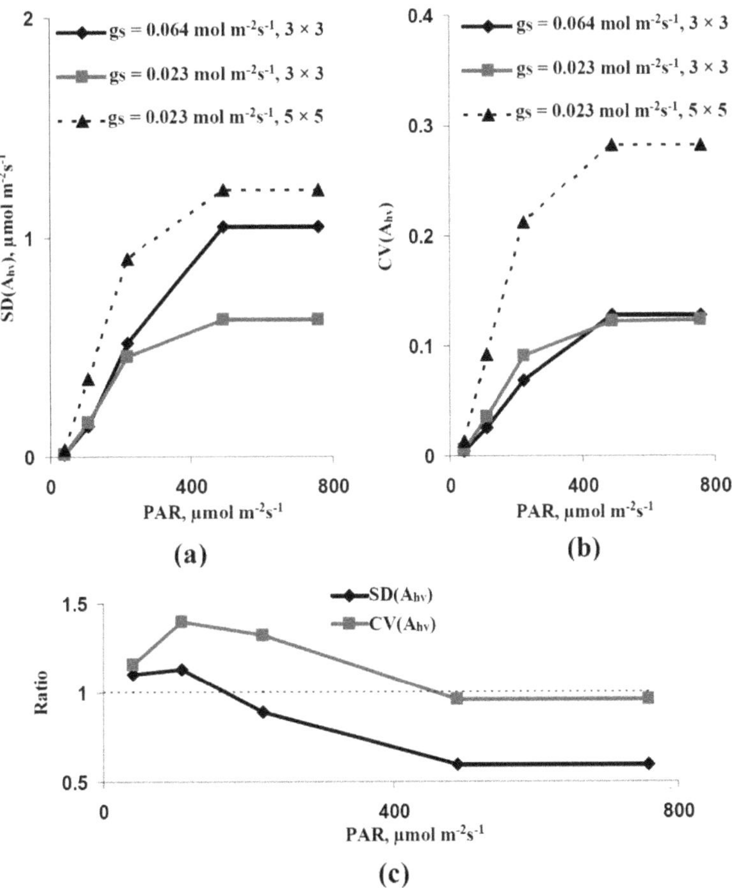

Figure 4. Dependences of parameters of the simulated spatial heterogeneity of the photosynthetic CO_2 assimilation rate (A_{hv}) on the intensity of the photosynthetic active radiation (PAR). (**a**) Dependence of the standard deviation of Ahv (SD(A_{hv})) on the PAR intensity. There were three variants of parameters. (i) The average g_S of the leaf was 0.064 mol m^{-2}s^{-1}, each stomata was located in the center of the 3 × 3 elements square. This variant was assumed as the control. (ii) The average g_S of the leaf was decreased to 0.023 mol m^{-2}s^{-1}. The CO_2 conductance in individual stomata was decreased; each stomata was located in the center of the 3 × 3 elements square. (iii) The average g_S of the leaf was decreased to 0.023 mol m^{-2}s^{-1}. The CO_2 conductance in individual stomata was not changed; each stomata was located in the center of the 5 × 5 elements square. (**b**) Dependence of the coefficient of variation of A_{hv} (CV(A_{hv})) on the PAR intensity. (**c**) Dependence of the ratio of the SD(Ahv) at g_S = 0.023 mol m^{-2}s^{-1} (3 × 3 elements) to the SD(Ahv) at g_S = 0.064 mol m^{-2}s^{-1} (3 × 3 elements) on the PAR intensity and the analogical dependence for CV(A_{hv}).

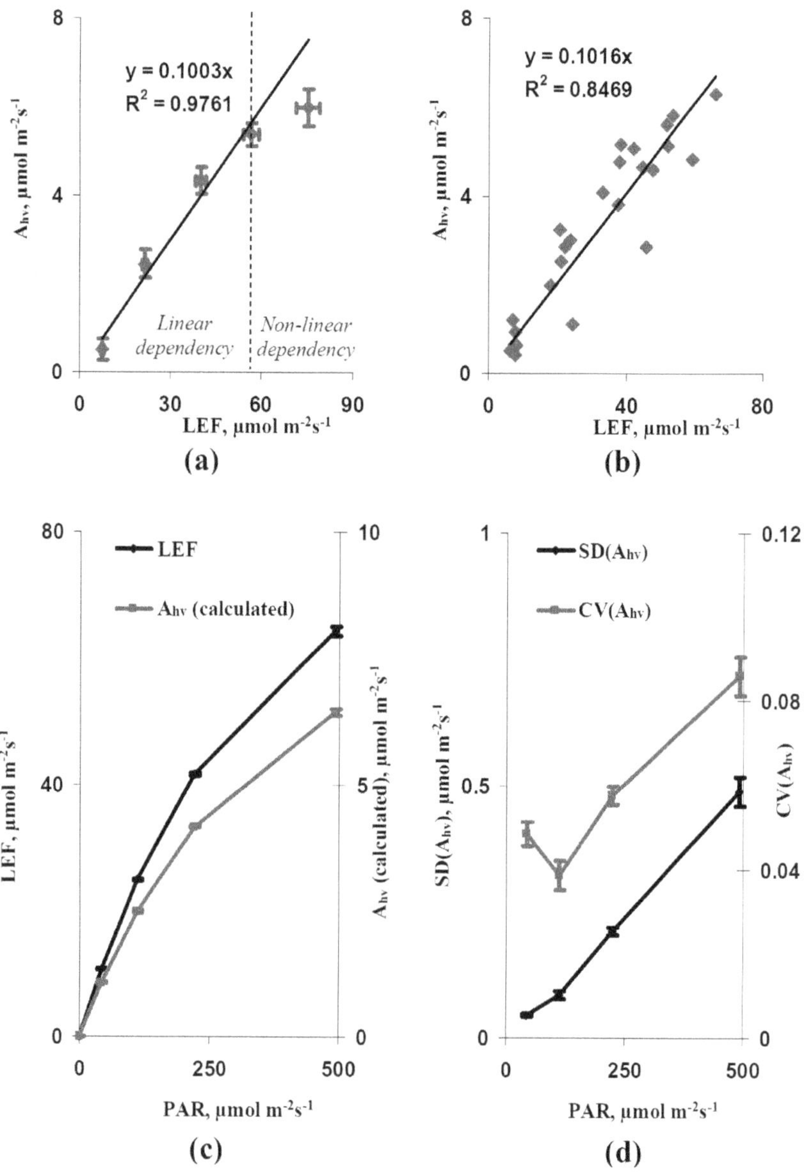

Figure 5. The dependence of average photosynthetic CO_2 assimilation rate (A_{hv}) on the average linear electron flow (LEF) at 34, 108, 239, 425, and 758 μmol m^{-2}s^{-1} intensities of actinic light (n = 5–7) and the linear calibration Equation (**a**), the dependence of individual A_{hv} on individual LEF at 34, 108, 239, and 425 μmol m^{-2}s^{-1} light intensities (n = 25) and the linear calibration Equation (**b**), dependences of LEF and A_{hv} (calculated) on the PAR intensity (n = 6) (**c**), and dependences of parameters of the spatial heterogeneity of Ahv (calculated) (SD(Ahv) and CV(A_{hv})) on the PAR intensity (n = 6) (**d**). R^2 is the determination coefficient. Ahv (calculated) was calculated based on LEF and the calibration Equation. A combination of Dual-PAM-300 and GFS-3000 was used for development of the calibration Equation. IMAGING-PAM M-Series MINI Version was used for analysis of the spatial heterogeneity of A_{hv}. Pea seedlings were used in all variants of experiments.

It is shown that the increase in light intensity increased the linear electron flow and calculated A_{hv} (Figure 5c). The experimental $SD(A_{hv})$ and $CV(A_{hv})$, which showed the spatial heterogeneity of the photosynthetic CO_2 assimilation rate in leaves, were also increased with increasing light intensity (Figure 5d). This result was in good accordance with the results of the simulation and supported the induction of the photosynthetic spatial heterogeneity under excess light conditions.

Finally, we experimentally checked the increase in $CV(A_{hv})$ at the decreased average g_S that was predicted by the developed model. It is shown that the short-term drought (1 day) decreased the g_S in pea leaves (Figure 6a), which was probably related to the stomata closing. $CV(A_{hv})$, calculated based on the variation coefficient of LEF, was significantly increased during the short-term drought (Figure 6b). This result experimentally supported the increase in photosynthetic spatial heterogeneity due to the stomata closing.

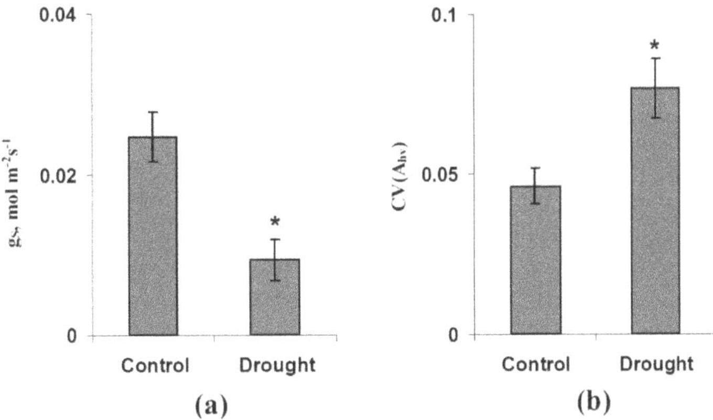

Figure 6. Influence of the short-term drought (1 day) on the leaf CO_2 conductance (g_S) (**a**) and the coefficient of variation of A_{hv} ($CV(A_{hv})$) showing the relative spatial heterogeneity of this parameter in the leaf (**b**) ($n = 6$). GFS-3000 was used for the gS measurement (averaged in the investigated area of the leaf) and IMAGING-PAM M-Series MINI Version was used for the analysis of the spatial heterogeneity of A_{hv} (based on the spatial heterogeneity of LEF and the calibration Equation). The moderate light intensity (249 μmol m^{-2}s^{-1}) was used in this experiment. Pea seedlings were irrigated in the control and were not irrigated under drought conditions. *, difference with the control was significant.

4. Discussion

Photosynthesis is a complex process [1,2] that can be strongly affected by numerous abiotic stressors [3,4,15,16]. The simulation of photosynthetic processes is an effective prediction tool of photosynthetic changes under the action of stressors [31]. There are photosynthetic models focusing on descriptions of the primary light absorption [32–34], photosynthetic light reactions [5,35–40], photosynthetic dark reactions, and CO_2 fluxes [41–44], etc. However, mathematical models of photosynthetic processes in the scale of the leaf surface, which can be used for revealing the spatial heterogeneity of the distribution of photosynthetic parameters on this surface, are weakly developed. Our current work—devoted to the solution of this problem—shows two important results.

First, the developed two-dimensional model of C_3 photosynthesis in the leaf, which is based on the FvCB model [42,49–51], descriptions of stomatal and transmembrane fluxes of CO_2 and lateral fluxes of CO_2 and HCO_3^- [73–75], and the simplified model of the H^+ and K^+ transport [70,76,77,80] can qualitatively simulate the experimental results, including the shape of dependence of the average A_{hv} in the leaf on the light intensity and the influence of the average g_S on the photosynthetic CO_2 assimilation rate (see Figures 2 and 3). It is important that this accordance between the experimental and the simulated results does not require additional adaptation of parameters of the photosynthetic description in the developed model because standard parameters of the FvCB model [50] are used (Table S1 in File S1). This result verifies the efficiency of the developed model for the simulation of the average A_{hv}. Furthermore, considering that this model can also describe the spatial heterogeneity of the A_{hv} distribution on the leaf surface, it is a potential tool for the investigation of the influence of stressors on this heterogeneity.

Second, the developed model predicts the increase in the A_{hv} spatial heterogeneity on the leaf surface with increasing light intensity (Figure 4). This effect is related to the stomatal CO_2 conductance and the quantity of open stomata supporting the CO_2 flux into the leaf, because the decrease in this conductance or the quantity of open stomata per leaf area increases the simulated photosynthetic spatial heterogeneity (especially at the weak and moderate light intensities). The results of analysis of the developed model are in good accordance with works showing the relations of the spatial heterogeneity and the dynamics of the stomata opening to the distribution of photosynthetic parameters in leaves [81–83]. Additionally, there are works [83–86] showing an increase in the spatial heterogeneity of photosynthetic parameters under drought conditions. The participation of the stomata closing due to this effect is a discussion question [85,86]; however, considering the influence of drought on stomata [87,88], this participation cannot be excluded.

Our experimental results support the prediction of the developed model: an increase in light intensity increases the variation coefficient of the photosynthetic CO_2 assimilation rate in pea leaves (Figure 5d) and a decrease in leaf CO_2 conductance, induced by the short-term drought, also increases this coefficient (Figure 6). These results, which are in good accordance with the noted experimental works by other authors showing the positive drought influence on the photosynthetic spatial heterogeneity in leaves [83–86] additionally verify the developed model.

A potential mechanism of the revealed light-induced increase in the A_{hv} spatial heterogeneity can be related to the heterogeneity of the stromal CO_2 concentration in the different cells. In accordance with the FvCB model [42,49–51], this concentration can strongly influence A_{hv} in cells. On the other hand, CO_2 is propagated from stomata through lateral diffusion [89,90] and is consumed by photosynthetic processes, which can be dependent on the light intensity. It means that an increase in this intensity and the stimulation of photosynthesis should increase the variability of the CO_2 concentration in different cells; i.e., the light intensity should influence the spatial heterogeneity of the stromal CO_2 concentration. The additional model analysis of the variation coefficient of this concentration shows that this coefficient is strongly increased by changes in the light intensity from 42 μmol m^{-2}s^{-1} to 221 μmol m^{-2}s^{-1} (from 0.013 to 0.100, respectively); thus, this mechanism can participate in an increase in the A_{hv} spatial heterogeneity under the excess light.

A decrease in the quantity of open stomata per leaf area should stimulate this effect by increasing the distance of the CO_2 diffusion. This supposition is supported by an increase in the variation coefficient of the simulated stromal CO_2 concentration from 0.100 to 0.180 by decreasing this quantity from one stomata per 9 cells to one stomata per 25 cells under the 221 μmol m^{-2}s^{-1} light intensity. In contrast, a decrease in the stomatal CO_2 conductance (without changes in the open stomata quantity) weakly influences this coefficient (data not shown). The last result shows that there are additional induction mechanisms of the A_{hv} spatial heterogeneity in the leaf. It cannot be excluded that these additional mechanisms also participate in influencing the light intensity on the A_{hv} heterogeneity.

The revealed stimulation of the A_{hv} spatial heterogeneity under excess light conditions and/or under the decreased leaf CO_2 conductance (imitation of the drought) can potentially modify the non-photochemical quenching of the chlorophyll fluorescence, including photodamage, state-transition in the light-harvesting complex, and energy-dependent quenching [3,4,18,19], because low A_{hv} in some parts of a leaf can strongly limit photosynthetic light reactions and can contribute to the induction of these processes. It means that this spatial heterogeneity can potentially modify the plant tolerance to the actions of the excess light. Particularly, cells with low CO_2 concentration in the stroma and weak activity of the photosynthetic CO_2 assimilation should have a low threshold for both photodamage and induction of protective changes in the photosynthetic machinery. It can be expected that these cells can influence damage and tolerance of whole leaves under the action of stressors (e.g., through the production and propagation of reactive oxygen species [71]); however, this supposition requires further development of the model (e.g., a description of the damage of photosynthetic machinery in the model can be included in the model) and the model-based investigations.

Additionally, the increased A_{hv} spatial heterogeneity and related changes in photosynthetic light reactions can be used for the development of methods of remote sensing plant stress changes under excess light or drought conditions. Particularly, it can be expected that these stressors should increase the heterogeneity of the spatial distribution of PRI because this reflectance index is strongly related to photosynthetic parameters [61,62,64,66,67]. Potentially, this effect can be used for the development of methods of remote sensing the actions of excess light and drought on plants (based on the measurements of the spatial heterogeneity of PRI); however, this possible stimulation of PRI under the action of stressors requires future model-based and experimental investigations.

Figure 7 summarizes the results of our work and their potential importance for understanding the ways of plant damage and tolerance under the action of stressors and the development of methods for plant remote sensing. It should be additionally noted that the developed model can be used for future analysis of the influence of the stochastic spatial heterogeneity of its parameters on photosynthetic processes; e.g., the influence of the stochastic heterogeneity of the activity of H^+-ATPases in the plasma membrane [31], which is related to the CO_2 flux into mesophyll cells [71], or the influence of the stochastic heterogeneity of the CO_2 conductance of individual stomata can be investigated. It is known that the stochastic spatial heterogeneity of biological objects (including plants) can influence their systemic parameters (e.g., through "diversity-induced resonance" or similar effects, [31,68,69]); thus, the analysis of this problem based on the developed model can be an important task.

Other interesting perspectives of the model development can be: description of stomata regulation mechanisms by light intensity and drought (and potential interactions between these mechanisms), description of the light damage to photosynthetic machinery (and relation of this damage with stomata opening, the plasma membrane and chloroplast envelope CO_2 conductance, and activity of the CO_2 carboxylation), and description of the influence of photosynthetic processes to leaf reflectance (this description can be important for the development of methods of remote sensing). Finally, the parameterization of the model for specific plant species (e.g., plant species that are widely used in agriculture) can be an additional important task for the future development of the model.

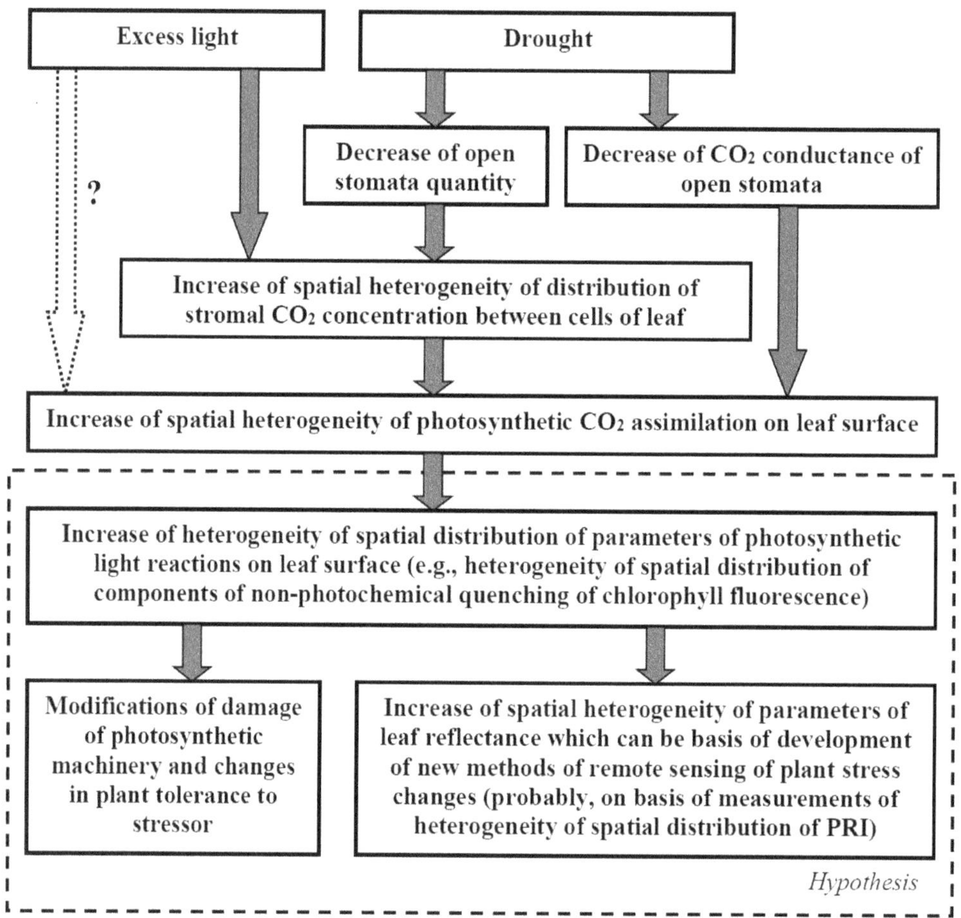

Figure 7. A scheme of potential ways the excess light and drought influencing the heterogeneity of the spatial distribution of photosynthetic parameters and the hypothetical importance of this heterogeneity for the plant tolerance and remote sensing of plant stress changes. The scheme is based on analysis of the developed model and experimental results (see Section 4 for details).

5. Materials and Methods

5.1. Experimental Procedure of Verification of Two-Dimensional Model of the C_3 Photosynthesis in Plant Leaves

We did not parameterize the two-dimensional model of C_3 photosynthesis in leaves for the specific plant, because using the standard parameters from earlier models, which were included in the current model, simplified parameterization and minimized potential errors in parameter values that were probable at the broad experimental search and could disrupt the model analysis.

Therefore, we could not expect a quantitative accordance between the simulated and the experimental photosynthetic parameters at verification. As a result, we analyzed the qualitive accordance between the results of the simulation and the results of the experimental investigation of the pea plant. Pea plants were selected based on our numerous early works, which investigated photosynthesis and its regulation in this plant object (e.g., [5,66,67,91]).

Thus, 2–3-week-old pea seedlings (*Pisum sativum* L., cultivar "Albumen") were used for verification of the two-dimensional model of C3 photosynthesis in plant leaves. The plants were cultivated in a sand substrate in a Binder KBW 240, with irrigation by the 50% Hoagland–Arnon medium (about 50 mL) performed every two days. Luminescent lamps FSL YZ18RR (Foshan Electrical And Lighting Co., Ltd., Foshan, China) were used for illumination (about 100 μmol m^{-2}s^{-1}). The weak water deficit (the short-term drought) was induced by an absence of irrigation of the experimental seedlings for 1 day.

A combination of a PAM-fluorometer Dual-PAM-100 and an infrared gas analyzer GFS-3000 (Heinz Walz GmbH, Effeltrich, Germany) was used for the investigation of the average photosynthetic parameters in the second mature leaves of the pea plant. A_{hv} was measured as the difference between the CO_2 assimilation rate after 10 min under the actinic blue light (Dual-PAM-100 was used as the source of this light) and this assimilation rate under dark conditions. The current CO_2 assimilation rate was measured by the gas analyzer GFS-3000. The leaf CO_2 conductance was calculated based on the leaf water conductance, which was measured by GFS-3000, in accordance with Cabrera et al. [92]. The GFS-3000 was also used for supporting the 360 ppm concentration of CO_2 and the 70% relative air humidity in the measuring cuvette.

A photosynthetic linear electron flow (LEF) was calculated based on the effective quantum yield of the photosystem II (Φ_{PSII}), the intensity of the actinic light (PAR), the fraction of absorbed light distributed to the photosystem II (dII = 0.42), and the fraction of PAR absorbed by the leaves (p = 0.88) in accordance with Equation (7) [91]:

$$\text{LEF} = p \cdot \text{dII} \cdot \Phi_{PSII} \cdot \text{PAR} \tag{7}$$

Φ_{PSII} was estimated after 10 min under the actinic light. This parameter was automatically calculated by the Dual-PAM-100 software based on the current levels of fluorescence (F) and the maximal fluorescence level after the preliminary illumination (F'_m), which were measured before initiation and before termination of the saturation pulse (300 ms, red light, 10,000 μmol m^{-2}s^{-1}), respectively, in accordance with the standard procedure of measurement by the PAM fluorometer. Equation (8) was used for the Φ_{PSII} calculation [93]:

$$\Phi_{PSII} = \frac{F'_m - F}{F'_m} \tag{8}$$

The blue light from the standard source of Dual-PAM-100 was used as the actinic light; its intensity was varied.

There were two variants of experiments combining the Dual-PAM-100 and the GFS-3000. First, we preliminary experimentally estimated the basic g_S that was used for the calculation of the stomatal CO_2 conductance in the model ($g_{S0} = g_S \cdot 9$ because one stomata per nine elements was used as the control variant in the model, Table S1 in File S1). Experiments were performed for 1 day; light curves were not analyzed. It was shown that g_S = 0.064 $\pm$ 0.04 mol m^{-2}s^{-1} (n = 6). As a result, g_S = 0.064 mol m^{-2}s^{-1} (and g_{S0} = 0.576 mol m^{-2}s^{-1}) was used as the basic leaf CO_2 conductance. In the model, the decreased g_S was provided by the decreased g_S^0 or the decreased quantity of stomata per leaf area (from one stomata per 3 $\times$ 3 elements square to one stomata per 5 $\times$ 5 elements square, see Section 2); both decreased g_S should be the same when compared. Thus, the decreased g_S was calculated as the multiplication between the basic g_S and 9/25 (the decreased g_{S0} was similarly calculated, Table S1 in File S1).

Second, we analyzed the experimental light curves, which were investigated for the long-time experimental series (about 2 weeks). In this case, the experimental g_S was more varied than the g_S in the first case (g_S = 0.058 $\pm$ 0.11 mol m^{-2}s^{-1}, n = 14). This variability was used for the additional verification of the model; all experimental records in this series were ranged and divided into two groups with the low (g_S < threshold value) and high (g_S > threshold value) CO_2 conductance. We found that using the 0.04 mol m^{-2}s^{-1} threshold value provided an average g_S which was similar to the leaf CO_2 conductance in the model: 0.069 $\pm$ 0.004 mol m^{-2}s^{-1} (n = 9) and 0.027 $\pm$ 0.007 mol m^{-2}s^{-1} (n = 5). After that, we separately statistically analyzed the light dependences in these two groups (with the low and high CO_2 conductance) to verify the developed model.

A system of PAM imaging IMAGING-PAM M-Series MINI Version (Heinz Walz GmbH, Effeltrich, Germany) was used for the measurements of the spatial distribution of photosynthetic parameters. The blue light from the standard source of this system was used as the actinic light; its intensity was varied. Φ_{PSII} was estimated at the saturation pulse (in accordance with Equation (8)) after 10 min under the actinic light.

The analysis of the spatial distributions of LEF was based on the analysis of grayscale images of the spatial distribution of the quantum yield of photosystem II, which were created by software of the IMAGING-PAM M-Series MINI Version. These grayscale images were analyzed using ImageJ 1.46r. The analysis showed the average value and the standard deviation of Φ_{PSII} in the standard round ROI in the center of the leaf. The coefficient of variation was calculated as the ratio of the standard deviation of the average value. The parameters of LEF (the average value, standard deviation, and coefficient of variation) were calculated based on Equation (7) as the simple proportion. These parameters were used for the estimation of the parameters of A_{hv} (the average value, standard deviation, and coefficient of variation) based on the calibration curve (see Section 3.2).

5.2. Statistics

Means and standard errors were used in the statistical analysis and Student's *t*-test was used for the estimation of significance. The spatial heterogeneity was estimated based on the standard deviation of A_{hv} (SD(A_{hv})) and the coefficient of variation of this photosynthetic parameter (CV(A_{hv})). Numbers of repetitions were shown in figures.

6. Conclusions

The work was devoted to the development of a two-dimensional model of C_3 photosynthesis in the plant leaf and further analysis of the induction of the spatial heterogeneity of the CO_2 assimilation rate under the excess light and a decrease in the leaf CO_2 conductance; this g_S decrease imitated the action of a short-term drought. First, it was shown that the developed two-dimensional model of C_3 photosynthesis in the leaf (based on the FvCB model, the descriptions of the fluxes of CO_2 and $HCO_3{}^-$, and the simplified model of the H^+ and K^+ transport) qualitatively simulated the experimental results. Second, the analysis of the developed model showed that the increase in the light intensity and the decrease in the average leaf CO_2 conductance should increase the spatial heterogeneity of the photosynthetic CO_2 assimilation rate on the leaf surface. Experimental investigations supported these theoretical results. Thus, the developed model can be used as a tool for theoretical investigations of the influence of environmental factors on the spatial heterogeneity of the distribution of photosynthetic parameters in the leaf. Finally, there are some potential ways to further develop the model, including its parameterization for specific plant species, additional description of stomata regulation by light and drought, description of light damage to photosynthetic machinery, description of relations between photosynthesis and leaf reflectance, analysis of influence of stochastic heterogeneity in photosynthetic and stomata parameters, and others.

Supplementary Materials: The following supporting information can be downloaded at: https://www.mdpi.com/article/10.3390/plants11233285/s1, File S1 "Equations and parameters of the two-dimensional photosynthetic model", Refs [50,51,70,72–80,94–98] have mentioned in Supplementary Materials.

Author Contributions: Conceptualization, E.S. and V.S.; methodology, E.S., D.R. and V.S.; software, E.S.; formal analysis, E.S.; investigation, E.S., D.R. and E.G.; writing—original draft preparation, E.S. and V.S.; writing—review and editing, V.S.; supervision, V.S.; project administration, E.S.; funding acquisition, E.S. All authors have read and agreed to the published version of the manuscript.

Funding: The investigation was funded by the Russian Foundation for Basic Research, project number 20-34-90086 Aspiranti.

Institutional Review Board Statement: Not applicable.

Informed Consent Statement: Not applicable.

Data Availability Statement: The data presented in this study are available upon request from the corresponding author.

Conflicts of Interest: The authors declare no conflict of interest. The funders had no role in the design of the study; in the collection, analyses, or interpretation of data; in the writing of the manuscript; or in the decision to publish the results.

References

1. Allen, J.F. Cyclic, pseudocyclic and noncyclic photophosphorylation: New links in the chain. *Trends Plant Sci.* **2003**, *8*, 15–19. [CrossRef] [PubMed]
2. Johnson, M.P. Photosynthesis. *Essays Biochem.* **2016**, *60*, 255–273. [CrossRef] [PubMed]
3. Ruban, A.V. Evolution under the sun: Optimizing light harvesting in photosynthesis. *J. Exp. Bot.* **2015**, *66*, 7–23. [CrossRef]
4. Ruban, A.V. Nonphotochemical chlorophyll fluorescence quenching: Mechanism and effectiveness in protecting plants from photodamage. *Plant Physiol.* **2016**, *170*, 1903–1916. [CrossRef] [PubMed]
5. Sukhova, E.; Khlopkov, A.; Vodeneev, V.; Sukhov, V. Simulation of a nonphotochemical quenching in plant leaf under different light intensities. *Biochim. Biophys. Acta Bioenerg.* **2020**, *1861*, 148138. [CrossRef]
6. Tikkanen, M.; Grieco, M.; Nurmi, M.; Rantala, M.; Suorsa, M.; Aro, E.-M. Regulation of the photosynthetic apparatus under fluctuating growth light. *Phil. Trans. R. Soc. B* **2012**, *367*, 3486–3493. [CrossRef] [PubMed]
7. Huang, W.; Hu, H.; Zhang, S.B. Photorespiration plays an important role in the regulation of photosynthetic electron flow under fluctuating light in tobacco plants grown under full sunlight. *Front. Plant Sci.* **2015**, *6*, 621. [CrossRef] [PubMed]
8. Retkute, R.; Smith-Unna, S.E.; Smith, R.W.; Burgess, A.J.; Jensen, O.E.; Johnson, G.N.; Preston, S.P.; Murchie, E.H. Exploiting heterogeneous environments: Does photosynthetic acclimation optimize carbon gain in fluctuating light? *J. Exp. Bot.* **2015**, *66*, 2437–2447. [CrossRef]
9. Kaiser, E.; Morales, A.; Harbinson, J. Fluctuating light takes crop photosynthesis on a rollercoaster ride. *Plant Physiol.* **2018**, *176*, 977–989. [CrossRef]
10. Flexas, J.; Medrano, H. Drought-inhibition of photosynthesis in C_3 plants: Stomatal and non-stomatal limitations revisited. *Ann. Bot.* **2002**, *89*, 183–189. [CrossRef] [PubMed]
11. Medrano, H.; Escalona, J.M.; Bota, J.; Gulías, J.; Flexas, J. Regulation of photosynthesis of C_3 plants in response to progressive drought: Stomatal conductance as a reference parameter. *Ann. Bot.* **2002**, *89*, 895–905. [CrossRef] [PubMed]
12. Zivcak, M.; Brestic, M.; Balatova, Z.; Drevenakova, P.; Olsovska, K.; Kalaji, H.M.; Yang, X.; Allakhverdiev, S.I. Photosynthetic electron transport and specific photoprotective responses in wheat leaves under drought stress. *Photosynth. Res.* **2013**, *117*, 529–546. [CrossRef] [PubMed]
13. Antolín, M.C.; Hekneby, M.; Sánchez-Díaz, M. Contrasting responses of photosynthesis at low temperatures in different annual legume species. *Photosynthetica* **2005**, *43*, 65–74. [CrossRef]
14. Bukhov, N.G.; Wiese, C.; Neimanis, S.; Heber, U. Heat sensitivity of chloroplasts and leaves: Leakage of protons from thylakoids and reversible activation of cyclic electron transport. *Photosynth. Res.* **1999**, *59*, 81–93. [CrossRef]
15. Allakhverdiev, S.I.; Kreslavski, V.D.; Klimov, V.V.; Los, D.A.; Carpentier, R.; Mohanty, P. Heat stress: An overview of molecular responses in photosynthesis. *Photosynth. Res.* **2008**, *98*, 541–550. [CrossRef] [PubMed]
16. Zhang, R.; Sharkey, T.D. Photosynthetic electron transport and proton flux under moderate heat stress. *Photosynth. Res.* **2009**, *100*, 29–43. [CrossRef] [PubMed]
17. Fischer, B.B.; Hideg, É.; Krieger-Liszkay, A. Production, detection, and signaling of singlet oxygen in photosynthetic organisms. *Antioxid. Redox Signal.* **2013**, *18*, 2145–2162. [CrossRef]
18. Müller, P.; Li, X.P.; Niyogi, K.K. Non-photochemical quenching. A response to excess light energy. *Plant Physiol.* **2001**, *125*, 1558–1566. [CrossRef] [PubMed]
19. Cruz, J.A.; Avenson, T.J.; Kanazawa, A.; Takizawa, K.; Edwards, G.E.; Kramer, D.M. Plasticity in light reactions of photosynthesis for energy production and photoprotection. *J. Exp. Bot.* **2005**, *56*, 395–406. [CrossRef]
20. Joliot, P.; Johnson, G.N. Regulation of cyclic and linear electron flow in higher plants. *Proc. Natl. Acad. Sci. USA* **2011**, *108*, 13317–13322. [CrossRef]
21. Alte, F.; Stengel, A.; Benz, J.P.; Petersen, E.; Soll, J.; Groll, M.; Bölter, B. Ferredoxin: NADPH oxidoreductase is recruited to thylakoids by binding to a polyproline type II helix in a pH-dependent manner. *Proc. Natl. Acad. Sci. USA* **2010**, *107*, 19260–19265. [CrossRef] [PubMed]
22. Benz, J.P.; Stengel, A.; Lintala, M.; Lee, Y.H.; Weber, A.; Philippar, K.; Gügel, I.L.; Kaieda, S.; Ikegami, T.; Mulo, P.; et al. Arabidopsis Tic62 and ferredoxin-NADP(H) oxidoreductase form light-regulated complexes that are integrated into the chloroplast redox poise. *Plant Cell* **2010**, *21*, 3965–3983. [CrossRef]
23. Kozaki, A.; Takeba, G. Photorespiration protects C_3 plants from photooxidation. *Nature* **1996**, *384*, 557–560. [CrossRef]

24. Davis, P.A.; Hangarter, R.P. Chloroplast movement provides photoprotection to plants by redistributing PSII damage within leaves. *Photosynth. Res.* **2012**, *112*, 153–161. [CrossRef] [PubMed]
25. Wada, M. Chloroplast movement. *Plant Sci.* **2013**, *210*, 177–182. [CrossRef]
26. Ptushenko, O.S.; Ptushenko, V.V.; Solovchenko, A.E. Spectrum of light as a determinant of plant functioning: A historical perspective. *Life* **2020**, *10*, 25. [CrossRef] [PubMed]
27. Miyake, C.; Yokota, A. Cyclic flow of electrons within PSII in thylakoid membranes. *Plant Cell Physiol.* **2001**, *42*, 508–515. [CrossRef]
28. Miyake, C.; Yonekura, K.; Kobayashi, Y.; Yokota, A. Cyclic electron flow within PSII functions in intact chloroplasts from spinach leaves. *Plant Cell Physiol.* **2002**, *43*, 951–957. [CrossRef] [PubMed]
29. Jajoo, A.; Mekala, N.R.; Tongra, T.; Tiwari, A.; Grieco, M.; Tikkanen, M.; Aro, E.M. Low pH-induced regulation of excitation energy between the two photosystems. *FEBS Lett.* **2014**, *588*, 970–974. [CrossRef] [PubMed]
30. Demmig-Adams, B.; Adams, W.W., III. The role of xanthophyll cycle carotenoids in the protection of photosynthesis. *Trends Plant Sci.* **1996**, *1*, 21–26. [CrossRef]
31. Sukhova, E.M.; Vodeneev, V.A.; Sukhov, V.S. Mathematical modeling of photosynthesis and analysis of plant productivity. *Biochem. (Moscow) Suppl. Ser. A Membr. Cell Biol.* **2021**, *15*, 52–72. [CrossRef]
32. Bernhardt, K.; Trissl, H.-W. Theories for kinetics and yields of fluorescence and photochemistry: How, if at all, can different models of antenna organization be distinguished experimentally? *Biochim. Biophys. Acta Bioenerg.* **1999**, *1409*, 125–142. [CrossRef] [PubMed]
33. Vredenberg, W.J. A three-state model for energy trapping and chlorophyll fluorescence in photosystem II incorporating radical pair recombination. *Biophys. J.* **2000**, *79*, 26–38. [CrossRef] [PubMed]
34. Bulychev, A.A.; Vredenberg, W.J. Modulation of photosystem II chlorophyll fluorescence by electrogenic events generated by photosystem I. *Bioelectrochemistry* **2001**, *54*, 157–168. [CrossRef] [PubMed]
35. Lazár, D. Chlorophyll a fluorescence rise induced by high light illumination of dark-adapted plant tissue studied by means of a model of photosystem II and considering photosystem II heterogeneity. *J. Theor. Biol.* **2003**, *220*, 469–503. [CrossRef] [PubMed]
36. Porcar-Castell, A.; Bäck, J.; Juurola, E.; Hari, P. Dynamics of the energy flow through photosystem II under changing light conditions: A model approach. *Func. Plant Biol.* **2006**, *33*, 229–239. [CrossRef]
37. Ebenhöh, O.; Houwaart, T.; Lokstein, H.; Schlede, S.; Tirok, K. A minimal mathematical model of nonphotochemical quenching of chlorophyll fluorescence. *Biosystems* **2011**, *103*, 196–204. [CrossRef] [PubMed]
38. Tikhonov, A.N.; Vershubskii, A.V. Computer modeling of electron and proton transport in chloroplasts. *Biosystems* **2014**, *121*, 1–21. [CrossRef] [PubMed]
39. Morales, A.; Yin, X.; Harbinson, J.; Driever, S.M.; Molenaar, J.; Kramer, D.M.; Struik, P.C. In silico analysis of the regulation of the photosynthetic electron transport chain in C_3 plants. *Plant Physiol.* **2018**, *176*, 1247–1261. [CrossRef] [PubMed]
40. Belyaeva, N.E.; Bulychev, A.A.; Riznichenko, G.Y.; Rubin, A.B. Analyzing both the fast and the slow phases of chlorophyll a fluorescence and P700 absorbance changes in dark-adapted and preilluminated pea leaves using a thylakoid membrane model. *Photosynth. Res.* **2019**, *140*, 1–19. [CrossRef]
41. Laisk, A.; Eichelmann, H.; Oja, V.; Eatherall, A.; Walker, D.A. A mathematical model of carbon metabolism in photosynthesis: Difficulties in explaining oscillations by fructose 2,6-bisphosphate regulation. *Proc. R. Soc. Lond. B Biol. Sci.* **1989**, *237*, 389–415.
42. Von Caemmerer, S. Steady-state models of photosynthesis. *Plant Cell Environ.* **2013**, *36*, 1617–1630. [CrossRef] [PubMed]
43. Zhu, X.-G.; Wang, Y.; Ort, D.R.; Long, S.P. E-photosynthesis: A comprehensive dynamic mechanistic model of C_3 photosynthesis: From light capture to sucrose synthesis. *Plant Cell Environ.* **2013**, *36*, 1711–1727. [CrossRef] [PubMed]
44. Berghuijs, H.N.; Yin, X.; Ho, Q.T.; Driever, S.M.; Retta, M.A.; Nicolaï, B.M.; Struik, P.C. Mesophyll conductance and reaction-diffusion models for CO_2 transport in C_3 leaves; needs, opportunities and challenges. *Plant Sci.* **2016**, *252*, 62–75. [CrossRef] [PubMed]
45. Wu, A.; Song, Y.; van Oosterom, E.J.; Hammer, G.L. Connecting biochemical photosynthesis models with crop models to support crop improvement. *Front. Plant Sci.* **2016**, *7*, 1518. [CrossRef] [PubMed]
46. Yin, X.; Struik, P.C. Can increased leaf photosynthesis be converted into higher crop mass production? A simulation study for rice using the crop model GECROS. *J. Exp. Bot.* **2017**, *68*, 2345–2360. [CrossRef] [PubMed]
47. Friend, A.D.; Geider, R.J.; Behrenfeld, M.J.; Still, C.J. Photosynthesis in global-scale models. In *Photosynthesis in Silico. Advances in Photosynthesis and Respiration*; Laisk, A., Nedbal, L., Govindjee, Eds.; Springer: Dordrecht, The Netherlands, 2009; Volume 29, pp. 465–497.
48. Pietsch, S.A.; Hasenauer, H. Photosynthesis within large-scale ecosystem models. In *Photosynthesis in Silico. Advances in Photosynthesis and Respiration*; Laisk, A., Nedbal, L., Govindjee, Eds.; Springer: Dordrecht, The Netherlands, 2009; Volume 29, pp. 441–464.
49. Farquhar, G.D.; von Caemmerer, S.; Berry, J.A. A biochemical model of photosynthetic CO_2 assimilation in leaves of C_3 species. *Planta* **1980**, *149*, 78–90. [CrossRef] [PubMed]
50. Von Caemmerer, S.; Farquhar, G.; Berry, J. Biochemical model of C_3 photosynthesis. In *Photosynthesis in Silico. Advances in Photosynthesis and Respiration*; Laisk, A., Nedbal, L., Govindjee, Eds.; Springer: Dordrecht, The Netherlands, 2009; Volume 29, pp. 209–230.

51. Bernacchi, C.J.; Rosenthal, D.M.; Pimentel, C.; Long, S.P.; Farquhar, G.D. Modeling the temperature dependence of C_3. In *Photosynthesis in Silico. Advances in Photosynthesis and Respiration*; Laisk, A., Nedbal, L., Govindjee, Eds.; Springer: Dordrecht, The Netherlands, 2009; Volume 29, pp. 231–246.

52. Niinemets, Ü.; Anten, N.P.R. Packing the photosynthetic machinery: From leaf to canopy. In *Photosynthesis in Silico. Advances in Photosynthesis and Respiration*; Laisk, A., Nedbal, L., Govindjee, Eds.; Springer: Dordrecht, The Netherlands, 2009; Volume 29, pp. 363–399.

53. Zhu, X.G.; Long, S.P. Can increase in Rubisco specificity increase carbon gain by whole canopy? A modeling analysis. In *Photosynthesis in Silico. Advances in Photosynthesis and Respiration*; Laisk, A., Nedbal, L., Govindjee, Eds.; Springer: Dordrecht, The Netherlands, 2009; Volume 29, pp. 401–416.

54. Song, Q.; Zhang, G.; Zhu, X.-G. Optimal crop canopy architecture to maximise canopy photosynthetic CO_2 uptake under elevated CO_2—A theoretical study using a mechanistic model of canopy photosynthesis. *Func. Plant Biol.* **2013**, *40*, 109–124. [CrossRef] [PubMed]

55. Ho, Q.T.; Berghuijs, H.N.; Watté, R.; Verboven, P.; Herremans, E.; Yin, X.; Retta, M.A.; Aernouts, B.; Saeys, W.; Helfen, L.; et al. Three-dimensional microscale modelling of CO_2 transport and light propagation in tomato leaves enlightens photosynthesis. *Plant Cell Environ.* **2016**, *39*, 50–61. [CrossRef]

56. Wu, A.; Doherty, A.; Farquhar, G.D.; Hammer, G.L. Simulating daily field crop canopy photosynthesis: An integrated software package. *Funct. Plant Biol.* **2018**, *45*, 362–377. [CrossRef] [PubMed]

57. Peñuelas, J.; Garbulsky, M.F.; Filella, I. Photochemical reflectance index (PRI) and remote sensing of plant CO_2 uptake. *New Phytol.* **2011**, *191*, 596–599. [CrossRef] [PubMed]

58. Zhang, C.; Filella, I.; Garbulsky, M.F.; Peñuelas, J. Affecting factors and recent improvements of the photochemical reflectance index (PRI) for remotely sensing foliar, canopy and ecosystemic radiation-use efficiencies. *Remote Sens.* **2016**, *8*, 677. [CrossRef]

59. Sukhova, E.; Sukhov, V. Connection of the Photochemical Reflectance Index (PRI) with the photosystem ii quantum yield and nonphotochemical quenching can be dependent on variations of photosynthetic parameters among investigated plants: A meta-analysis. *Remote Sens.* **2018**, *10*, 771. [CrossRef]

60. Kior, A.; Sukhov, V.; Sukhova, E. Application of reflectance indices for remote sensing of plants and revealing actions of stressors. *Photonics* **2021**, *8*, 582. [CrossRef]

61. Gamon, J.A.; Peñuelas, J.; Field, C.B. A narrow-waveband spectral index that tracks diurnal changes in photosynthetic efficiency. *Remote Sens. Environ.* **1992**, *41*, 35–44. [CrossRef]

62. Evain, S.; Flexas, J.; Moya, I. A new instrument for passive remote sensing: 2. Measurement of leaf and canopy reflectance changes at 531 nm and their relationship with photosynthesis and chlorophyll fluorescence. *Remote Sens. Environ.* **2004**, *91*, 175–185. [CrossRef]

63. Kováč, D.; Veselovská, P.; Klem, K.; Večeřová, K.; Ač, A.; Peñuelas, J.; Urban, O. Potential of photochemical reflectance index for indicating photochemistry and light use efficiency in leaves of European beech and Norway spruce trees. *Remote Sens.* **2018**, *10*, 1202. [CrossRef]

64. Sukhova, E.; Sukhov, V. Analysis of light-induced changes in the photochemical reflectance index (PRI) in leaves of pea, wheat, and pumpkin using pulses of green-yellow measuring light. *Remote Sens.* **2019**, *11*, 810. [CrossRef]

65. Kohzuma, K.; Tamaki, M.; Hikosaka, K. Corrected photochemical reflectance index (PRI) is an effective tool for detecting environmental stresses in agricultural crops under light conditions. *J. Plant Res.* **2021**, *134*, 683–694. [CrossRef] [PubMed]

66. Yudina, L.; Sukhova, E.; Gromova, E.; Nerush, V.; Vodeneev, V.; Sukhov, V. A light-induced decrease in the photochemical reflectance index (PRI) can be used to estimate the energy-dependent component of non-photochemical quenching under heat stress and soil drought in pea, wheat, and pumpkin. *Photosynth. Res.* **2020**, *146*, 175–187. [CrossRef]

67. Sukhov, V.; Sukhova, E.; Khlopkov, A.; Yudina, L.; Ryabkova, A.; Telnykh, A.; Sergeeva, E.; Vodeneev, V.; Turchin, I. Proximal imaging of changes in photochemical reflectance index in leaves based on using pulses of green-yellow light. *Remote Sens.* **2021**, *13*, 1762. [CrossRef]

68. Tessone, C.J.; Mirasso, C.R.; Toral, R.; Gunton, J.D. Diversity-induced resonance. *Phys. Rev. Lett.* **2006**, *97*, 194101. [CrossRef] [PubMed]

69. Liang, X.; Zhang, X.; Zhao, L. Diversity-induced resonance for optimally suprathreshold signals. *Chaos* **2020**, *30*, 103101. [CrossRef] [PubMed]

70. Sukhova, E.; Ratnitsyna, D.; Sukhov, V. Stochastic spatial heterogeneity in activities of H^+-ATP-ases in electrically connected plant cells decreases threshold for cooling-induced electrical responses. *Int. J. Mol. Sci.* **2021**, *22*, 8254. [CrossRef] [PubMed]

71. Sukhova, E.; Sukhov, V. Electrical signals, plant tolerance to actions of stressors, and programmed cell death: Is interaction possible? *Plants* **2021**, *10*, 1704. [CrossRef] [PubMed]

72. Winter, H.; Robinson, D.G.; Heldt, H.W. Subcellular volumes and metabolite concentrations in spinach leaves. *Planta* **1994**, *193*, 530–535. [CrossRef]

73. Tholen, D.; Zhu, X.-G. The mechanistic basis of internal conductance: A theoretical analysis of mesophyll cell photosynthesis and CO_2 diffusion. *Plant Physiol.* **2011**, *156*, 90–105. [CrossRef] [PubMed]

74. Evans, J.R.; Kaldenhoff, R.; Genty, B.; Terashima, I. Resistances along the CO_2 diffusion pathway inside leaves. *J. Exp. Bot.* **2009**, *60*, 2235–2248. [CrossRef] [PubMed]

75. Sukhova, E.M.; Sukhov, V.S. Dependence of the CO_2 uptake in a plant cell on the plasma membrane H^+-ATPase activity: Theoretical analysis. *Biochem. Mosc. Suppl. Ser. A* **2018**, *12*, 146–159. [CrossRef]

76. Sukhov, V.; Vodeneev, V. A mathematical model of action potential in cells of vascular plants. *J. Membr. Biol.* **2009**, *232*, 59–67. [CrossRef]

77. Sukhova, E.; Akinchits, E.; Sukhov, V. Mathematical models of electrical activity in plants. *J. Membr. Biol.* **2017**, *250*, 407–423. [CrossRef]

78. Kinoshita, T.; Shimazaki, K. Blue light activates the plasma membrane H^+-ATPase by phosphorylation of the C-terminus in stomatal guard cells. *EMBO J.* **1999**, *18*, 5548–5558. [CrossRef] [PubMed]

79. Gradmann, D. Impact of apoplast volume on ionic relations in plant cells. *J. Membr. Biol.* **2001**, *184*, 61–69. [CrossRef] [PubMed]

80. Sukhov, V.; Nerush, V.; Orlova, L.; Vodeneev, V. Simulation of action potential propagation in plants. *J. Theor. Biol.* **2011**, *291*, 47–55. [CrossRef] [PubMed]

81. Cardon, Z.G.; Mott, K.A.; Berry, J.A. Dynamics of patchy stomatal movements, and their contribution to steady-state and oscillating stomatal conductance calculated using gas-exchange techniques. *Plant Cell Environ.* **1994**, *17*, 995–1007. [CrossRef]

82. Siebke, K.; Weis, E. Assimilation images of leaves of *Glechoma hederacea*: Analysis of non-synchronous stomata related oscillations. *Planta* **1995**, *196*, 155–165. [CrossRef]

83. Schurr, U.; Walter, A.; Rascher, U. Functional dynamics of plant growth and photosynthesis–from steady-state to dynamics–from homogeneity to heterogeneity. *Plant Cell Environ.* **2006**, *29*, 340–352. [CrossRef]

84. Sharkey, T.D.; Seemann, J.R. Mild water stress effects on carbon-reduction-cycle intermediates, ribulose bisphosphate carboxylase activity, and spatial homogeneity of photosynthesis in intact leaves. *Plant Physiol.* **1989**, *89*, 1060–1065. [CrossRef]

85. Meyer, S.; Genty, B. Heterogeneous inhibition of photosynthesis over the leaf surface of *Rosa rubiginosa* L. during water stress and abscisic acid treatment: Induction of a metabolic component by limitation of CO_2 diffusion. *Planta* **1999**, *210*, 126–131. [CrossRef] [PubMed]

86. Osmond, C.B.; Kramer, D.; Lüttge, U. Reversible, water stress-indiced non-uniform chlorophyll fluorescence quenching in wilting leaves of *Potentilla reptans* may not be due to patchy stomatal responses. *Plant Biol.* **1999**, *1*, 618–624. [CrossRef]

87. Kim, T.H.; Böhmer, M.; Hu, H.; Nishimura, N.; Schroeder, J.I. Guard cell signal transduction network: Advances in understanding abscisic acid, CO_2, and Ca^{2+} signaling. *Annu. Rev. Plant Biol.* **2010**, *61*, 561–591. [CrossRef] [PubMed]

88. Christmann, A.; Grill, E.; Huang, J. Hydraulic signals in long-distance signaling. *Curr. Opin. Plant Biol.* **2013**, *16*, 293–300. [CrossRef]

89. Pieruschka, R.; Schurr, U.; Jahnke, S. Lateral gas diffusion inside leaves. *J. Exp. Bot.* **2005**, *56*, 857–864. [CrossRef]

90. Pieruschka, R.; Chavarría-Krauser, A.; Schurr, U.; Jahnke, S. Photosynthesis in lightfleck areas of homobaric and heterobaric leaves. *J. Exp. Bot.* **2010**, *61*, 1031–1039. [CrossRef] [PubMed]

91. Sukhov, V.; Surova, L.; Sherstneva, O.; Katicheva, L.; Vodeneev, V. Variation potential influence on photosynthetic cyclic electron flow in pea. *Front. Plant Sci.* **2015**, *5*, 766. [CrossRef] [PubMed]

92. Cabrera, J.C.B.; Hirl, R.T.; Schäufele, R.; Macdonald, A.; Schnyder, H. Stomatal conductance limited the CO_2 response of grassland in the last century. *BMC Biol.* **2021**, *19*, 50. [CrossRef]

93. Maxwell, K.; Johnson, G.N. Chlorophyll fluorescence–A practical guide. *J. Exp. Bot.* **2000**, *51*, 659–668. [CrossRef] [PubMed]

94. Flexas, J.; Barbour, M.M.; Brendel, O.; Cabrera, H.M.; Carriquí, M.; Díaz-Espejo, A.; Douthe, C.; Dreyer, E.; Ferrio, J.P.; Gago, J.; et al. Mesophyll diffusion conductance to CO_2: An unappreciated central player in photosynthesis. *Plant Sci.* **2012**, *193*, 70–84. [CrossRef]

95. Day, T.A.; Vogelmann, T.C. Alterations in photosynthesis and pigment distributions in pea leaves following UV-B exposure. *Physiol. Plant.* **1995**, *94*, 433–440. [CrossRef]

96. Antal, T.K.; Kovalenko, I.B.; Rubin, A.B.; Tyystjärvi, E. Photosynthesis-related quantities for education and modeling. *Photosynth Res.* **2013**, *117*, 1–30. [CrossRef]

97. Roeske, C.A.; Chollet, R. Role of metabolites in the reversible light activation of pyruvate, orthophosphate dikinase in *Zea mays* mesophyll cells in Vivo. *Plant Physiol.* **1989**, *90*, 330–337. [CrossRef] [PubMed]

98. Wang, Y.; Wu, W.H. Plant sensing and signaling in response to K^+-deficiency. *Mol. Plant.* **2010**, *3*, 280–287. [CrossRef] [PubMed]

Article

Metabolic Integration of Spectral and Chemical Cues Mediating Plant Responses to Competitors and Herbivores

Alexander Chautá and André Kessler *

Department of Ecology and Evolutionary Biology, Cornell University, E445 Corson Hall, Ithaca, NY 14850, USA
* Correspondence: ak357@cornell.edu

Abstract: Light quality and chemicals in a plant's environment can provide crucial information about the presence and nature of antagonists, such as competitors and herbivores. Here, we evaluate the roles of three sources of information—shifts in the red:far red (R:FR) ratio of light reflected off of potentially competing neighbors, induced metabolic changes to damage by insect herbivores, and induced changes to volatile organic compounds emitted from herbivore-damaged neighboring plants—to affect metabolic responses in the tall goldenrod, *Solidago altissima*. We address the hypothesis that plants integrate the information available about competitors and herbivory to optimize metabolic responses to interacting stressors by exposing plants to the different types of environmental information in isolation and combination. We found strong interactions between the exposure to decreased R:FR light ratios and damage on the induction of secondary metabolites (volatile and non-volatile) in plants. Similarly, the perception of VOCs emitted from neighboring plants was altered by the simultaneous exposure to spectral cues from neighbors. These results suggest that plants integrate spectral and chemical environmental cues to change the production and perception of volatile and non-volatile compounds and highlight the role of plant context-dependent metabolic responses in mediating population and community dynamics.

Keywords: competition; herbivory; induced defenses; plant communication; red:far red; volatiles; plant resistance; secondary metabolites

Citation: Chautá, A.; Kessler, A. Metabolic Integration of Spectral and Chemical Cues Mediating Plant Responses to Competitors and Herbivores. *Plants* **2022**, *11*, 2768. https://doi.org/10.3390/plants11202768

Academic Editors: Frantisek Baluska and Gustavo Maia Souza

Received: 12 September 2022
Accepted: 17 October 2022
Published: 19 October 2022

Publisher's Note: MDPI stays neutral with regard to jurisdictional claims in published maps and institutional affiliations.

1. Introduction

The ability to perceive, process, and integrate information from the environment is essential for any kind of behavioral or phenotypically plastic response, and thus for the fitness of any organism [1]. Plants are not an exception and, much like animals, have been shown to perceive and process information coded in light [2], sound [3], infochemicals (e.g., volatile organic compounds (VOCs)) [4,5], and touch [6]. Out of these, the perception of light and VOCs have received increased recent attention as they can encode information about the most important antagonistic interactions plants can have with other organisms; competition and herbivory/pathogen attack, respectively [7]. Plants perceive light with several specialized pigment molecules, such as chlorophylls and phytochromes. Among them, phytochromes regulate different processes such as germination, etiolation, shade avoidance, floral induction, induction of bud dormancy, tuberization, tropic orientation, and proximity perception [8,9]. Phytochromes are present in two interconvertible forms: P_r and P_{fr}, and the relative cytosolic concentrations of P_r and P_{fr} are determined by the ratio between red (λ max = 615–720 nm) and far-red light (λ max = 725–755 nm) [10]. A low ratio of red:far-red (R:FR) light transforms phytochrome into its inactive form (P_r), which attenuates the degradation of phytochrome-interacting factors (PIF), which, in turn, leads to different physiological changes in the plant. The absorption of red light by photosynthesizing plants, increases the relative amount of far-red light reflected off of their leaves, allowing neighboring plants to perceive the presence of other plants in their vicinity. Neighboring plants, in turn, are potential competitors, which is why plants are thought to preferentially allocate resources to competition by increasing elongating growth when exposed to lower R:FR light ratios [11]. Moreover, experiments on tobacco and tomato have demonstrated that this phytochrome-mediated perception of changes of the R:FR light ratio is also associated with a simultaneous reduced allocation of resources into direct resistance to herbivores and an attenuation of induced chemical resistance [12,13]. Likely underlying these attenuated metabolic responses to herbivory with corresponding effects on pathogen as well as herbivore resistance are apparent alterations of jasmonic acid (JA)- and salicylic acid (SA)-mediated gene expression in plants exposed to elevated FR light ratios [9,14]. Although the direction of the effect of FR light on individual VOCs [15–17] and non-volatile compounds [18,19], and so the expression of different types of resistance (e.g., direct vs. indirect resistance) may vary, the plants' ability to perceive changes in the R:FR light ratio seems to allow for a fine-tuning of the allocation of resources into competition or anti-herbivore defenses [20].

Like the perception of differences in light quality, the ability of plants to produce and perceive chemical environmental cues, such as VOCs, seems to play an important role in coping with multiple environmental challenges. VOCs are crucial in mediating plant direct and indirect resistance against herbivores [21]. After damage by herbivores, plants emit increased and attacker-specific blooms of VOCs, often called herbivory-induced plant volatiles (HIPV) [22]. These HIPVs are often repellent to foraging herbivores (direct repellence/resistance) and can also function as effective cues that attract natural enemies of herbivores, such as predators and parasitoids (information-mediated indirect defenses) [22,23]. Moreover, HIPVs can also be perceived by other plants, which respond by priming or directly inducing increased production of defense-related secondary metabolites, and thus increased resistance in anticipation of oncoming herbivores [24–28]. The mechanisms of plant VOC perception are debated to this date [29] and may include direct alteration of membrane potentials [30], specific receptors [31], and the transformation of VOCs into direct defensive compounds by the receiver plants [32]. However, very much like shifts in R:FR light ratios encode potential competition with neighbors, HIPVs provide a reliable cue associated with the probability of future herbivory. The fact, that plants have these different abilities to adaptively respond to changed light quality and HIPVs from neighbors, in turn, raises the question of how plants integrate these two different types of information to optimize responses to two of the most fitness-impacting environmental factors, competition, and herbivory.

Solidago altissima L. (Asteraceae) dominates early succession habitats in northeastern North America [33,34]. This species grows in dense patches where it competes for light with a diverse Asteracea-dominated plant community. Additionally, this species is attacked by a large diversity of insect herbivores [35,36]. Most importantly, however, plant community composition [37], as well as population genetic composition [38,39], are driven by a strong interaction between competition and insect herbivory on the dominant species *S. altissima*. Moreover, previous studies have demonstrated that *S. altissima* plants strongly respond to HIPVs from neighboring plants by priming and directly inducing changes in secondary metabolism and resistance [27]. Moreover, HIPV-mediated plant-to-plant information transfer affects herbivore distribution [40] and, in consequence, is under strong herbivory-mediated natural selection [28]. The particularly strong interaction observed between competitive ability and herbivory in determining the persistence of *S. altissima* plants in a population and community raises the more general question of how plants can utilize environmental information to adjust their phenotypes to varying environmental conditions while minimizing the combined, often synergistic impact of antagonistic biotic factors. The hypothesis that we are addressing here is that plants can integrate the information available on future herbivory and competition to induce metabolic changes that minimize the negative fitness effects of multiple interacting antagonists. Here, we test two major predictions associated with this hypothesis: (A) Secondary metabolite responses to herbivory should be altered in the presence of a neighbor (i.e., perception of lower R:FR ratio). (B) Perception of oncoming herbivory (i.e., HIPVs from damaged neighbors) should be altered by the presence of a potentially competitive neighbor (i.e., perception of lower R:FR ratio). This hypothesis seems particularly relevant in the study system we chose for this project, the tall goldenrod *S. altissima* (Figure 1a), where herbivory can be the major factor mediating competition with neighbors [37,39], and more nuanced and integrated responses to the combined perception of competitors and herbivores may maximize plant fitness. Here, we use *S. altissima* in factorial manipulative experiments to address the above-mentioned hypothesis and further our understanding of how plants integrate two different sources of biotic environmental information (light and HIPVs, Figure 1).

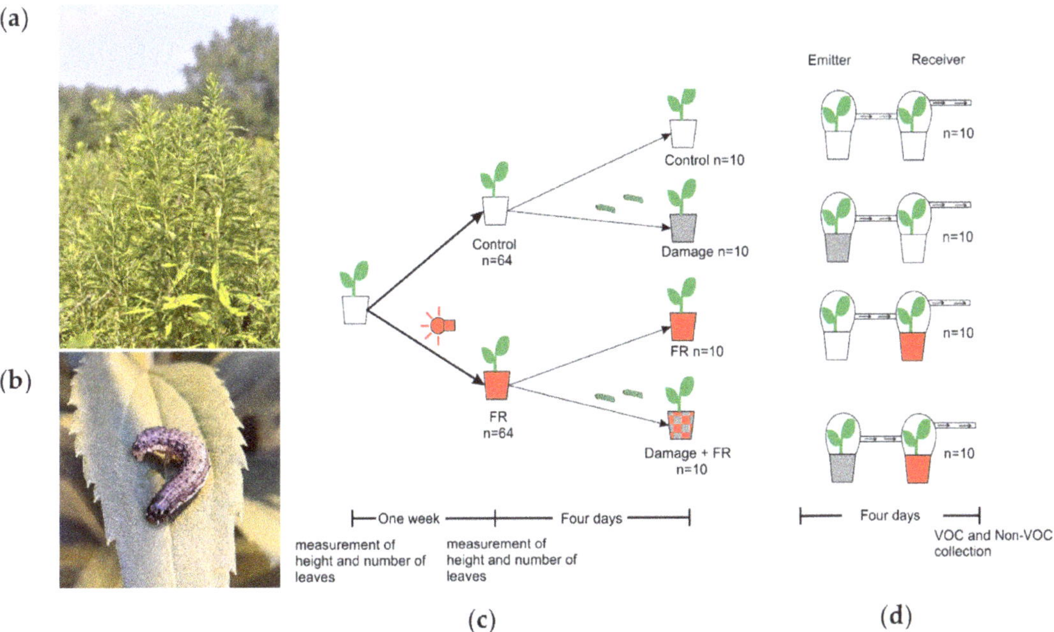

Figure 1. Experimental system. (**a**) Tall goldenrod, *Solidago altissima*, in dense early succession stands. (**b**) Fall Armyworm, *Spodoptera frugiperda* (Noctuidae: Lepidoptera), a generalist herbivore on *S. altissima*. (**c**) Scheme showing the sequence of treatments of *Solidago altissima* plants in the experiments. Plants were divided into two groups; one was exposed to supplemental far-red (FR) light and the other was kept as a control (ambient light). Then, half of the plants in each treatment were damaged for four days by two *S. frugiperda* larvae in L3. (**d**) Plants in the supplemented FR light and control treatments were set up to receive volatile organic compounds (VOC) from plants that were damaged by *S. frugiperda* for four days or from control plants that received no damage.

2. Results

2.1. Effect of FR Light on Plant Growth

Exposure to supplemental FR light resulted in increased stem elongation relative to plants under normal light conditions (t = −10.264, df = 63, $p < 0.001$, Figure 2a); however, there was no effect on the number of new leaves that grew between measurements (t = −0.93743, df = 63, $p = 0.3521$, Figure 2b).

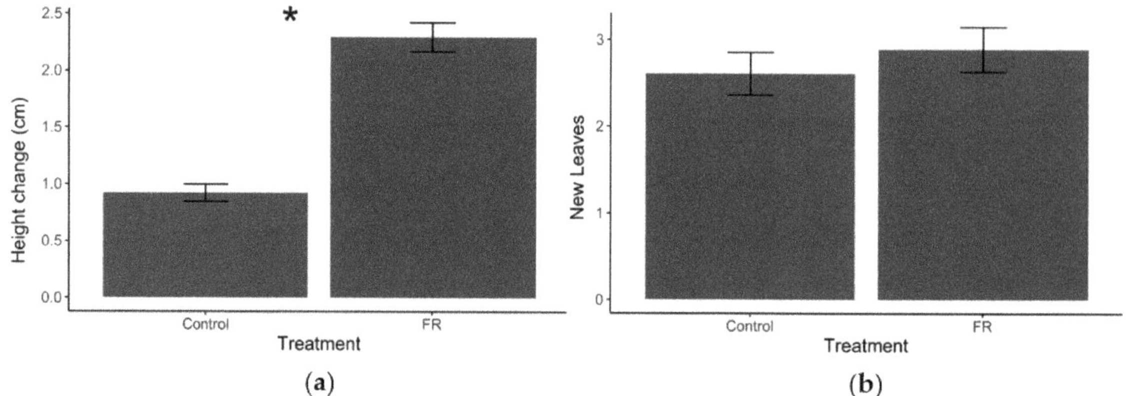

(a)   (b)

Figure 2. Plant growth responses to supplemental far-red (FR) light exposure. (**a**) Mean (±SEM) stem height change and (**b**) mean (±SEM) number of new leaves produced by *Solidago altissima* plants growing under regular light conditions supplemented with FR light (reduced red:far-red ratio) or under regular (control) light conditions over one week. * Represent statistical differences ($p < 0.001$) based on a Student's *t*-test.

2.2. Effect of Herbivory and FR Radiation on Plant Chemistry

Both FR exposure and herbivory by *Spodoptera frugiperda* larvae (Figure 1b) affected VOC production (PERMANOVA, $F_{1,36} = 9.5016$, $p = 0.001$ and $F_{1,36} = 4.3728$, $p = 0.007$, respectively), and we identified an interaction between both factors (FR x herbivore damage, $F_{1,36} = 7.2251$, $p = 0.001$). The post hoc analyses identified differences in VOC bouquet compositions between all the treatments except between plants damaged by herbivores and plants that were exposed to both enhanced FR light and herbivore damage (FR + Damage) (Table 1).

Table 1. VOC composition as a function of light quality and herbivore damage. Pairwise (post hoc) comparison of the VOC bouquets emitted from *Solidago altissima* plants that had been exposed to control light (Control) or regular light supplemented with FR light (FR), while being not damaged or damaged by two larvae of *S. frugiperda* in L3 instar (Damage).

Treatment	Control	Damage	FR
Damage	$F_{1,18} = 3.872$, $p = 0.001$ *		
FR	$F_{1,18} = 19.12$, $p = 0.001$ *	$F_{1,18} = 7.85$, $p = 0.001$ *	
FR + Damage	$F_{1,18} = 5.70$ $p = 0.001$ *	$F_{1,18} = 0.86$, $p = 0.283$	$F_{1,18} = 7.7928$, $p = 0.001$ *

* Represent statistical differences ($p < 0.05$) after false discovery rate (FDR) adjustment.

These general changes in compound composition in response to the exposure to supplemented FR light and herbivory are also evident in an NMDS analysis (Figures 3a and S1a, stress value = 0.1150099). Random Forest Analyses and post hoc ANOVAs revealed 30 individual VOCs whose emissions varied with treatment and predominantly increased in response to herbivore damage or the combination of FR exposure and damage (Figure 3c). Nonvolatile compound compositions were also affected by supplemented FR exposure (PERMANOVA, $F_{1,36} = 3.2915$, $p = 0.011$) but only marginally by damage ($F_{1,36} = 2.0827$, $p = 0.060$); however, we observed a strong interaction between both factors (FR × herbivore damage, $F_{1,36} = 5.4194$, $p = 0.001$).

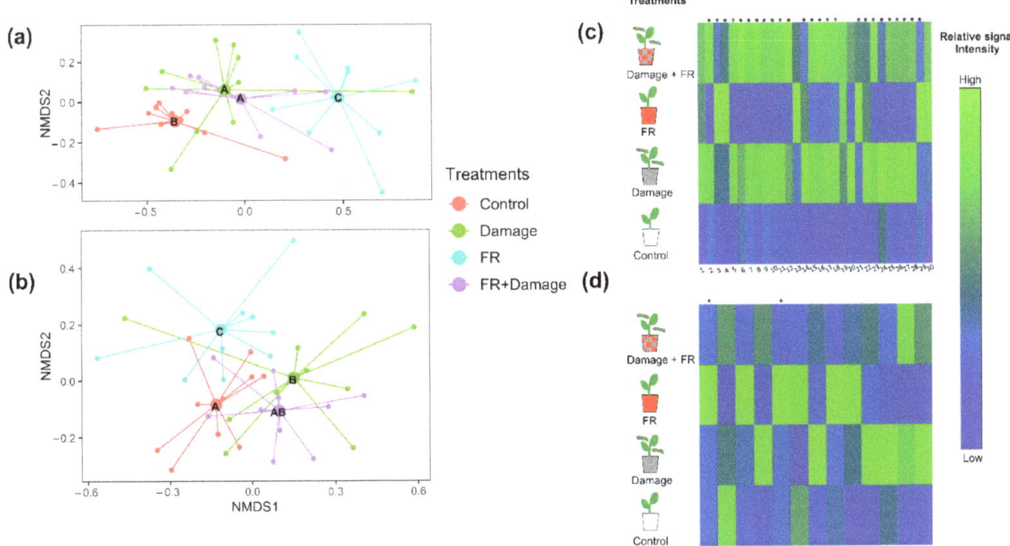

Figure 3. Plant secondary metabolite production in response to supplemented far-red (FR) light exposure and herbivore damage. Non-metric multidimensional scaling (NMDS) of (**a**) volatile organic compound emissions (stress value = 0.115) and (**b**) non-volatile secondary metabolite production (stress value = 0.163) of *Solidago altissima* plants, growing under reduced red:far red light ratios (FR), with damage by larvae of *Spodoptera frugiperda* (Damage), the combination of FR supplementation and damage (FR + Damage), or under control light (control) conditions. Different letters in the NMDS centroids indicate significant differences based on a post hoc test of a PERMANOVA. Heat map of (**c**) the emission of volatile organic compounds (VOCs) with tentative identification (1. Cymene, 2. Dimethyl-1,3,7-nonatriene, 3. (3Z)-Hexenyl acetate, 4. 1,3,8 -Menthatriene, 5. Unknown 1, 6. Unknown 2, 7. Unknown 3, 8. p-Copaene, 9. β-Bourbonene, 10. α-Cubebene, 11. Linalyl isobutyrate, 12. Unknown 4, 13. Bornyl acetate, 14. 1,5-Cyclodecadiene, 15. β-Caryophyllene, 16. Unknown 5, 17. Unknown Sesquiterpene 1, 18. α-Humulene, 19. Unknown Sesquiterpene 2, 20. γ-Muurolene 21. Unknown sesquiterpene 3, 22. Methyl salicylate, 23. Unknown 6, 24. Unknown aliphatic compound, 25. Unknown 7, 26. Unknown 8, 27. Unknown 9, 28. α-Phellandrene, 29. β-pinene, and 30. Limonene), and (**d**) the production of non-volatile compounds (1. Unknown 10, 2. Diterpene 1, 3. Coumaric acid 1, 4. Coumaric acid 2, 5. Chlorogenic acid, 6. Coumaric acid 3, 7. Flavonoid 1, 8. Diterpene 2, 9. Diterpene 3, 10. Diterpene 4, 11. Diterpene 5, 12. Diterpene 6, and 13. Diterpene 7) whose production is significantly varying with treatment ($p < 0.05$). The different treatments include untreated controls, plants exposed to increased FR radiation (FR), plants damaged by *S. frugiperda* caterpillars (Damage), and plants that received both treatments (FR + Damage). The heat maps include those compounds that explain most of the variation between treatments based on Random Forest Analyses. Different shades of color represent different signal intensity based on individual ANOVAs. * Indicate significance ($p < 0.05$) after false discovery rate adjustment.

The post hoc analyses of non-volatile secondary metabolite composition revealed differences between all the treatments except for the comparison between damage vs. FR + Damage and Control vs. FR + Damage (Table 2), which is also reflected in the NMDS analysis (Figure 3b and S1b, stress value = 0.2587829). Random Forest Analyses and subsequent ANOVAs identified 13 non-volatile compounds that show a pronounced increase in two treatments (Damage and FR), while their production tended to be lower in the combined FR + Damage treatment (Figure 3d). Of those compounds, there are seven diterpeneoids, two coumaric acid derivatives, one flavonoid, one chlorogenic acid derivative, and one currently unknown compound.

Table 2. Non-volatile secondary metabolite composition as a function of light quality and herbivore damage. Pairwise (post hoc) comparison of the non-volatile secondary metabolite composition *Solidago altissima* plants that had been exposed to control light (Control), or control light supplemented with FR light (FR), while being undamaged or damaged by two larvae of *S. frugiperda* in L3 instar (Damage).

Treatment	Control	Damage	FR
Damage	$F_{1,18} = 2.5411, p = 0.03$ *		
FR	$F_{1,18} = 7.851, p = 0.003$ *	$F_{1,18} = 3.75, p = 0.001$ *	
FR + Damage	$F_{1,18} = 1.33, p = 0.153$	$F_{1,18} = 0.887, p = 0.461$	$F_{1,18} = 4.7507, p = 0.002$ *

* Represent statistical differences ($p < 0.05$) after false discovery rate (FDR) adjustment.

2.3. Effect of FR Light on the Perception of VOCs

The overall composition of VOCs emitted from plants exposed to VOCs from neighboring plants did not change with exposure to increased FR light ratios ($F_{1,36} = 1.444, p = 0.095$) or with the exposure to VOCs from damaged plants ($F_{1,36} = 1.459, p = 0.106$), nor was there an interaction between both factors ($F_{1,36} = 1.543, p = 0.086$; Figure 4a and Figure S2a). However, Random Forest Analyses and ANOVAs performed on individual compounds identified differential response patterns for five compounds. Four of those compounds were stronger emitted from plants that were exposed to increased FR light and received VOCs from control plants. However, those compounds showed lower emission rates when simultaneously exposed to increased FR light and VOCs from damaged neighbors, indicating an integration of the two types of environmental information (Figure 4c). The fifth compound was emitted in higher amounts from plants exposed to FR light while also receiving VOCs from damaged neighbors (Figure 4c).

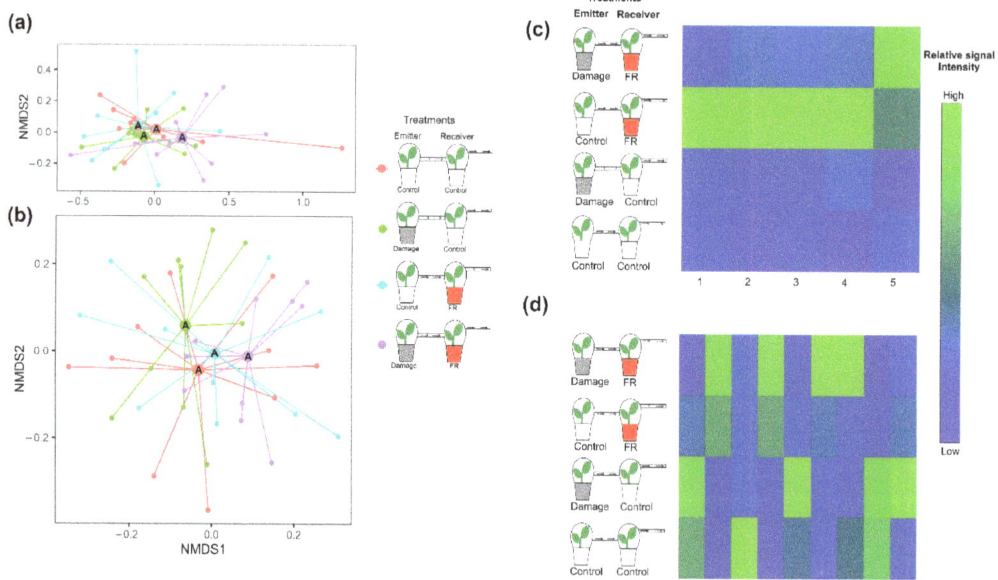

Figure 4. Plant secondary metabolite production in response to supplemented far-red (FR) light exposure and volatile organic compounds (VOCs) from neighbors. Plant secondary metabolite

production of plants under supplemented far-red light (FR, red pots) or normal light (control, white pots) that were exposed to VOCs from control plants or plants that were damaged by two larvae of *S. furgiperda* (Damage, grey pots). Non-metric multidimensional scaling (NMDS) of (**a**) VOC emissions (stress value = 0.08728402) and (**b**) non-volatile secondary metabolite production (stress value = 0.2796113) of *Solidago altissima* plants under the different exposure treatments. Different letters in the NMDS centroids indicate significant differences based on a post hoc test of a PERMANOVA. Heat map of (**c**) the emission of VOCs with tentative identification (1. Unknown 1, 2. α-Ylangene, 3. Bornyl acetate, 4. Unknown 2, and 5. β-Phellandrene) and (**d**) the production of non-volatile compounds (1. Coumaric acid, 2. Caffeic acid, 3. Flavonol 1, 4. Diterpene 1, 5. Flavonol 2, 6. Unknown, 7. Chlorogenic acid, 8. Diterpene 2, and 9. Diterpene 3) whose production significantly varied with treatment ($p < 0.05$). The different treatments include plants exposed to supplemented FR radiation or regular light and plants exposed to VOCs from undamaged plants or VOCs from plants damaged by *Spodoptera frugiperda* caterpillars, and plants that were exposed to both FR light supplementation and VOCs from damaged plants. The heat maps include those compounds that explain most of the variation between treatments based on Random Forest Analyses. Different shades of color represent different signal intensity based on individual ANOVAs.

Like the VOC responses, the overall non-volatile compound composition from plants that were exposed to increased FR radiation or were exposed to VOCs from herbivore-damaged plants did not change (PERMANOVA, $F_{1,36} = 1.50148$, $p = 0.119$ and $F_{1,36} = 0.85814$, $p = 0.497$, respectively) relative to those from controls. Moreover, perception of neighbors and herbivory on neighboring plants (FR and Damage VOCs) did not interact to affect non-volatile secondary metabolite production ($F_{1,36} = 0.96695$, $p = 0.380$; Figure 4b and Figure S2b). However, ANOVA analyses of individual compounds showed significant changes for some of them (Figure 4d).

3. Discussion

Plants can perceive neighbors by the shift in the R:FR ratio of light reflected off of green leaves and are commonly observed to respond with accelerated stem elongation [41]. In confirmation of these earlier findings, we found *S. altissima* plants responding in a very similar way. The exposure of the plants to increased FR light radiation resulted in stem elongation but did not alter the number of leaves produced. At the same time, plant secondary metabolism as well as plant metabolic responses to herbivory and to VOCs from neighboring plants were strongly affected by FR perception. Specifically, *S. altissima* plants exposed to both supplemented FR light and herbivory induced differences in secondary metabolite production (volatiles and non-volatiles) with both factors interacting, suggesting that both constitutive and induced secondary metabolite production are strongly affected by the exposure to spectral cues from neighboring plants. This coordinated change in growth and metabolism in response to perceived potential competitors has recently been interpreted as a differential allocation of resources into plant competitive ability and defensive functions [13,20]. Similarly, but to a somewhat smaller extent, the ability of plants to perceive volatiles coming from neighboring plants (with and without damage) was affected by the exposure to FR light. These findings are significant as they suggest that plants process information about actual herbivory as well as potential future herbivory differently when exposed to neighboring plants, and plants differentially integrate information about different types of antagonists (e.g., herbivores and competitors) to induce metabolic responses.

3.1. FR Light and Plant Growth

In shade-intolerant plants, a low ratio of R:FR light induces differential growth, such as stem elongation [42], that can provide a competitive advantage over neighbors in the natural habitats [41]. This stem elongation is regulated by gibberellin A1- and Indole-3-acetic acid (IAA)-mediated cell expansion rather than cell propagation [43,44], and thus results in an increase in the internode distance rather than an increase in the number of internodes [41]. *Solidago altissima* plants respond to FR supplementation in a similar way despite the fact that this species seems well adapted to growth in dense, high-competition environments. Similar plant-endogenous signaling mechanisms are thought to also mediate the correlated changes in plant secondary metabolism and the changes in the inducibility of metabolic responses to other environmental cues and stressors, such as herbivory [13]. Thus, the functional question for why plants induce changes to competition, indicating light quality has to be answered on two levels. On one hand, we need to explain why secondary metabolism changes in response to different light quality in the first place (i.e., the potential benefit of altered constitutive defenses). On the other hand, one has to probe the potential effects of light-quality-mediated changes on the perception of other stressors, such as herbivory (i.e., integration of environmental information from different sources).

3.2. Effect of Increased FR Light on Constitutive and Herbivory-Induced Secondary Metabolism

Effects of increased FR light ratios on secondary metabolite production, specifically volatile compounds, have been shown in plants as different as *Petunia* × *hybrida* [16], *Hordeum vulgare* [15], *Ocimum basilicum* [17], *Nicotiana sylvestris*, *Solanum lycopersicon* [12,13], and now *S. altissima*. The previous studies suggest the involvement of a wider range of phytohormones that had been found important for the growth responses. For example, a low R:FR ratio causes downregulation in the jasmonic acid (JA) pathway in shade-intolerant plants [14,20]. This pathway is crucial in the induced production of defensive compounds in plants but commonly reduces plant growth [45]. In *Arabidopsis thaliana*, JA is repressed by low R:FR light ratios [20], suggesting a priority of growth over defenses. Interestingly, the increased expression of IAA signaling, induced by a low R:FR ratio, has long been known as an inhibitor of JA responses and, on a mechanistic level, may explain the differential allocation of resources into cell elongation and away from secondary metabolism [46–49]. Moreover, the inhibition of wound- or herbivory-induced JA signaling will also significantly impair the herbivory-mediated induction of defense-related secondary metabolites [50]. This would certainly explain the aforementioned findings on the FR-induced allocation into growth away from constitutive and induced secondary metabolite production and resistance in tobacco, *Nicotiana sylvestris*, and tomato, *S. lycopersicon* [12,13]. From a functional perspective, this kind of response is likely adaptive in plant systems where competition with neighbors is substantially more impacting on plant fitness than herbivory. In systems like *S. altissima*, where herbivory can be the major factor mediating competition with neighbors [37,39], more nuanced and integrated responses to the combined perception of competitors and herbivores may suit the plants better. While this project focused on the evidence for such an integration of different types of information, it goes beyond the scope of this paper to investigate the actual resistance and plant fitness effects.

However, in light of both the wider functional hypothesis as well as the objective of this study, there are several remarkable induction patterns in *S. altissima's* response to FR light and herbivory. Previous studies have shown that FR light and damage affect the production of VOCs in plants [15,16,22]. When *S. altissima* plants are exposed to a combination of herbivore damage and higher FR light ratios, the chemical profile becomes more like one of the plants with damage. This suggests that, different from the previous studies, in the case of *S. altissima*, secondary metabolism and its induction by herbivores are not suppressed by reduced R:FR light ratios. More importantly, the fact that the combination of FR light supplementation (i.e., perception of a potential competitor) and herbivory-induced different volatile and non-volatile secondary metabolite profiles is strong support for the information integration hypothesis. Interestingly, compounds that were mostly upregulated by higher FR light ratios in *S. altissima* plants were the ones that are downregulated with damage or the combined exposure to supplemented FR light and herbivory and vice versa (Figure 3c). Non-volatile compounds have been observed to change with increased FR light exposure in other study systems [18,19]. Similar to VOCs, the changes in non-volatiles could be related to changes in the JA pathway. In how far these differential inductions of plant secondary metabolism affect subsequent interactions with other organisms, such as herbivores and neighboring plants, remains to be determined. Interestingly, some of the compounds upregulated by FR light supplementation in *S. altissima* are diterpenes (Diterpene 2 and 3, Figure 3d), which are known for having functions as anti-feedants and growth inhibitors for *Solidago* herbivores [51,52]. While previous studies have found downregulation of defenses in response to the exposure to FR light [13], this upregulation of defense metabolites in *S. altissima* indicates a differently regulated response to competition and herbivory. This is particularly remarkable when we consider that in *S. altissima* the defensive function of induced resistance and VOC-mediated information transfer [28] are only realized when plants are in close proximity, so that herbivores can move freely from plant to plant and thus spread the risk of damage among all members of the plant population [40]. In conclusion, our data suggest that *S. altissima* can integrate both types of signals (spectral and chemical) and responds with a stronger induction of defenses and clearer information encoded in HIPV emissions when exposed to cues from competitors and herbivores simultaneously.

3.3. Effect of Ingreased FR Light on the Perception of HIPVs from Neighbors

The second prediction in the information integration hypothesis went one step further and suggested that, if plants can integrate the information of a perceived neighbor with the information provided by an actively feeding herbivore, plants may also be able to integrate the perceived neighbor with cues that indicate future herbivory (i.e., HIPV emitted from damaged neighboring plants). Overall, *S. altissima* secondary metabolite profiles did not differ in response to the exposure to VOCs from control plants or plants exposed to FR light supplementation (Figure 4a,b). This is not necessarily surprising as exposure to VOCs alone has rarely been found to induce significant metabolic changes without additional damage to the leaf tissue (e.g., priming of plant responses: [53]). For example, in *S. altissima*, only 19 compounds of the non-volatile fraction of the recorded secondary metabolites were directly inducible by VOCs from neighboring plants without additional herbivore damage [27]. In accordance with these earlier findings, we also found several individual compounds induced by the simple exposure of the plant to neighbor HIPVs. Most surprisingly, plants under supplemented FR light receiving VOCs from an undamaged control plant increase the production of four VOCs dramatically (Figure 4c). This suggests that increased FR light ratios make plants more perceptive of VOCs from neighbors. Recent studies have suggested that reduced R:FR light ratios emitted from neighboring plants are used by *Arabidopsis thaliana* as a signal for kin recognition that mediates interactions among kin neighbors, reducing competition for resources [54]. The FR-induced VOC emission as well as the FR-mediated differential perception of VOCs can provide an alternative and likely more specific mechanism of kin recognition. In *S. altissima*, one individual can be surrounded by several clonal ramets, which is why the availability of information about the neighbor's genetic relatedness would be beneficial to the receiver plant because it would allow for a reduced investment into resources for competition against itself or close relatives [55].

From the nine non-volatile compounds that were affected by the treatments, the greatest increment was evident in plants that received VOCs from damaged neighbors (Figure 4d), indicating that *S. altissima* can detect and respond to HIPVs. However, the compounds that are upregulated are not the same in the different exposure treatments. On the one hand, plant exposure to HIPVs directly induces the production of a coumaric acid derivative, a flavonoid, and two diterpene acids; on the other hand, plants simultaneously exposed to FR and HIPVs increased the production of a different set of compounds, including a chlorogenic acid derivative, one diterpene acid, a caffeic acid derivative, and an unknown compound. This ultimately indicates that plants exposed to light reflected off of potentially competitive neighbors perceived and processed the information encoded in HIPVs from herbivore-attacked neighbors differently from plants that stand isolated without neighbors. More generally, our data suggest that VOCs can provide specific information about the presence as well as about the identity and herbivory status of a plant neighbor. Moreover, the data support the hypothesis that *S. altissima* plants integrate the information encoded in VOCs and herbivore damage with that of spectral information from neighbors to induce distinct changes in their metabolism. While the ecological outcomes as well as the detailed molecular mechanisms of this integration of environmental information remain to be revealed, previous studies observing herbivore-induced changes of similar magnitudes reported significant effects on population and community dynamics [56]. Most importantly, the implied population and species-specific differential integration of information concluded from this study may provide an explanation for plant responses that do not have outcomes predicted by the observed phytohormonal signaling [57] and can explain local adaptation [38] and associational resistance dynamics [58].

4. Materials and Methods

4.1. Plant Material

Seeds of *S. altissima* were bulk collected in winter 2020 from plants around Bebe Lake, Ithaca, NY and then stored in a freezer at $-20\ °C$. After a month, the seeds were put into LM1 germination mix soil (Lambert) for germination at Cornell University's greenhouse with a photoperiod of 16:8 h light:dark. Once the plants had germinated, they were repotted into individual clear polyethylene terephthalate plastic cups of 500 mL capacity, and an initial measurement of the length of the plant and the number of leaves was taken. All plants (n = 128) were grown under natural light supplemented with high-pressure sodium lamps that produce 200 $\mu mol/m^2/s$ of white light to complete a photoperiod of 16:8 h day:night. In addition, half of the plants were under supplemental far-red (FR) light, using a FR LED strip (Forever Green Indoors, $\lambda = 730$ nm) of 114 cm length with 32 LED bulbs. The FR lamps were covered with a blue filter (Roscolux, Supergel, Cinegel no. 83 Medium Blue) to remove residual red light following the protocol of [14]. The lamp was located at 15 cm to the side of the plant and 10 cm from the ground to simulate the angle of light and the intensity coming from a neighboring plant. After one week, the second measurement of height and number of leaves was recorded to assess the effect of increased ratios of FR light on plant growth. After these measurements, two larvae of *Spodoptera frugiperda* in their third instar were added to each of 10 plants in each light treatment (FR and control), completing four groups of plants: Control, Damage, FR, and FR + Damage (Figure 1c). After another four days with the larvae actively feeding, 10 plants in the damage treatment and 10 plants in the control treatment were used as emitter plants in a plant VOC-exposure experiment. Their VOC emissions were pulled into receiver plant chambers that included either control plants under normal light conditions or plants supplemented with FR light (Figure 1d). The chambers of both the emitter and receiver plants were connected through 0.7 cm diameter silicon tubing (BIO-RAD, Hercules, CA, USA), and the chamber of the receiver was connected to an active air sampling vacuum pump (IONTIK) pulling air at about 450 mL/min. The pumps generate a constant flow of air from the emitter to the receiver plants (Figure 1d).

The pumps were changed twice a day, to ensure that there would be at least 22 h of flow per day. After four days of VOC exposure, we collected VOCs using adsorbent traps and leaf material to analyze non-volatile metabolites. Volatile samples of each plant were taken by enclosing the plant into 500 mL polyethylene cups that were connected to an ORBO-32 charcoal adsorbent tube (Supelco®, SIGMA-ALDRICH, Inc. St Louise, MO, USA). The air was pulled through the charcoal traps using an active air sampling vacuum pump (IONTIK) pulling air at about 450 mL/min. Additionally, leaf samples were collected and flash-frozen in liquid nitrogen and later stored at $-80\ °C$ until further analysis. To understand if the chemical response to damage is affected by the presence of a neighbor, we compared the volatile and non-volatile chemical profiles of the emitter plants (Control, Damage, FR, and FR + Damage). To understand if the perception of volatiles is affected by FR exposure, we compared the volatile and non-volatile metabolites produced by the receiver plants.

4.2. Secondary Metabolite Analysis

For the analysis of VOCs emitted from the experimental plants, each of the ORBO-32 charcoal traps that were used in the collection was spiked with 5 µL of tetraline (90 ng/mL) as an internal standard. The charcoal traps were washed with 400 µL of dichloromethane, which was then injected in a Varian CP-3800 gas chromatograph (GC) coupled with a Saturn 2200 mass spectrometer (MS) and equipped with a CP-8400 autosampler. The GC-MS was fitted with a DB-WAX column, (Agilent, J&W Scientific, Santa Clara, CA, USA) of 60 m × 0.25 mm id capillary column coated with polyethylene glycol (0.25 mm film thickness). The temperature program began with an injection temperature of 225 °C, heated from 45 °C to 130 °C at 10 °C/minute, then from 130 °C to 180 at 5 °C/min, and finally from 180 °C to 250 °C at 20 °C/minute with a 5 min hold at 230 and 250 °C. The samples were standardized by expressing signal intensity (peak area) of each peak relative to that of the internal standard. Compound and compound class identity were determined comparing mass spectra and retention time indices with NIST library records and previously published VOC data of *S. altissima* [27,59].

For the high-performance liquid chromatography (HPLC) analysis of non-volatile compounds, leaf samples (150–250 mg/sample) were homogenized and extracted in 1 mL of 90% methanol using a FastPrep® tissue homogenizer (MP Biomedicals®, Irvine, CA, USA) at 6 m/s for 90 s using 0.9 g grinding beads (Zirconia/Silica 2.3 mm, Biospec®, Bartlesville, OK, USA). The samples were then centrifuged at 4 °C for 15 min at 14,000 rpm, and we analyzed 15 µL of the by HPLC on an Agilent® 1100 series HPLC. We used 99.9% acetonitrile and 0.25% H_3PO_4 as the mobile phase. The elution system consisted of aqueous 0.25% H_3PO_4 and acetonitrile (ACN), which were pumped through a Gemini C18 reverse-phase column (3 µm, 150 × 4.6 mm, Phenomenex, Torrance, CA, USA) at a rate of 0.7 mL/min with increasing concentrations of ACN: 0–5 min, 0–20% ACN; 5–35 min, 20–95% ACN; and 35–45 min, 95% ACN. The area of each peak was standardized by the mass of the leaf tissue extracted. The individual compound or class identity was determined based on the UV spectra and retention times of authentic standards.

4.3. Statistical Analysis

The differences in growth between FR and control plants were analyzed using a Student's *t*-test. The overall composition of volatile and non-volatile plant secondary metabolites was inspected using nonmetric multidimensional scaling (Bray–Curtis distance matrix; metaMDS in *vegan* package) and tested for the effects of the FR light exposure and herbivory on the composition with a PERMANOVA with 999 permutations using the adonis2 function in the *vegan* package using the trials as strata. For the emitters, we used the exposure to FR and damage as independent factors and the relative abundance of compounds as dependent factors. For the receivers, we used the exposure to FR and the damage on the emitter plant as independent factors, and the relative abundance of compounds as dependent factors. If PERMANOVAs showed significant results, a post hoc test was run using the function *pairwise.adonis2* from the library *pairwise.adonis* [60]. We adjusted the *p* values for the multiple corrections using the false discovery rate (FDR) adjustment (*p. adjust* in package *stats*). Additionally, individual ANOVAs were run for each volatile and non-volatile compound from emitter and receiver plants that were identified as those compounds most explaining the variation between treatments in a Random Forest Analysis using the *randomForest* package in R. Results from these compounds were included in a heat map analysis for easier visualization of the complex differences. FDR adjustments to the ANOVA results are reflected in the figures. All statistical analyses were performed using the R program [61].

Supplementary Materials: The following supporting information can be downloaded at: https://www.mdpi.com/article/10.3390/plants11202768/s1, Figure S1: Plant secondary metabolite production in response to supplemented far-red (FR) light exposure and herbivore damage; Figure S2:

Plant secondary metabolite production in response to supplemented far-red (FR) light exposure and volatile organic compounds (VOCs) from neighbors.

Author Contributions: A.C. and A.K. conceived the ideas and designed the methodology; A.C. collected the data; A.C. and A.K. analyzed the data; A.C. and A.K. led the writing of the manuscript. All authors have read and agreed to the published version of the manuscript.

Funding: The research was funded by a grant to A.C. by Fundación CEIBA (Centro de Estudios Interdisciplinarios Básicos y Aplicados) and a grant from NIFA Multistate NE-1501 (2021-22-194) to A.K.

Data Availability Statement: Raw data are available on Cornell's eCommons (https://hdl.handle.net/1813/111891 (accessed on 17 October 2022)) data repository under the title of this publication.

Acknowledgments: We thank Anurag Agrawal and Robert Raguso for comments on an earlier draft of this paper and Aino Kalske for insights and preliminary experiments that generated the ideas for this project. This study was conducted at Cornell University, which is located on the traditional homelands of the Gayogo̱hó꞉nǫ' (the Cayuga Nation) of the Haudenosaunee Confederacy. We acknowledge the painful history of Gayogo̱hó꞉nǫ' dispossession and honor the ongoing connection of Gayogo̱hó꞉nǫ' people, past and present, to these lands and waters.

Conflicts of Interest: The authors declare no conflict of interest.

References

1. Chen, M.; Chory, J.; Fankhauser, C. Light Signal Transduction in Higher Plants. *Annu. Rev. Genet.* **2004**, *38*, 87–117. [CrossRef] [PubMed]
2. Carvalho, R.F.; Takaki, M.; Azevedo, R.A. Plant Pigments: The Many Faces of Light Perception. *Acta Physiol. Plant.* **2011**, *33*, 241–248. [CrossRef]
3. Khait, I.; Obolski, U.; Yovel, Y.; Hadany, L. Sound Perception in Plants. *Semin. Cell Dev. Biol.* **2019**, *92*, 134–138. [CrossRef] [PubMed]
4. Heil, M.; Ton, J. Long-Distance Signalling in Plant Defence. *Trends Plant Sci.* **2008**, *13*, 264–272. [CrossRef] [PubMed]
5. Kessler, A. The Information Landscape of Plant Constitutive and Induced Secondary Metabolite Production. *Curr. Opin. Insect Sci.* **2015**, *8*, 47–53. [CrossRef]
6. Bae, H.; Mishra, R.C. Plant Cognition: Ability to Perceive 'Touch' and 'Sound'. In *Sensory Biology of Plants*; Sopory, S., Ed.; Springer Nature: Singapore, 2019; pp. 137–162. ISBN 978-981-13-8921-4.
7. Karban, R. Plant Behaviour and Communication. *Ecol. Lett.* **2008**, *11*, 727–739. [CrossRef]
8. Smith, H. Physiological and Ecological Function within the Phytochrome Family. *Annu. Rev. Plant Physiol. Plant Mol. Biol.* **1995**, *46*, 289–315. [CrossRef]
9. De Wit, M.; Spoel, S.H.; Sanchez-Perez, G.F.; Gommers, C.M.M.; Pieterse, C.M.J.; Voesenek, L.A.C.J.; Pierik, R. Perception of Low Red:Far-Red Ratio Compromises Both Salicylic Acid- and Jasmonic Acid-Dependent Pathogen Defences in Arabidopsis. *Plant J.* **2013**, *75*, 90–103. [CrossRef]
10. Rockwell, N.C.; Su, Y.S.; Lagarias, J.C. Phytochrome Structure and Signaling Mechanisms. *Annu. Rev. Plant Biol.* **2006**, *57*, 837–858. [CrossRef]
11. Casal, J.J.; Sánchez, R.A.; Deregibus, V.A. The Effect of Light Quality on Shoot Extension Growth in Three Species of Grasses. *Ann. Bot.* **1987**, *59*, 1–7. [CrossRef]
12. Cortés, L.E.; Weldegergis, B.T.; Boccalandro, H.E.; Dicke, M.; Ballaré, C.L. Trading Direct for Indirect Defense? Phytochrome B Inactivation in Tomato Attenuates Direct Anti-Herbivore Defenses Whilst Enhancing Volatile-Mediated Attraction of Predators. *New Phytol.* **2016**, *212*, 1057–1071. [CrossRef] [PubMed]
13. Izaguirre, M.M.; Mazza, C.A.; Biondini, M.; Baldwin, I.T.; Ballaré, C.L. Remote Sensing of Future Competitors: Impacts on Plants Defenses. *Proc. Natl. Acad. Sci. USA* **2006**, *103*, 7170–7174. [CrossRef]
14. Fernández-Milmanda, G.L.; Crocco, C.D.; Reichelt, M.; Mazza, C.A.; Köllner, T.G.; Zhang, T.; Cargnel, M.D.; Lichy, M.Z.; Fiorucci, A.S.; Fankhauser, C.; et al. A Light-Dependent Molecular Link between Competition Cues and Defence Responses in Plants. *Nat. Plants* **2020**, *6*, 223–230. [CrossRef] [PubMed]
15. Kegge, W.; Ninkovic, V.; Glinwood, R.; Welschen, R.A.M.; Voesenek, L.A.C.J.; Pierik, R. Red:Far-Red Light Conditions Affect the Emission of Volatile Organic Compounds from Barley (Hordeum Vulgare), Leading to Altered Biomass Allocation in Neighbouring Plants. *Ann. Bot.* **2015**, *115*, 961–970. [CrossRef] [PubMed]
16. Colquhoun, T.A.; Schwieterman, M.L.; Gilbert, J.L.; Jaworski, E.A.; Langer, K.M.; Jones, C.R.; Rushing, G.V.; Hunter, T.M.; Olmstead, J.; Clark, D.G.; et al. Light Modulation of Volatile Organic Compounds from Petunia Flowers and Select Fruits. *Postharvest Biol. Technol.* **2013**, *86*, 37–44. [CrossRef]

17. Carvalho, S.D.; Schwieterman, M.L.; Abrahan, C.E.; Colquhoun, T.A.; Folta, K.M. Light Quality Dependent Changes in Morphology, Antioxidant Capacity, and Volatile Production in Sweet Basil (Ocimum Basilicum). *Front. Plant Sci.* **2016**, *7*, 1–14. [CrossRef]
18. Tegelberg, R.; Julkunen-Tiitto, R.; Aphalo, P.J. Red:Far-Red Light Ratio and UV-B Radiation: Their Effects on Leaf Phenolics and Growth of Silver Birch Seedlings. *Plant Cell Environ.* **2004**, *27*, 1005–1013. [CrossRef]
19. Kuo, T.C.; Chen, C.; Chen, S.; Lu, I.; Chu, M.; Huang, L.; Lin, C.; Chen, C.; Lo, H.; Jeng, S.; et al. The Effect of Red Light and Far-Red Light Conditions on Secondary Metabolism in Agarwood. *BMC Plant Biol.* **2015**, *15*, 1–10. [CrossRef]
20. Leone, M.; Keller, M.M.; Cerrudo, I.; Ballaré, C.L. To Grow or Defend? Low Red: Far-Red Ratios Reduce Jasmonate Sensitivity in Arabidopsis Seedlings by Promoting DELLA Degradation and Increasing JAZ10 Stability. *New Phytol.* **2014**, *204*, 355–367. [CrossRef]
21. Dudareva, N.; Klempien, A.; Muhlemann, J.K.; Kaplan, I. Biosynthesis, Function and Metabolic Engineering of Plant Volatile Organic Compounds. *New Phytol.* **2013**, *198*, 16–32. [CrossRef]
22. Becker, C.; Desneux, N.; Monticelli, L.; Fernandez, X.; Michel, T.; Lavoir, A.V. Effects of Abiotic Factors on HIPV-Mediated Interactions between Plants and Parasitoids. *Biomed Res. Int.* **2015**, *2015*, 1–18. [CrossRef] [PubMed]
23. Dicke, M.; Baldwin, I.T. The Evolutionary Context for Herbivore-Induced Plant Volatiles: Beyond the "Cry for Help". *Trends Plant Sci.* **2010**, *15*, 167–175. [CrossRef] [PubMed]
24. Karban, R.; Shiojiri, K.; Ishizaki, S. Plant Communication—Why Should Plants Emit Volatile Cues? *J. Plant Interact.* **2011**, *6*, 81–84. [CrossRef]
25. Karban, R. Plant Communication System. *Annu. Rev. Ecol. Evol. Syst.* **2021**, *52*, 1–24. [CrossRef]
26. Okada, K.; Abe, H.; Arimura, G.I. Jasmonates Induce Both Defense Responses and Communication in Monocotyledonous and Dicotyledonous Plants. *Plant Cell Physiol.* **2015**, *56*, 16–27. [CrossRef]
27. Morrell, K.; Kessler, A. Plant Communication in a Widespread Goldenrod: Keeping Herbivores on the Move. *Funct. Ecol.* **2017**, *31*, 1049–1061. [CrossRef]
28. Kalske, A.; Shiojiri, K.; Uesugi, A.; Sakata, Y.; Morrell, K.; Kessler, A. Insect Herbivory Selects for Volatile-Mediated Plant-Plant Communication. *Curr. Biol.* **2019**, *29*, 3128–3133.e3. [CrossRef] [PubMed]
29. Erb, M. Volatiles as Inducers and Suppressors of Plant Defense and Immunity—Origins, Specificity, Perception and Signaling. *Curr. Opin. Plant Biol.* **2018**, *44*, 117–121. [CrossRef]
30. Heil, M. Herbivore-Induced Plant Volatiles: Targets, Perception and Unanswered Questions. *New Phytol.* **2014**, *204*, 297–306. [CrossRef]
31. Gallie, D.R. Ethylene Receptors in Plants—Why so Much Complexity? *F1000Prime Rep.* **2015**, *7*, 1–12. [CrossRef]
32. Sugimoto, K.; Matsui, K.; Iijima, Y.; Akakabe, Y.; Muramoto, S.; Ozawa, R.; Uefune, M.; Sasaki, R.; Alamgir, K.M.; Akitake, S.; et al. Intake and Transformation to a Glycoside of (Z)-3-Hexenol from Infested Neighbors Reveals a Mode of Plant Odor Reception and Defense. *Proc. Natl. Acad. Sci. USA* **2014**, *111*, 7144–7149. [CrossRef] [PubMed]
33. Etterson, J.R.; Delf, D.E.; Craig, T.P.; Ando, Y. NOTE/NOTE Parallel Patterns of Clinal Variation in *Solidago altissima* in Its Native Range in Central USA and Its Invasive Range in Japan. *Botany* **2008**, *97*, 91–97. [CrossRef]
34. Howard, M.M.; Kalske, A.; Kessler, A. Eco-Evolutionary Processes Affecting Plant–Herbivore Interactions during Early Community Succession. *Oecologia* **2018**, *187*, 547–559. [CrossRef]
35. Maddox, G.D.; Root, R.B. Resistance to 16 Diverse Species of Herbivorous Insects within a Population of Goldenrod, *Solidago altissima*: Genetic Variation and Heritability. *Oecologia* **1987**, *72*, 8–14. [CrossRef] [PubMed]
36. Maddox, G.D.; Root, R.B. Structure of the Encounter between Goldenrod (*Solidago altissima*) and Its Diverse Insecr Fauna. *Ecology* **1990**, *71*, 2115–2124. [CrossRef]
37. Carson, W.P.; Root, R.B. Herbivory and Plant Species Coexistence: Community Regulation by an Outbreaking Phytophagous Insect. *Ecol. Monogr.* **2000**, *70*, 73–99. [CrossRef]
38. Bode, R.F.; Kessler, A. Herbivore Pressure on Goldenrod (*Solidago altissima* L., Asteraceae): Its Effects on Herbivore Resistance and Vegetative Reproduction. *J. Ecol.* **2012**, *100*, 795–801. [CrossRef]
39. Uesugi, A.; Kessler, A. Herbivore Exclusion Drives the Evolution of Plant Competitiveness via Increased Allelopathy. *New Phytol.* **2013**, *198*, 916–924. [CrossRef]
40. Rubin, I.N.; Ellner, S.P.; Kessler, A.; Morrell, K.A. Informed Herbivore Movement and Interplant Communication Determine the Effects of Induced Resistance in an Individual-Based Model. *J. Anim. Ecol.* **2015**, *84*, 1273–1285. [CrossRef]
41. Demotes-Mainard, S.; Péron, T.; Corot, A.; Bertheloot, J.; Le Gourrierec, J.; Pelleschi-Travier, S.; Crespel, L.; Morel, P.; Huché-Thélier, L.; Boumaza, R.; et al. Plant Responses to Red and Far-Red Lights, Applications in Horticulture. *Environ. Exp. Bot.* **2016**, *121*, 4–21. [CrossRef]
42. Fankhauser, C.; Batschauer, A. Shadow on the Plant: A Strategy to Exit. *Cell* **2016**, *164*, 15–17. [CrossRef]
43. Kurepin, L.V.; Emery, R.J.N.; Pharis, R.P.; Reid, D.M.; Kurepin, L.V.; Emery, R.J.N.; Pharis, R.P.; Reid, D.M. Uncoupling Light Quality from Light Irradiance Effects in Helianthus Annuus Shoots: Putative Roles for Plant Hormones in Leaf and Internode Growth. *J. Exp. Bot.* **2007**, *58*, 2145–2157. [CrossRef] [PubMed]
44. Pierik, R.; Ballaré, C.L.; Dicke, M. Ecology of Plant Volatiles: Taking a Plant Community Perspective. *Plant Cell Environ.* **2014**, *37*, 1845–1853. [CrossRef]

45. Cipollini, D.; Lieurance, D.M. Expression and Costs of Induced Defense Traits in Alliaria Petiolata, a Widespread Invasive Plant. *Basic Appl. Ecol.* **2012**, *13*, 432–440. [CrossRef]
46. Mason, H.S.; Mullet, J.E. Expression of Two Soybean Vegetative Storage Protein Genes during Development and in Response to Water Deficit, Wounding, and Jasmonic Acid. *Plant Cell* **1990**, *2*, 569–579. [CrossRef] [PubMed]
47. Mason, H.S.; DeWald, D.B.; Creelman, R.A.; Mullet, J.E. Coregulation of Soybean Vegetative Storage Protein Gene Expression by Methyl Jasmonate and Soluble Sugars. *Plant Physiol.* **1992**, *98*, 859–867. [CrossRef]
48. Dewald, D.B.; Sadka, A.; Mullet, J.E. Sucrose Modulation of Soybean Vsp Gene Expression Is Inhibited by Auxin. *Plant Physiol.* **1994**, *104*, 439–444. [CrossRef]
49. Thornburg, R.W.; Li, X. Wounding Nicotiana Tabacum Leaves Causes a Decline in Endogenous Indole-3-Acetic Acid. *Plant Physiol.* **1991**, *96*, 802–805. [CrossRef] [PubMed]
50. Baldwin, I.T.; Zhang, Z.P.; Diab, N.; Ohnmeiss, T.E.; McCloud, E.S.; Lynds, G.Y.; Schmelz, E.A. Quantification, Correlations and Manipulations of Wound-Induced Changes in Jasmonic Acid and Nicotine in Nicotiana Sylvestris. *Planta* **1997**, *201*, 397–404. [CrossRef]
51. Cooper-Driver, G.A.; Le Quesne, P.W. Diterpenoids as Insect Antifeedants and Growth Inhibitors: Role in Solidago Species. In *Allelochemicals: Role in Agriculture and Forestry*; Waller, G.R., Ed.; American Chemical Society: Washington, DC, USA, 1987; pp. 534–550. ISBN 9780841209923.
52. Uesugi, A.; Kessler, A. Herbivore Release Drives Parallel Patterns of Evolutionary Divergence in Invasive Plant Phenotypes. *J. Ecol.* **2016**, *104*, 876–886. [CrossRef]
53. Kessler, A.; Halitschke, R.; Diezel, C.; Baldwin, I.T. Priming of Plant Defense Responses in Nature by Airborne Signaling between Artemisia Tridentata and Nicotiana Attenuata. *Oecologia* **2006**, *148*, 280–292. [CrossRef]
54. Crepy, M.A.; Casal, J.J. Photoreceptor-Mediated Kin Recognition in Plants. *New Phytol.* **2015**, *205*, 329–338. [CrossRef]
55. Semchenko, M.; Saar, S.; Lepik, A. Plant Root Exudates Mediate Neighbour Recognition and Trigger Complex Behavioural Changes. *New Phytol.* **2014**, *204*, 631–637. [CrossRef] [PubMed]
56. Kessler, A.; Kalske, A. Plant Secondary Metabolite Diversity and Species Interactions. *Annu. Rev. Ecol. Evol. Syst.* **2018**, *49*, 115–138. [CrossRef]
57. Mertens, D.; Boege, K.; Kessler, A.; Koricheva, J.; Thaler, J.S.; Whiteman, N.K.; Poelman, E.H. Predictability of Biotic Stress Structures Plant Defence Evolution. *Trends Ecol. Evol.* **2021**, *36*, 444–456. [CrossRef]
58. Barbosa, P.; Hines, J.; Kaplan, I.; Martinson, H.; Szczepaniec, A.; Szendrei, Z. Associational Resistance and Associational Susceptibility: Having Right or Wrong Neighbors. *Annu. Rev. Ecol. Evol. Syst.* **2009**, *40*, 1–20. [CrossRef]
59. Lawson, S.K.; Sharp, L.G.; Powers, C.N.; McFeeters, R.L.; Satyal, P.; Setzer, W.N. Volatile Compositions and Antifungal Activities of Native American Medicinal Plants: Focus on the Aster0aceae. *Plants* **2020**, *9*, 126. [CrossRef]
60. Martinez Arbizu, P. *PairwiseAdonis: Pairwise Multilevel Comparison Using Adonis*, Version 0.4; 2020. Available online: https://github.com/pmartinezarbizu/pairwiseAdonis#readme (accessed on 11 September 2022).
61. R Team Core. *R: A Language and Environment for Statistical Computing*, Version 2.6.2; R Foundation for Statistical Computing: Vienna, Austria, 2021.

MDPI AG
Grosspeteranlage 5
4052 Basel
Switzerland
Tel.: +41 61 683 77 34

Plants Editorial Office
E-mail: plants@mdpi.com
www.mdpi.com/journal/plants